MTP International Review of Science

Surface Chemistry and Colloids

MTP International Review of Science

Publisher's Note

The MTP International Review of Science is an important new venture in scientific publishing, which we present in association with MTP Medical and Technical Publishing Co. Ltd. and University Park Press, Baltimore. The basic concept of the Review is to provide regular authoritative reviews of entire disciplines. We are starting with chemistry because the problems of literature survey are probably more acute in this subject than in any other. As a matter of policy, the authorship of the MTP Review of Chemistry is international and distinguished; the subject coverage is extensive, systematic and critical; and most important of all, new issues of the Review will be published every two years.

In the MTP Review of Chemistry (Series One), Inorganic, Physical and Organic Chemistry are comprehensively reviewed in 33 text volumes and 3 index volumes, details of which are shown opposite. In general, the reviews cover the period 1967 to 1971. In 1974, it is planned to issue the MTP Review of Chemistry (Series Two), consisting of a similar set of volumes covering the period 1971 to 1973. Series Three is planned for 1976, and so on.

The MTP Review of Chemistry has been conceived within a carefully organised editorial framework. The over-all plan was drawn up, and the volume editors were appointed, by three consultant editors. In turn, each volume editor planned the coverage of his field and appointed authors to write on subjects which were within the area of their own research experience. No geographical restriction was imposed. Hence, the 300 or so contributions to the MTP Review of Chemistry come from many countries of the world and provide an authoritative account of progress in chemistry.

To facilitate rapid production, individual volumes do not have an index. Instead, each chapter has been prefaced with a detailed list of contents, and an index to the 13 volumes of the MTP Review of Physical Chemistry (Series One) will appear, as a separate volume, after publication of the final volume. Similar arrangements will apply to the MTP Review of Organic Chemistry (Series One) and to subsequent series.

Butterworth & Co. (Publishers) Ltd.

**Physical Chemistry
Series One**
Consultant Editor
A. D. Buckingham
*Department of Chemistry
University of Cambridge*

Volume titles and Editors

1 THEORETICAL CHEMISTRY
Professor W. Byers Brown, *University of Manchester*

2 MOLECULAR STRUCTURE AND PROPERTIES
Professor G. Allen, *University of Manchester*

3 SPECTROSCOPY
Dr. D. A. Ramsay, F.R.S.C.,
National Research Council of Canada

4 MAGNETIC RESONANCE
Professor C. A. McDowell, F.R.S.C.,
University of British Columbia

5 MASS SPECTROMETRY
Professor A. Maccoll, *University College, University of London*

6 ELECTROCHEMISTRY
Professor J. O'M Bockris, *University of Pennsylvania*

7 SURFACE CHEMISTRY AND COLLOIDS
Professor M. Kerker, *Clarkson College of Technology, New York*

8 MACROMOLECULAR SCIENCE
Professor C. E. H. Bawn, F.R.S.,
University of Liverpool

9 CHEMICAL KINETICS
Professor J. C. Polanyi, F.R.S.,
University of Toronto

10 THERMOCHEMISTRY AND THERMODYNAMICS
Dr. H. A. Skinner, *University of Manchester*

11 CHEMICAL CRYSTALLOGRAPHY
Professor J. Monteath Robertson, F.R.S.,
University of Glasgow

12 ANALYTICAL CHEMISTRY — PART 1
Professor T. S. West, *Imperial College, University of London*

13 ANALYTICAL CHEMISTRY — PART 2
Professor T. S. West, *Imperial College, University of London*

INDEX VOLUME

**Inorganic Chemistry
Series One**
Consultant Editor
H. J. Emeléus, F.R.S.
*Department of Chemistry
University of Cambridge*

Volume titles and Editors

1 **MAIN GROUP ELEMENTS—
HYDROGEN AND GROUPS I–IV**
Professor M. F. Lappert, *University of Sussex*

2 **MAIN GROUP ELEMENTS—
GROUPS V AND VI**
Professor C. C. Addison, F.R.S. and
Dr. D. B. Sowerby, *University of Nottingham*

3 **MAIN GROUP ELEMENTS—
GROUP VII AND NOBLE GASES**
Professor Viktor Gutmann, *Technical University of Vienna*

4 **ORGANOMETALLIC DERIVATIVES
OF THE MAIN GROUP
ELEMENTS**
Dr. B. J. Aylett, *Westfield College, University of London*

5 **TRANSITION METALS—PART 1**
Professor D. W. A. Sharp, *University of Glasgow*

6 **TRANSITION METALS—PART 2**
Dr. M. J. Mays, *University of Cambridge*

7 **LANTHANIDES AND ACTINIDES**
Professor K. W. Bagnall, *University of Manchester*

8 **RADIOCHEMISTRY**
Dr. A. G. Maddock, *University of Cambridge*

9 **REACTION MECHANISMS IN
INORGANIC CHEMISTRY**
Professor M. L. Tobe, *University College, University of London*

10 **SOLID STATE CHEMISTRY**
Dr. L. E. J. Roberts, *Atomic Energy Research Establishment, Harwell*

INDEX VOLUME

**Organic Chemistry
Series One**
Consultant Editor
D. H. Hey, F.R.S.
*Department of Chemistry
King's College, University of London*

Volume titles and Editors

1 **STRUCTURE DETERMINATION
IN ORGANIC CHEMISTRY**
Professor W. D. Ollis, F.R.S., *University of Sheffield*

2 **ALIPHATIC COMPOUNDS**
Professor N. B. Chapman,
Hull University

3 **AROMATIC COMPOUNDS**
Professor H. Zollinger, *Swiss Federal Institute of Technology*

4 **HETEROCYCLIC COMPOUNDS**
Dr. K. Schofield, *University of Exeter*

5 **ALICYCLIC COMPOUNDS**
Professor W. Parker, *University of Stirling*

6 **AMINO ACIDS, PEPTIDES AND
RELATED COMPOUNDS**
Professor D. H. Hey, F.R.S. and
Dr. D. I. John,
King's College, University of London

7 **CARBOHYDRATES**
Professor G. O. Aspinall, *Trent, University, Ontario*

8 **STEROIDS**
Dr. W. D. Johns, *G. D. Searle & Co., Chicago*

9 **ALKALOIDS**
Professor K. F. Wiesner, F.R.S.,
University of New Brunswick

10 **FREE RADICAL REACTIONS**
Professor W. A. Waters, F.R.S.,
University of Oxford

INDEX VOLUME

Physical Chemistry
Series One

Consultant Editor
A. D. Buckingham

MTP International Review of Science

Volume 7
Surface Chemistry and Colloids

Edited by **M. Kerker**
Clarkson College of Technology, New York

Butterworths · London
University Park Press · Baltimore

THE BUTTERWORTH GROUP

ENGLAND
Butterworth & Co (Publishers) Ltd
London: 88 Kingsway, WC2B 6AB

AUSTRALIA
Butterworths Pty Ltd
Sydney: 586 Pacific Highway 2067
Melbourne: 343 Little Collins Street, 3000
Brisbane: 240 Queen Street, 4000

NEW ZEALAND
Butterworths of New Zealand Ltd
Wellington: 26 28 Waring Taylor Street, 1

SOUTH AFRICA
Butterworth & Co (South Africa) (Pty) Ltd
Durban: 152–154 Gale Street

ISBN 0 408 70268 0

UNIVERSITY PARK PRESS

U.S.A. and CANADA
University Park Press Inc
Chamber of Commerce Building
Baltimore, Maryland, 21202

> Library of Congress Cataloging in Publication Data
>
> Kerker, Milton.
>   Surface chemistry and colloids.
>
>   (Physical chemistry, series one, v. 7) (MTP international review of science)
>   1. Surface chemistry. 2. Colloids. I. Title.
> [DNLM: 1. Colloids. 2. Surface properties.
> W1 PH683K v. 7 1972 XNLM: [QD 506 S961 1972]]
> QD453.2.P58 vol. 7 [QD506]   541'.3'08s [541'.345]
> ISBN 0–8391–1021–0   72–5216

First Published 1972 and © 1972
MTP MEDICAL AND TECHNICAL PUBLISHING CO. LTD.
Seacourt Tower
West Way
Oxford, OX2 OJW
and
BUTTERWORTH & CO. (PUBLISHERS) LTD.

Filmset by Photoprint Plates Ltd., Rayleigh, Essex
Printed in England by Redwood Press Ltd., Trowbridge, Wilts
and bound by R. J. Acford Ltd., Chichester, Sussex

Consultant Editor's Note

The MTP International Review of Science is designed to provide a comprehensive, critical and continuing survey of progress in research. The difficult problem of keeping up with advances on a reasonably broad front makes the idea of the Review especially appealing, and I was grateful to be given the opportunity of helping to plan it.

This particular 13-volume section is concerned with Physical Chemistry, Chemical Crystallography and Analytical Chemistry. The subdivision of Physical Chemistry adopted is not completely conventional, but it has been designed to reflect current research trends and it is hoped that it will appeal to the reader. Each volume has been edited by a distinguished chemist and has been written by a team of authoritative scientists. Each author has assessed and interpreted research progress in a specialised topic in terms of his own experience. I believe that their efforts have produced very useful and timely accounts of progress in these branches of chemistry, and that the volumes will make a valuable contribution towards the solution of our problem of keeping abreast of progress in research.

It is my pleasure to thank all those who have collaborated in making this venture possible – the volume editors, the chapter authors and the publishers.

Cambridge   A. D. Buckingham

Preface

Colloid Chemistry and its extension to Surface Chemistry emerged as a somewhat autonomous discipline at the turn of this century in association with the New Physical Chemistry. Even then Colloid Chemistry had a strong component of precise measurement and mathematical analysis, yet major components of this science dealt qualitatively and descriptively with poorly, or at best difficultly, defined systems. A mere perusal of this volume should assure the reader that this is no longer true.

The topics under review are not exhaustive, and we hope that the next issue of this continuing series will fill the gaps. The omission of electrical phenomena may be the most glaring. And of course, within each chapter, the authors have had to concentrate on only parts of the subject. Their charge was to review the advances of the past five years with the expectation that there would be an opportunity on a biennial basis to round out the picture.

This science is in its early middle age and its state of health is excellent. The problems are clearly formulated and the models are well articulated. Perhaps we still cannot unambiguously describe real membranes and solid surfaces and micelles; yet the picture is clear enough to set up good experiments and to interpret them clearly. One can do good science in Colloid and Surface Chemistry and have the gratification that this science is central to an understanding of some of the most important problems in technology and in nature—both animate and inanimate.

The editor's job has been a simple one, thanks to the zeal and the genius of the authors.

Potsdam, New York   Milton Kerker

Contents

Insoluble monolayers   1
G. L. Gaines, Jr., *General Electric Research and Development Centre, U.S.A.*

Membranes: their interfacial chemistry and biophysics   25
H. Ti Tien, *Michigan State University*

Recent advances in the study of solid surfaces   79
W. P. Ellis, *University of California, Los Alamos Scientific Laboratory*

Aggregation in surfactant systems   97
Per Ekwall, *Stockholm, Sweden,* and Ingvar Danielsson and Per Stenius, *Åbo Akademi, Finland*

Theory of homogeneous nucleation from the vapour   147
G. M. Pound, *Stanford University, California,*
Kazumi Nishioka and Jens Lothe, *Blindern University, Oslo*

The physical adsorption of gases   189
S. J. Gregg, *Brunel University, Uxbridge*

Chemisorption on tungsten crystal planes   225
H. Wise, *Stanford Research Institute, Menlo Park, California*

Hydrosols   241
D. H. Napper and R. J. Hunter, *University of Sydney*

1
Insoluble Monolayers

G. L. GAINES, Jr.
General Electric Research and Development Center, U.S.A.

1.1	INTRODUCTORY REMARKS	1
1.2	MONOLAYER PROPERTIES	2
	1.2.1 *Film spreading and stability*	3
	1.2.2 *Surface pressure and potential*	4
	1.2.3 *Surface viscosity*	5
	1.2.4 *Wave damping*	6
	1.2.5 *Miscellaneous experimental techniques*	6
	1.2.6 *Equations of state and thermodynamic properties*	7
1.3	SELECTED MONOLAYER-FORMING MATERIALS	8
	1.3.1 *'Small' molecules*	8
	1.3.1.1 *Counter-ion effects in ionised monolayers*	9
	1.3.1.2 *Phospholipids*	9
	1.3.1.3 *Porphyrins and chlorophyll*	10
	1.3.1.4 *Miscellaneous small molecules and sub-phase effects*	10
	1.3.2 *Macromolecules*	11
	1.3.2.1 *Proteins and polypeptides*	11
	1.3.2.2 *Synthetic polymers*	12
1.4	MIXED FILMS	13
1.5	MONOLAYERS ON SOLIDS; MULTILAYERS	14
1.6	REACTIONS IN MONOLAYERS	16
1.7	PERMEATION THROUGH MONOLAYERS; EVAPORATION CONTROL	17
1.8	SOME ADDITIONAL IMPLICATIONS AND APPLICATIONS	18
	NOTE ADDED IN PROOF	19

1.1 INTRODUCTORY REMARKS

Insoluble monolayers—films a single molecule thick spread at a liquid—gas interface—continue to attract scientific attention. As models for studying a host of more complex interfacial phenomena, from lubrication to life processes occurring at or within biological membranes, their apparent simplicity is most appealing. Yet there are many aspects of the behaviour of molecules in monolayers which are not well understood. This review attempts to report progress in understanding and developments in technique reflected in the literature during the five-year period up to mid-1971. While mention is made of a variety of substances studied in monolayers, especially at the air–water interface, and a number of applications of monolayer techniques are cited, no attempt is made to provide a comprehensive catalogue of all monolayer studies. Only spread monolayers are discussed; except for films transferred to solids from liquid surfaces ('built-up' films), monolayers on solids are not considered, nor are liquid–liquid interfaces or adsorbed films ('soluble monolayers'). Work prior to 1966 has been treated in the author's monograph[1], standard texts such as Adamson's[2], and in a review by Mingins and Standish[3].

Several recent reviews of aspects of monolayer research have appeared. Lange[4] has summarised work on non-ionic surfactants of the ethylene oxide adduct type. Some materials of this class may be studied as either spread or adsorbed films; Lange's review covers both kinds of studies, and indicates the similarities and differences between them. Shah[5] provided a broad review of monolayers of lipids, including simple long chain compounds such as acids, but emphasising phospholipids, mixed monolayers, and chemical reactions in lipid monolayers. Cadenhead[6] reviewed the narrower field of monolayers of synthetic phospholipids in depth. Monolayer studies on synthetic polymers were summarised in an article on 'Surface Chemistry and Polymers' by Rosoff[7]. Pilpel[8] included spread monolayer studies in a review on heavy-metal soaps. A brief discussion of monolayers deposited on solids, with particular attention to their relevance to lubrication phenomena, was given by Popiel[9]. Gershfeld's summary[10] of energetics and thermodynamic relationships in monolayers emphasises recent experimental developments in the measurement of very low surface pressures as well as their interpretation; it also includes a useful, though brief, review of more general aspects of lipid monolayer properties. Although the author has not seen them, other general reviews have appeared in Japan[11] and Spain[12]. Joly[13] has treated monolayer viscosity, but this summary may be somewhat obsolete in view of very recent developments in this area (cf. Section 1.2.3). Note should also be made of the Fifth International Congress of Surface Activity, held in Barcelona in 1969[14], which included a number of publications on monolayers, as had the earlier Congresses.

Three papers by Giles and Forrester[15] provide a wealth of fascinating historical details on the origins of scientific study of films on water. These cover the period from Benjamin Franklin's observations of the spreading of an oil film (which Giles describes as 'the first recorded scientific experiment in surface chemistry') through Agnes Pockels' development of monolayer experimental techniques, with briefer mention of more recent developments.

1.2 MONOLAYER PROPERTIES

The study of monolayers demands considerable attention to experimental detail, especially because minute traces of impurities can vitiate the significance of careful measurements, and there are many possible experimental artifacts. Accordingly, the recent literature includes continuing development of standard methods, as well as new techniques which have been applied to spread films. Attention has been given to some spreading difficulties and film loss or degradation processes. At the same time, further attempts have been made to interpret several of the properties of single-component films in a quantitative way.

1.2.1 Film spreading and stability

Whether spreading solvents are retained in or alter the properties of spread monolayers continues to be a matter of concern. Mingins, Owens and Iles[16] found considerable effects on the surface pressure of octadecyltrimethylammonium bromide films. Barnes, Elliot and Grigg[17], however, made direct measurements of the amount of radioactivity tagged decane retained in an octadecanol monolayer, and concluded that this hydrocarbon, if pure, evaporated completely. Pagano and Gershfeld[18] also found that petroleum ether evaporated rapidly, and did not alter the surface vapour pressure of an oleyl alcohol film (0.085 ± 0.001 dyn cm^{-1}). In view of their relevance to the spreading solvent question, recent studies of the solubility in water of various hydrocarbons are noted[19].

Spreading from the vapour phase and rate of spreading studies have been primarily related to the question of evaporation control; cf. Section 1.7.

The instability of monolayers with respect to desorption has been considered by Gershfeld and Patlak[20]. If the desorbing film is in equilibrium with a thin unstirred layer, either in the liquid subphase for a solution process or in the vapour for evaporation from the monolayer, the activity coefficients in the monolayer can be estimated from the desorption rate. Gershfeld[21] has also estimated cohesive energies from desorption rate studies on long-chain sulphates, phosphonates and fatty acids.

Heikkila *et al.*[22] have made a detailed study of collapse and film losses from fatty acid monolayers. They obtained values of equilibrium spreading pressure (Π_e) for the C_{16}–C_{22} saturated compounds, oleic and linoleic acids, in substantial agreement with earlier work, and demonstrated that slow collapse to these surface pressures occurred when the films were overcompressed. When the monolayers were maintained at surface pressures below Π_e, film losses as a function of pH and temperature strongly suggested a solution process. Motomura *et al.*[23] derived an equation to allow for the effect of varying compression rate on the observed surface pressure – area (Π–A) curve, when the monolayer is slowly dissolving. With this relationship, they were able to correct for the solution process in measurements on myristic acid films. Compression rate effects have also been reported occasionally in other systems[24], but in few studies have such kinetic effects been examined in detail.

Chemical degradation may also be a problem when susceptible molecules are spread in monolayers. Simple calculations[25] suggest that protective agents dissolved in the sub-phase may be ineffective against even traces of reactive species (e.g. oxygen, for a readily oxidisable monolayer) in the gas phase, because of the relatively slow diffusion in the liquid, compared to gaseous molecule collision rates. In the case of cholesterol, direct evidence of slow oxidation in monolayers has been obtained by thin-layer chromatography of collected films[26]. While very short monolayer exposure to air produced no detectable oxidation, leaving the monolayer spread for 45 min led to considerable amounts of several different oxidation products. In this case, both inhibition and enhancement effects from reagents dissolved in the subphase were found.

1.2.2 Surface pressure and potential

The most commonly measured physical property of a monolayer is the reduction of surface tension, or surface pressure (Π), which the film produces. Measured together with the area of the film, usually expressed as area occupied per molecule, it defines the familiar $\Pi-A$ curve which characterises the mechanical properties of the monolayer. The Langmuir float method and the Wilhelmy plate continue to be the 'standard' methods for surface pressure measurement.

Considerable attention has been paid recently to the Wilhelmy plate method. Both end corrections[27, 28] and roughness of the plate[29, 30] have been treated in detail. These papers indicate details of technique and corrections which can be important in absolute surface tension determinations, but the use of the Wilhelmy method for studies of monolayers generally minimises errors from these sources, because of the differential nature of the measurement. Another difficulty, that of adsorption on the plate leading to faulty readings, has been re-emphasised (with regard to amines on platinum)[31]. Refinements in the Langmuir float method have mainly involved improved sensing devices (some of which could therefore be applied to other methods as well); one detector, in particular, has been claimed to have a sensitivity of 10^{-4} dyn cm^{-1} [32]. Many reports have described film balance and trough systems which provide for increased sensitivity, speed, atmosphere control, or other particular advantages; typical references are listed[18, 33-39].

Note must also be taken of the possibility of deriving $\Pi-A$ curves from ripple-damping measurements, discussed in more detail in Section 1.2.4 of this review.

The change in Volta potential at the liquid-gas interface produced by a spread monolayer – the 'surface potential' of the film – is a useful experimental parameter, although its significance in molecular terms remains cloudy. Jarvis and Scheiman[40] have measured the effect of added electrolytes on the Volta potential at the air-water interface; at salt concentrations below 1M, the potential changes did not exceed ± 30 mV, but high concentrations of certain salts [NaI, NaSCN, Ca(NO$_3$)$_2$] produce ΔVs of -150 mV or more. Generally, it appears that the anions have the predominant effect in controlling the sign and magnitude of the potential changes; no quantitative interpretation is available, however.

Two methods for direct differential measurement of monolayer surface potentials using ionising electrodes have been described. In one, a single electrode (using ^{241}Am as the radioactive source) is moved alternately over the clean and film-covered surface in a Langmuir trough[41]. In the second procedure[42] two ionising electrodes are mounted on a rotating frame, and the potential difference between them is measured as each is alternately positioned over the clean and film-covered surface; no trough electrode is required. Both techniques minimise the problem of drift of the air–electrode Volta potential; the latter method also eliminates any trough electrode problems, and is useful for studying adsorbed films as well as spread monolayers.

1.2.3 Surface viscosity

A major development in the past few years has been detailed study of surface rheological phenomena. Several analyses have shown that proper allowance for the interaction between a surface film and the sub-phase is critically dependent on the experimental geometry. For this reason, meaningful absolute measurements are only possible if the experimental apparatus conforms closely in design to that considered in a particular analysis. Even comparative estimates made with incompletely analysed equipment are highly suspect; Mannheimer and Burton[43], for example, have shown that conventional calculations from results with torsional viscometers could indicate large temperature effects even if the absolute surface (shear) viscosity were in fact independent of temperature.

In an important series of papers[44–48] Mannheimer et al. have considered the viscous traction viscometer in detail. In such an instrument, the film is spread in a stationary annular canal, and motion is imparted by rotating the (concentric) liquid container; flow at the interface is detected by the movement of floating particles in the film[49]. In the latest embodiment of this method[46], the canal is formed by concentric cylinders which reach nearly to the bottom of the liquid container, and the interface is kept flat by the presence of a non-wettable step at the proper level. The ratio of canal width to depth is critical, and must be determined accurately. A complete analysis for Newtonian surface films is available, and the surface yield stress for rigid films can be closely approximated. Mannheimer and Schechter[47] have also analysed this system for the case of a film which behaves as a Bingham plastic i.e. is rigid up to some critical yield stress, then exhibits Newtonian behaviour. In addition they have considered relaxation measurements and the frequency response of the surface when the impressed motion is periodic[48]. For these cases, a two dimensional Maxwell model (spring and dashpot) was assumed in the analysis.

Hegde and Slattery[50] have extended Mannheimer's analysis by a more general consideration of the dependence of surface shear stress on shear rate. Osborne[51] has provided an extension to include films at liquid–liquid interfaces; he also discussed torsional viscometers, and indicated one configuration (with stationary guard cylinders above and below the interface) which might make such instruments capable of rigorous analysis.

In the same period, Goodrich and his co-workers[52–54] have examined more conventional torsional viscometer designs (in which the torsional element just touches or floats at the interface) in detail. They find[53] that the rotating disk element, because of mathematical difficulties, cannot be analysed exactly in real systems. A rotating ring making knife-edge contact with a semi-infinite liquid surface, however, is subject to quantitative analysis[54].

To date, no experimental studies of spread insoluble monolayers by these exact methods have been reported. Such measurements are clearly very desirable.

1.2.4 Wave damping

As already remarked, the 'first recorded scientific experiment' on a monolayer was Franklin's wave damping observation. Yet only in the past few years have detailed studies of the phenomenon become available. In addition to providing quantitative analysis of this natural phenomenon, they suggest interesting experimental techniques for the laboratory study of spread films. An excellent review of work through 1968 has been published[55].

Several experimental techniques have been used to study wave damping. Mann's combination of mechanical generator and sensor in the surface[56] has been modified in various ways by Hansen and his co-workers[57-59]. New optical detection methods have also been described by Garrett and Zisman[60] and Battezzati[61]. In the former, a movable screen above the trough is used to find the point of focus of light diffracted by the ripples, which act as lenses. The latter device makes use of the interference pattern produced when monochromatic light is used; it is claimed to be rapid and simple, but is presently less sensitive than other methods. While the mechanical technique so far seems more sensitive, the presence of a solid object (the sensor) in the surface can perturb the ripple field; Mann and Hansen attempted to correct for this effect, but the correction has been criticised[62].

Essentially complete solutions of the wave damping equations relating the observed wavelengths and amplitudes to the surface tension, its gradients (i.e. film compressibility), and surface rheological parameters have been obtained[55, 62]. Unfortunately, because of mathematical complexities as well as experimental difficulties already noted, it is generally rather difficult to evaluate the surface parameters from the ripple measurements, although the validity of the procedure has been demonstrated[57]. Developments to date have largely been concerned with verifying the equations for various kinds of surface layers, particularly adsorbed films where diffusional interchange between surface and bulk is important.

On the other hand, damping measurements on poly(dimethyl siloxane)[60] and poly(oxyethylene)[63] monolayers show complex sequences of maxima and minima. Presumably these changes in damping coefficients with film compression reflect changes in molecular orientation or conformation in the film, which are not readily interpretable in terms of the hydrodynamic theory (cf. Section 1.3.2.2).

1.2.5 Miscellaneous experimental techniques

Smith[37, 64] has described the use of an ellipsometer incorporated into a Langmuir trough arrangement to study monolayers on mercury. Fluorescence from chlorophyll in monolayers has been studied in a new apparatus by Trosper, Park and Sauer[65]. In this device, the exciting light enters the trough vertically from below, and filters are used to isolate the emitted fluorescence; polarisation of the emission can be measured.

Bernett et al.[66] have described a new evaporation rate measuring system, in which two identical liquid surfaces, one clean and the other film covered, are exposed to flowing helium. Thermal conductivity detectors and flowmeters provide a measure of differences in evaporation rate between the two surfaces. Diffusion of film molecules within a monolayer has now been measured with the aid of carbon-14 labelled substances; the results indicate that a two-dimensional form of Fick's law adequately describes the diffusion process[67].

Brief reports have appeared describing two methods of observation which, while not yet well developed, might form the basis for useful techniques. One is the flow pattern developed in monolayer-covered surfaces by a cylinder rotating in the surface[68]. The second involves heat-flow measurements in a Langmuir trough during film compression[69].

A number of additional techniques have been applied to built-up films on solids; they are noted in Section 1.5.

1.2.6 Equations of state and thermodynamic properties

A number of attempts to interpret monolayer measurements in fundamental terms have been reported; relations applicable to single-component films are mentioned here, while consideration of mixed films is noted in Section 1.4.

Smith[70] has derived an equation for the liquid-expanded state in monolayers of long-chain compounds, based on a model of the chain as a stack of hard disks. He tested his equation with Π–A curves for several fatty acids at various temperatures, and found agreement within about 15%. An alternative expression was obtained by Motomura and Matuura[71] on the basis of a lattice model in which —CH_2— segments were assumed to occupy sites in a quasi-lattice; the statistical treatment is analogous to the Flory–Huggins polymer-solution theory. This equation was compared only to data for the C_{14}–C_{16} fatty acids, but agreement was good.

The 'intermediate state', characterised by the region of high compressibility in the Π–A isotherm lying between the condensed and expanded film regions, has been re-evaluated by Cadenhead and Demchak[72]. On the basis of studies of monolayers of 4-amino-p-terphenyl, which closely approximates a 'rigid rotor', they conclude that hindered rotation is not a major cause of the occurrence of the intermediate state, but that this state is in some way related to internal motions of paraffinic chains.

Motomura[73] has suggested that the intermediate state is a kinetic, rather than an equilibrium, phenomenon. In the same paper, he also presented a general thermodynamic discussion of monolayer states, spreading and collapse, including non-equilibrium terms. Since it is not obvious how to evaluate the irreversible variables, the affinity and extent of reaction, no application of this analysis to experiment has yet been made.

The equation of state of ionised monolayers continues to be a topic of discussion. Three major problems which remain unresolved are (i) the detailed analysis of electrostatic effects, (ii) the proper form of the contribution to Π by the uncharged film, and (iii) whether at very large areas the $\Pi-A$ product for an ionised monolayer approaches kT or $2\,kT$. With regard to the last point, an experimental study[74] of monolayers of docosyltriethylammonium bromide has supported the values $2\,kT$ on pure water and kT on salt solutions, as predicted by Haydon[75]. Experimental evidence on the other problems is very limited, and in fact the specific counter-ion effects which have been discovered (discussed in Section 1.3.1.1) raise questions about the quantitative application of any simple theory to experiment. Theoretical papers, however, have been numerous[76–81].

The proper thermodynamic treatment of monolayer compression data has been the subject of extended discussion by Gershfeld and his co-workers (cf. his review[10]). The evaluation of a free energy of compression, he correctly notes, requires surface pressure measurements to large enough areas that the film-forming molecules have negligible interaction with one another, i.e. to the ideal gas or effectively infinitely dilute surface solution region of the $\Pi-A$ isotherm[82]. With such data, the free energy of compression, ΔFc, can be evaluated from

$$\Delta Fc = F - F_i = -\int_{A_i}^{A_c} \Pi \alpha A,$$

where F is the free energy of the film in a coherent state characterised by molecular area A_c, and F_i is the free energy in the ideal gas state at area A_i. From the variation of measured free energies of compression with chain-length[82], it was concluded that these free energies could be decomposed into additive contributions due to hydrocarbon chains (which part was found to be characteristic of the coherent state—whether condensed or expanded) and the polar groups. An important check on the hydrocarbon chain interaction part of this free energy is provided by measurement of desorption rates, which may also be used to estimate the average cohesive energy per CH_2 group[20, 21]. In a very recent paper[83], Gershfeld and Pagano have applied these concepts to new measurements on several fatty acids, alcohols and other lipids, having both saturated and unsaturated chains. They have also calculated the enthalpies and entropies of compression from the temperature dependence of ΔF_c. They conclude that the hydrocarbon regions of the films are energetically similar to bulk hydrocarbon liquids, with the expanded films being slightly less condensed and the condensed films slightly more so than the bulk liquid phase. Polar group contributions to ΔF_c (which are opposite in sign to the hydrocarbon chain contributions) are significant, in most cases examined being larger in absolute magnitude than the non-polar part of the free energy. The entropy changes suggest that changes in water structure at the interface occur when the coherent state is reached.

1.3 SELECTED MONOLAYER-FORMING MATERIALS

1.3.1 'Small' molecules

Major interest in lipid and pigment monolayers has continued to be stimulated by the biological importance of such molecules in interfacial films. Certain interesting recent developments are discussed here.

1.3.1.1 Counter-ion effects in ionised monolayers

While it has long been recognised that specific counter-ion binding may occur with charged films, this phenomenon has been the subject of detailed study in several laboratories in the past few years. Goddard and co-workers[84] have studied the Π–A and surface potential effects of lithium, sodium, potassium, and tetralkylammonium cations on fatty acid (C_{18}–C_{22}) monolayers. Rosano et al.[85] extended these studies to include Rb^+, Cs^+ and NH_4^+. Patil, Matthews and Cornwell[86] have also studied Na^+ and K^+ ions at various pH and ionic strength. Shah[87] noted that certain buffers, notably tris–HCl, give anomalous results, and this effect was confirmed by Patil et al. On the other hand, unbuffered solutions lead to problems because of absorption of atmospheric CO_2 [84].

While there are some discrepancies in detail in these reports, perhaps due to these differences of pH control and co-ion effects, they broadly agree on certain features. On alkaline sub-phases (pH 9 and above), all of the fatty acid films are expanded at low pressures, the carboxyl groups presumably being completely ionised. On compression, a plateau develops; the surface pressure and area for the plateau depend on pH, ionic strength, and the nature of the specific counter-ion. This behaviour has been ascribed to a decrease in degree of ionisation as the film is compressed; experiments in which un-ionised stearyl alcohol was added to a stearic acid film support this interpretation, as do the ionic strength effects observed[86]. The cation condensing effect generally decreases with increasing (crystallographic) ionic radius – i.e. $Li^+ > Na^+ > K^+$, etc. – although exceptions have been noted.

In the case of the alkaline earth stearates, calcium appears to be anomalous, producing the greatest condensation, but the condensing effect of the other ions also showed a decrease with ionic size[88]. These effects, for both alkali and alkaline earth cations, have been discussed in terms of their similarity to the ion sequences observed with ion exchange resin or glass electrode selectivity, soap micelle ion binding, and other phenomena.

Counter-ion selectivities have also been reported for long-chain amines and sulphates[89], nonadecyl benzene sulphonate[90], and alkyl phosphates[91–93]. The benzene sulphonate exhibited the same cation sequence as the fatty acids, but that for docosyl sulphate monolayers is reversed.

Smith and Serrins[94] have made a detailed study of the reaction of Cu^{2+} with erucic acid monolayers; they were particularly interested in the process of monolayer collapse, which is augmented in the presence of cupric ions.

1.3.1.2 Phospholipids

Studies of monolayers of phospholipids are numerous. Careful examination of synthetic single-component films in the past decade has enhanced our understanding of these materials, although some discrepancies between results obtained in different laboratories[95] still suggest that the purity of these preparations may be difficult to control. Because of their biological implications, a large number of studies of mixed monolayers involving phospholipids have been carried out; quantitative interpretation of such work, unfortunately, remains rather uncertain (cf. Section 1.4).

Aside from their biological importance, interest in the monolayer properties of phospholipids stems from the zwitterionic nature of the polar headgroups in the lecithins and cephalins. While the orientation of these zwitterionic groups has been the subject of much discussion[96-98], no firm conclusions can yet be drawn on how they are arrayed at the air–water interface. As might be expected, the presence of zwitterions produces a variety of pH and ionic strength effects which are not well understood. In view of the specific ion effects noted with molecules having only one ionisable group (Section 1.3.1.1), it may well be that sorting out these effects with zwitterionic surfactants will require much more study, although several papers have given an indication of the specificities which exist[91, 99, 100]. Cadenhead's review[6] describes recent studies in considerable detail.

1.3.1.3 Porphyrins and chlorophyll

The methyl esters of several different metal-free porphyrins having different distributions of polar groups around the porphyrin ring form monolayers with particularly interesting spectroscopic properties[101]. The strong interactions between the chromophores in these rigid, close-packed films produce large distortions of the monolayer visible absorption spectra, as compared with the spectra of solutions. A particularly striking case is provided by two isomers of co-proporphyrin tetramethyl ester, which in bulk are spectroscopically indistinguishable, but whose monolayer spectra differ so much that the films even appear different in colour to the naked eye.

Chemical reactions of chlorophyll[102] and its magnesium-free derivative, pheophytin[103], in monolayers have been studied. These results provide further evidence of the considerable care required in handling these labile compounds. An interesting study of mixed monolayers of chlorophyll with other chloroplast lipids has been reported[65, 104]. While $\Pi-A$ data do not indicate any specific interactions, fluorescence polarisation studies suggest that certain lipids can affect the orientation of the chromophore.

1.3.1.4 Miscellaneous small molecules and sub-phase effects

Most studies of monolayers of non-polymeric materials continue to be concerned with long-chain substituted compounds. Some additional examples include alkyl phosphates[92, 93] quinones[25], quaternary salts[211], phosphines[105], and long-chain esters of dihydroxyacetone and phenols[106]. However, there is also interest in other types of structures. Sterols and porphyrins have been remarked on elsewhere in this review. In addition relatively rigid cyclic structures which have been examined or re-examined include 4-amino-p-terphenyl[72] (cf. Section 1.2.6), abietic acid[107], cyclopentadecanone[108], and the cyclic polypeptide antibiotic, alamethicin[109]. Smith[110] has reported studies of monolayers of naphthalene, anthracene, cholesterol, and several adamantane derivatives on mercury surfaces.

Cadenhead and his co-workers have examined the effect of mixed sub-phases on monolayer characteristics. The addition of glycerol[111] to the aqueous sub-phase has a relatively small effect on the tension of the clean surface. Condensed monolayers spread on aqueous glycerol solutions show little difference in Π–A characteristics from those observed on water, but the measured surface potentials are lower. Expanded films, on the other hand, become much more expanded on glycerol-containing sub-phases. Ethanol additions[112], which change the surface tension of water appreciably, show comparable effects, with a superimposed surface pressure reduction; the surface pressure change also displaces the observed Π–A curves for condensed films such as stearic acid.

Strong (4 M) salt solutions have also been reported[113] to have significant effects on the Π–A curves of octadecanol.

1.3.2 Macromolecules

Major problems still exist in the definition of materials of this class. Preparations of both biological polymers and synthetic macromolecules are seldom sufficiently well characterised that artifacts due to the presence of small molecule fragments, polymerisation catalysts, or other contaminants can be easily ruled out. Kinetic effects (i.e. partial collapse, incomplete spreading, or irreversibility of compression) are often not examined; this reviewer remains sceptical, therefore, of much of the interpretation which is attempted. Nevertheless, much work has been reported, and a selection of interesting results is catalogued here.

1.3.2.1 *Proteins and polypeptides*

A valuable review by Loeb[114] summarised work in this area up to mid-1965, and presented a detailed critique of the state of knowledge at that time. Since then, combinations of techniques have shed further light on the behaviour of monolayers of both synthetic and natural polypeptides.

Malcolm[115] has studied a number of systems, combining conventional Π–ΔV–A measurements with infrared spectroscopy of collapsed films and hydrogen–deuterium exchange studies. Loeb[116] has used multiple internal reflection infrared spectoscopy to examine monolayers deposited (without collapse) on germanium prisms. These studies show that certain polypeptides form ordered arrays of α-helices at the air–water interface, and that these structures are resistant to compression, persisting in the collapsed film. On the other hand, some compounds, such as β-lactoglobulin[116c], exhibit a change in conformation on compression. Changes of temperature and sub-phase composition may also lead to such conformational changes[117]. Surface potential measurements[118] have been used to provide interpretations of these effects. It then appears that 'surface denaturation' of proteins – long invoked to interpret monolayer behaviour – may be a complicated combination of processes, different for different molecules. A variety of forms of protein 'denaturation' in bulk are also now recognised[119].

Despite ambiguities about monolayers of these compounds, their great biological importance prompts further study of many of their properties.

For example, a recent study has compared elasticity (in cyclic small amplitude compression–expansion) of monolayers of bovine serum albumin and ovalbumin with the behaviour predicted from the Π–A isotherm of a slowly compressed film[120]. At pressures below 5 dyn cm^{-1}, the two observations are concordant if the amplitude of the cyclic compression is low, but at higher pressures differences appear. Some other work on such films is noted in Section 1.8.

1.3.2.2 Synthetic polymers

In addition to polypeptides, synthetic polymers unrelated to biological systems continue to provoke interest. Poly(siloxanes), in particular, have been extensively studied. Noll, Steinbach and Sucker[121] have extended their earlier studies on methyl siloxanes to polymers in which the silicon atoms bear a variety of groups, including n-alkyl, vinyl, halogenated methyl or epoxide-containing chains. Bernett and Zisman[122] have reported surface potential measurements as well as Π–A curves for several methyl-, n-alkyl- and methyltrifluoropropyl- siloxanes. Characteristics of the monolayers on water fall into several classes, depending on the length, structure, and polarity of the side-chain substituents. In nearly all cases, however, it appears that at low surface pressures the siloxane chains spread out completely at the air–water interface. On compression, various reorientations, of both side chains and the siloxane backbone, can account for the observed behaviour. Some siloxanes also form insoluble monolayers on various organic liquid sub-phases[123].

Wave-damping by siloxane films spread on water shows most interesting successions of maxima and minima as the films are compressed[60]. These results seem qualitatively consistent with the sequence of changes (helix formation and reorientation) which have been postulated to account for inflections in the Π–A curves[124], but no quantitative analysis is available.

Willis[125] has carried out a detailed study of films of poly(dimethylsiloxane)-substituted 11-undecanoic acids. These compounds orient horizontally at low surface pressures, but on compression the siloxane chains are forced out of the interface, the carboxyl groups remaining in the water.

Block copolymers of dimethyl siloxane with organic polymers which are themselves not spreadable on water form monolayers in which the siloxane chains spread, but the organic portion appears to occupy no area at the interface[126].

Other well-known spreadable polymers which have recently been investigated in more detail are poly(vinyl acetate) and poly(methyl methacrylate). For the latter, both compression rate effects[24] and tacticity-dependent phenomena[127] have been reported. Branched and linear poly(vinyl acetates) have been examined in monolayers[128]. A variety of copolymers, graft-modified non-polar polymers, and mixtures involving these types of structures have been studied[129–131]. Shuler and Zisman[63] have reported that poly(oxyethylene) of high molecular weight forms stable monolayers, and they examined wave-damping by these films, as well as their Π–ΔV–A characteristics. Cellulose[132] and some of its derivatives[133] have also been spread. The extent of ionisation of monolayers of polyelectrolytes such as poly(methacrylic acid) has been inferred from measurements of surface potential and pressure[134].

The polymerisation of monomers spread at a liquid–gas interface to produce two-dimensional polymer molecules continues to be a goal[135, 136]. Until someone is willing to take the trouble to collect the 'polymerised' monolayer and analytically demonstrate the existance of macromolecular species derived from the spread monolayer, however, such studies can be considered suggestive, rather than definitive.

The behaviour of polymers which bear long-chain substituents along the backbone chain, when spread as monolayers, may be intermediate between those of conventional long-chain lipids and unsubstituted polymers. Two interesting papers[137] have described the rate of two-dimensional crystallisation (due to the side chains) of poly(n-octadecyl methacrylate) monolayers.

The possibility of determining the molecular weight of a polymer from Π-A measurements at sufficiently low surface pressures[138] continues to be interesting. Romeo and Rosano[139] have reported measurements on monolayers of a lipo-polysaccharide at pressures of 0.01–0.08 dyn cm^{-1} which seem to give satisfactory results, as compared with ultracentrifuge data.

1.4 MIXED FILMS

Mixtures of different film-forming molecules in spread monolayers are important for a number of reasons. In applications such as evaporation control, multi-component films may be used either accidentally (because commercial surfactant preparations may be poly-disperse) or deliberately (for example, to enhance the rate of spreading). Biological membranes, of course, contain a variety of chemical species, and model studies of mixed monolayers have a long history of relevance to biological problems. Many technologically important systems involve complex mixtures at interfaces, for which monolayers also provide models. (An interesting example is provided by comparisons of foam stability with monolayer interactions[140].)

The fundamental interpretation of effects observed when films containing more than one component are spread, however, has remained a matter of controversy. Two types of limiting models – one in which essentially enthalpic interactions between molecular species in the film are invoked, the other athermal in nature and related solely to ordering or entropic effects – have been much discussed. There is, of course, no *a priori* reason that both types of phenomena may not be involved in particular systems.

A basic thermodynamic analysis of mixed film behaviour was developed by Crisp[141] and Goodrich[142]; the author has summarised and indicated some extensions of this treatment[143]. While there seems no reason to doubt the basic validity of the analysis, some details of its application have been subject to question. (In fact, most of these points were made by the original authors, but have been ignored by some subsequent workers.) In the first place, the calculation of a free energy of mixing by integration of the Π-A curves of the mixed and single component films requires that equilibrium data be obtained to low enough surface pressures that the film molecules exert no influence on one another. This point has been re-emphasised recently[144], and Pagano and Gershfeld[83, 145] have demonstrated its importance experimentally by careful measurements of Π-A data out to areas corresponding to gaseous film behaviour. These results strongly suggest that estimates of free energies of mixing which have been made without such very low pressure data may be grossly in error. Pagano and Gershfeld[145] have also demonstrated the utility of the coherent–gaseous surface phase transition, where the surface vapour pressure, Π_v, is observed, to demonstrate miscibility in spread films. By an argument analogous to Crisp's, they show that Π_v for a 'mixture' will be independent of composition where the components are immiscible, but will vary with composition for truly mixed monolayers.

Another difficulty which has not been treated relates to the definition of 'ideality' in two-dimensional mixtures. All treatments to date have considered ideal mixtures as those which obey mole fraction additivity relations; thermodynamic analysis, in particular, has been based on the assumption that in ideal mixtures the chemical potentials of the components are given by $\mu_i = \mu_i^0 + RT\ln N_i$, where N_i is the mole fraction of component i. Such a definition, however, is based on a random statistical analysis of molecules of identical size. It is well known[146] that for mixtures of molecules of different sizes, athermal mixing does not lead to mole fraction average behaviour. (While the difference is not large for molecules whose sizes differ only slightly, it becomes very important, for example, in mixtures of polymers with small molecules.)

In view of these problems, much of the interpretation of mixed monolayer studies remains somewhat uncertain. Another difficulty which reflects on much work which has been reported relates to the instability of cholesterol when spread, noted in Section 1.2.1 (cholesterol, of course, is a favourite component in mixed-film studies because of its biological importance and its interesting 'condensing effect' on many lipid films). Gershfeld and Pagano[147] have attempted to deal with the homogeneity and equilibration problems by using lipid-saturated solutions as sub-phases; unfortunately, they did not recognise the chemical degradation difficulty, so their quantitative results are of uncertain significance. Chemical instability in monolayers has also been observed for cholesteryl acetate[26] and esters of cholesterol with unsaturated fatty acids[148]. Chapman et al.[149] and Cadenhead, in his review article[6], have summarised the large number of studies of mixed phospholipid–cholesterol monolayers.

Joos[150] has extended Crisp's analysis of mixed monolayer collapse pressures to include cases where interactions or immiscibility in the bulk collapsed material occur, and where the two components have not very different collapse pressures. He has used a similar analysis to calculate the equilibrium spreading pressure of mixed crystals[151].

1.5 MONOLAYERS ON SOLIDS; MULTILAYERS

Studies on films transferred from liquid–gas interfaces to solid surfaces by some variant of the Langmuir–Blodgett technique continue to abound. Such work may be intended either to elucidate the nature of the monolayer itself, or to form structures (generally multilayers) which have interesting properties themselves. In the former case, there is always some uncertainty about how exactly the deposited film represents the monolayer which existed on the liquid surface; in the latter, details of the deposition process may affect the resulting structure in many ways.

Spink[152] examined the deposition of stearic acid on several different solids, including mica, glass and metals, using autoradiographic, electron microscopy and electron diffraction techniques. In the case of silver, he found significant differences between crystal faces; the deposition ratio was as low as 0.5 on 111 faces under certain conditions. In another paper[153], he examined the rate of vacuum desorption of the same deposited monolayers. With mica substrates, most monolayer desorption occurred with an activation energy similar to that for the sublimation of bulk stearic acid; a much lower activation energy, and lower desorption rate, was observed from silver surfaces. He re-emphasised the fact that details of the deposition and subsequent removal are affected by a very large number of experimental variables. Petzny and Quinn[154] reported that when barium stearate monolayers were deposited on mica sheets with small pores through them, the soap molecules diffused into the pores, thereby reducing their radii in a controlled manner. In a related study[155], the gas permeability through thin polymer membranes covered with stearate multilayers was measured. Both stearate and polymer [e.g. poly(γ-benzyl glutamate)] monolayers have been deposited on wet gel cellulose[156], and the resulting membranes appear to have reduced permeability to salts in aqueous solution. The structure of these composite films has not been characterised in detail, however.

A number of newer analytical techniques have been applied to built-up films in recent years. Photoelectron spectroscopy has been shown to be capable of elemental analysis in such layers, and is sensitive to small composition variation, as for example, when two layers of stearic acid were deposited on top of 200 layers of α-bromostearic acid[157]. Attenuated total reflection (ATR) of infrared radiation has been used to examine a variety of monolayers built up on appropriate prisms; substances examined have included alkyl phosphates[93], polypeptides[116, 158], and stearates[159].

The use of multilayers as gratings for x-ray spectrometry continues. Frans and Davidson[160] have discussed the optimum thickness of such films. Charles[161] has examined both the use of x-ray methods for studying multilayer structure and the effect of varying amounts of soap formation in lead stearate films on the x-ray reflection characteristics.

An extensive and important series of papers from H. Kuhn and his co-workers has been concerned with optical and electrical properties of built-up films. (In addition, reviews of the work have been given periodically[162], and a popular review has also appeared[163].) These workers have found that cadmium arachidate (C_{20} saturated fatty acid) gives the best balance of deposition ease and stability properties to make stable 'spacer' layers; a detailed account of their multilayer deposition techniques has been published[164]. In the same paper, the miscibility of arachidic acid with a cyanine dye and a merocyanine dye, both with C_{18} chains, was studied. It was found that the former gave mixed films with the fatty acid, while the latter did not.

The use of cadmium arachidate layers as spacers has permitted quantitative study of energy-transfer processes occurring with a variety of dye molecules; an interesting recent study has been concerned with energy transfer from dye molecules into silver bromide (leading to photographic sensitisation)[165]. Mixed monolayers of the stearyl substituted cyanine dyes with octadecane can also be formed; apparently the dye chromophores assume a regular packing which optimises interactions between the added hydrocarbon and the side chains of the dye[166].

Photo-induced structural changes in some dye monolayers on the water surface have also been inferred from studies of $\Pi-\Delta V-A$ properties on illumination[167].

Molecular motion through deposited monolayers has been assessed in experiments using both radioactive stearic acid and ^{45}Ca to label the counter ion[168]. Several techniques involving the deposition of protein monolayers have been used to permit collection of nucleic acids from solution for visualisation by electron microscopy[169]. Such procedures can also be used to estimate diffusion constants in solution[170]; this application is reminiscent of one described by Trurnit in 1954[171].

1.6 REACTIONS IN MONOLAYERS

In earlier sections, note has been made of certain reactions, especially oxidative degradation, which may present problems with spread monolayers of labile molecules. Such phenomena as solution from a spread film[20-22] or slow phase transitions[137] may also be considered 'reactions'. In the same class are spreading rate studies, which have largely been performed in connection with practical interest in evaporation control (cf. Section 1.7). The interaction of counter ions with ionised monolayers can also be examined as a rate process, although little work of this sort has been reported.

The hydrolysis of ester[172] and glyceride[173] monolayers has been studied with the aid of radioactive tracers. For the glyceride, of course, there are successive steps in the reaction as the three ester links in each molecule are broken; paper chromatography of material collected from the spread film permits determination of the various reaction products. A somewhat more complex problem is the hydrolysis of a polymer with ester side chains spread in a monolayer; a brief report on the hydrolysis of poly(methyl methacrylate) has appeared[174].

Enzyme-catalysed hydrolyses are also of great interest. In the case of glyceride hydrolysis by pancreatic lipase, identification of reaction products by thin-layer chromatography has been used to verify conclusions based on surface pressure measurements[175]; excellent linear rate plots were obtained. The hydrolysis of a lecithin monolayer by phospholipase has also been studied, using a tritium-labelled lecithin preparation[176]. Shah and Schulman[177] have reported on the hydrolysis of various lecithin monolayers by snake venom; a variety of cation and buffer composition specificities were observed.

Studies of the 'penetration' of spread monolayers by dissolved surface active materials continue to be reported. Typical studies include those of psychoactive drugs into lipid films[178], polyene antibiotics into cholesterol monolayers[179], serum proteins into sphingolipid films[180], bile salts into monoglyceride films[181], and polypeptide hormones into cholesterol and stearic acid monolayers[182].

When fatty acid and alcohol monolayers on water or various salt solutions are irradiated with ultraviolet light (wavelength ~ 2400 Å) in air, a variety of effects are observed[183]. Stearyl alcohol, which does not absorb at this wavelength, is not degraded. Stearic acid is oxidised, and Cu^{2+} ions in solution, for example, are catalytic and alter the observed kinetics.

1.7 PERMEATION THROUGH MONOLAYERS; EVAPORATION CONTROL

The practical use of monolayers to reduce evaporation from water storage reservoirs continues to encourage a wide variety of studies related to this application. In addition to more conventional powder spreading techniques or the use of solutions in volatile solvents, it has been shown that monolayers of cetyl alcohol (hexadecanol) can be spread from the vapour phase[184, 185]. Such a process, of course, also makes it possible to deposit monolayers on water droplets in a fog or aerosol, and some attention has been paid to monolayers on droplets and their effect on evaporation[185–188].

The rate at which a film spreads at the air–water interface is of major importance in evaporation control applications, and spreading rate enhancement is a desirable goal. Saylor and Barnes[189] have found that the incorporation of n-decane into solid spreading sources of fatty alcohols increases the spreading rate; the hydrocarbon evaporates from the mixed film after spreading has occurred[17]. The removal of spread cetyl alcohol molecules by a drop of solvent has been examined by Avetisyan and Trapeznikov[190]; they[191] have also studied the effect of waves on evaporation retardation.

In addition to the fatty alcohols, their derivatives such as alkoxy ethanols show promise for evaporation control. A series of papers by Katti et al.[192] has reported on the properties of monolayers of these substances. The rates of spreading and equilibrium spreading pressures of the alkoxy propanols and butanols are higher than those of the corresponding n-alcohols, but less than those of the alkoxy ethanols. As evaporation retardants, it was concluded that the alkoxy ethanols are the most efficient.

The importance of various mechanisms by which monolayers affect the overall evaporation rate under practical conditions continues to be a subject of controversy. MacRitchie[193] has suggested that the major effect, for cetyl alcohol monolayers, is not the inherent resistance of the monolayer but an increase in the thickness of the diffusion boundary layer in the air near the surface. His analysis and experiments, however, have been criticised[194, 195], and whether this effect is important under practical field conditions remains an open question.

Barnes et al.[196] have shown, on the other hand, that a statistical calculation based on an energy barrier model can give reasonable agreement with experimental values. In their calculation, it is assumed that holes in the monolayer, produced by thermal agitation, permit evaporation of water at the clean-surface rate; the absolute evaporation rate, therefore, is proportional to the (instantaneous) area fraction of holes.

Other phenomena related to evaporation reduction which have been studied include concentration of solutes near the surface[197], and the effect of surfactant molecules on super heating of water held near its normal boiling point[198]. The latter phenomenon appears to be extremely sensitive in certain cases; some long-chain fatty acids seem to have appreciable effects even at concentrations of one molecule per 5000 Å^2.

The possibility of reducing evaporation of volatile organic liquids has been re-examined by Bernett et al.[66]. A variety of partially fluorinated compounds, either adsorbed or spread on surfaces of toluene, 2,2,4-trimethylpentane, and nitromethane, were studied; none had any effect on evaporation rate when present as monomolecular films.

The transport of CO_2 through monolayers of long-chain alcohols[199] as well as ionised monolayers[200], has been studied by Hawke and his co-workers, using radio-tracer methods. These results also support the idea that the monolayers provide an intrinsic energy barrier to transport of small molecules.

1.8 SOME ADDITIONAL IMPLICATIONS AND APPLICATIONS

A variety of studies related to biological problems have been remarked throughout this review; it is clear that spread monolayers continue to provide important models for living systems. As detailed knowledge of the composition and structure of the natural units becomes available, it is to be expected that such model studies will become increasingly useful. It is interesting to note in this connection recent results from other disciplines[201,202] which support the idea, long ago inferred from monolayer studies, that lipids in living membrane systems have an essentially liquid-like character. It is obvious, however, that the value of any model studies will depend on combining sound physical chemistry with detailed analytical information.

An interesting biological problem[203,204] which has been the subject of much study by monolayer techniques is the nature and function of the lining of the alveolar surfaces of the mammalian lung. When extracted from the lung and spread at the air–water interface, the materials which constitute the lining have been found to exhibit compression–expansion hysteresis which bears some resemblance to the pressure hysteresis during respiration which maintains the integrity of the lung. Another question, whose origins date to 1925[205], relates to the area occupied by lipids in the surface of red blood cells; this matter has recently been re-examined[206]. Unfortunately, there is still no way to account quantitatively for the protein components of the membrane.

Monolayers on very large water surfaces, e.g. the ocean, have been of recent interest. The presence of such films, often from natural sources or accidental contamination, may affect a variety of processes occurring at the surface[207]. An interesting example is the effect of monolayers on salt nuclei produced in the air from breaking bubbles in sea water; up to three-fold increases in nuclei concentration were observed in the presence of insoluble films[208]. It has also been suggested that deliberately spread monolayers may be useful in controlling oil pollution from spills on water surfaces[209].

Analytical applications of monolayer techniques continue to be developed. An example is provided by Robb's determination of the solubility of vinyl stearate in water[210]. The equilibrated aqueous solution was extracted with n-hexane, and the organic solute estimated quantitatively by spreading.

Note added in proof

Several significant papers appeared after the completion of this manuscript, and are noted briefly here.

Eriksson ((1971). *J. Colloid Interface Sci.,* **37,** 659) has presented a new thermodynamic analysis of insoluble monolayers (both one- and two-component). His formalism uses the condition of equilibrium spreading (at Π_e) as the reference state, and the thermodynamic properties of the film are evaluated by integration of the $\Pi-A$ curve from that point.

An important symposium on 'Rheology at Interfaces' has contributed to the understanding of surface viscosity and ripple damping. Goodrich and Allen ((1971). *J. Colloid Interface Sci.,* **37,** 68) have provided an analysis of the double ring viscometer, in which one ring rotates in the surface while the concentric stationary ring is the torque sensor. Lifshutz, Hegde, and Slattery ((1971). *J. Colloid Interface Sci.,* **37,** 73) analysed the double knife-edge viscous traction instrument. Pintar, Israel and Wasan ((1971). *J. Colloid Interface Sci.,* **37,** 52) extended the Burton and Mannheimer analysis of the deep-channel viscometer to include a curved interface, gap between the channel walls and floor, and non-Newtonian flow. Three papers considered the effects of additional terms in the interfacial boundary conditions on the ripple damping equations (Mann, J. A. and Du, G. (1971). *J. Colloid Interface Sci.,* **37,** 2; Mann, J. A., Baret, J. F., Dechow, F. J. and Hansen, R. S., ibid., **37,** 14; Maver, E. and Eliassen, J. D., ibid., **37,** 228). The second of these also describes the rudiments of a new technique for studying interfacial viscoelastic properties at high frequencies by laser beat frequency spectroscopy.

A most elegant demonstration of the success of careful multilayer deposition techniques in building spacer layers of cadmium fatty acid soaps has been reported. Mann and Kuhn ((1971). *J. Appl. Phys.,* **42,** 4398) obtained good agreement between theory and measured values for tunnelling currents and metal–dielectric work functions for such layers between metal electrodes; these measurements provide an extremely sensitive test for the perfection of the built-up soap layers.

References

1. Gaines, G. L., Jr. (1966). *Insoluble Monolayers at Liquid-Gas Interfaces.* (New York: Wiley-Interscience)
2. Adamson, A. W. (1967). *The Physical Chemistry of Surfaces,* Second Edn. Chap. 3. (New York: Wiley-Interscience)
3. Mingins, J. and Standish, M. M. (1966). *Ann. Rep. Chem. Soc. (London),* **63,** 91
4. Lange, H. (1967). in *Non-Ionic Surfactants,* M. J. Schick, Ed., 443. (New York: Marcel Dekker)
5. Shah, D. O. (1970). *Advan. Lipid Res.,* **8,** 347

6. Cadenhead, D. A. (1970). in *Recent Progress in Surface Science,* Danielli, J. F., Riddiford, A. C., Rosenberg, M. D., Eds, **3,** 169 (New York: Academic Press)
7. Rosoff, M. (1969). in *Physical Methods in Macromolecular Chemistry,* Carroll, B., Ed., **1,** 1 (New York: Marcel Dekker)
8. Pilpel, N. (1969). *Advan. Colloid and Interface Sci.,* **2,** 261
9. Popiel, W. J. (1970). *Sci. Progr. (Oxford),* **58,** 237
10. Gershfeld, N. L. (1972), in *Techniques of Surface and Colloid Chemistry,* Good, R. J., Patrick, R. L. and Stromberg, R. R., Eds. (New York: Marcel Dekker)
11. Matuura, R. and Motomura, K. (1967). *Kagaku To Kogyo (Tokyo),* **20,** 87; through *Chem. Abstr.,* **66,** 62925
12. Castillo, M. (1968). *Rev. Real Acad. Farm. Barcelona,* No. 12, 89; through *Chem. Abs.,* **73,** 134162
13. Joly, M. (1966). *Abh. Deut. Akad. Wiss. Berlin, Kl. Chem., Geol., Biol.,* 683; through *Chem. Abstr.,* **67,** 91118
14. *Proc. Fifth Int. Cong. of Surface Activity* (1969). (Barcelona: Ediciones Unidas, S. A.)
15. Giles, C. H. (1969). *Chem. and Ind. (London),* 1616; Giles, C. H. and Forrester, S. D., ibid., (1970), 80; (1971), 43
16. Mingins, J., Owens, N. F. and Iles, D. H. (1969). *J. Phys. Chem.,* **73,** 2118
17. Barnes, G. T., Elliot, A. J. and Grigg, E. C. M. (1968). *J. Colloid Interface Sci.,* **26,** 230
18. Pagano, R. E. and Gershfeld, N. L. (in press). *J. Colloid Interface Sci.*
19. McAuliffe, C. (1966). *J. Phys. Chem.,* **70,** 1267; (1969). *Science,* **163,** 478
20. Gershfeld, N. L. and Patlak, C. S. (1966). *J. Phys. Chem.,* **70,** 286; Patlak, C. S. and Gershfeld, N. L. (1967). *J. Colloid Interface Sci.,* **25,** 503
21. Gershfeld, N. L. (1968). *Advan. Chem. Series,* **84,** 115
22. Heikkila, R. E., Kwong, C. and Cornwell, D. G. (1970). *J. Lipid Res.,* **11,** 190; Heikkila, R. E., Deamer, D. W. and Cornwell, D. G. (1970), ibid., **11,** 195
23. Motomura, K., Shibata, A., Nakamura, M. and Matuura, R. (1969). *J. Colloid Interface Sci.,* **29,** 623
24. Sutherland, J. E. and Miller, M. L. (1970). *J. Colloid Interface Sci.,* **32,** 184
25. Gaines, G. L., Jr. (1968). *J. Colloid Interface Sci.,* **28,** 334
26. Kamel, A. M., Weiner, N. D. and Felmeister, A. (1971). *J. Colloid Interface Sci.,* **35,** 163
27. Kawanishi, T., Seimiya, T. and Sasaki, T. (1970). *J. Colloid Interface Sci.,* **32,** 622
28. Pike, F. P. and Bonnet, J. C. (1970). *J. Colloid Interface Sci.,* **34,** 597
29. Princen, H. M. (1970). *Aust. J. Chem.,* **23,** 1789
30. Lane, J. E. and Jordan, D. O. (1970). *Aust. J. Chem.,* **23,** 2153; (1971), ibid., **24,** 1297
31. Rogeness, G. A. (1968). *J. Colloid Interface Sci.,* **26,** 131
32. Gershfeld, N. L., Pagano, R. E., Friauf, W. S. and Fuhrer, J. (1970). *Rev. Sci. Inst.,* **41,** 1356
33. Frommer, M. A. and Miller, I. R. (1965). *Rev. Sci. Inst.,* **36,** 707
34. Poulsen, B. J. and Lemberger, A. P. (1965). *J. Pharm. Sci.,* **54,** 875
35. Cadenhead, D. A., Demchak, R. J. and Phillips, M. C. (1967). *Koll. Z.,* **220,** 59
36. Suzuki, A., Ikeda, S. and Isemura, T. (1967). *Ann. Rep. Biol. Works, Fac. Sci., Osaka Univ.,* **15,** 83; through *Chem. Abstr.* **69,** 89879
37. Smith, T. (1968). *J. Colloid Interface Sci.,* **26,** 509
38. Vroman, L., Kanor, S. and Adams, A. L. (1968). *Rev. Sci. Inst.,* **39,** 278
39. Mendenhall, R. M. (1971). *Rev. Sci. Inst.,* **42,** 878
40. Jarvis, N. L. and Scheiman, M. A. (1968). *J. Phys. Chem.,* **72,** 74
41. Bergeron, J. A. and Gaines, G. L., Jr. (1967). *J. Colloid Interface Sci.,* **23,** 292
42. Plaisance, M. and Ter-Minassian Saraga, L. (1970). *Compt. Rend. Acad. Sci Paris,* **270,** 1269
43. Mannheimer, R. J. and Burton, R. A. (1970). *J. Colloid Interface Sci.,* **32,** 73
44. Burton, R. A. and Mannheimer, R. J. (1967). *Advan. Chem. Series,* **63,** 315
45. Mannheimer, R. J. and Schechter, R. S. (1967). *J. Colloid Interface Sci.,* **25,** 434; (1968), ibid., **27,** 324
46. Mannheimer, R. J. and Schechter, R. S. (1970). *J. Colloid Interface Sci.,* **32,** 195
47. Mannheimer, R. J. and Schechter, R. S. (1970). *J. Colloid Interface Sci.,* **32,** 212
48. Mannheimer, R. J. and Schechter, R. S. (1970). *J. Colloid Interface Sci.,* **32,** 225
49. Ewers, W. E. and Sack, R. A. (1954). *Aust. J. Chem.,* **7,** 40; Davies, J. T. (1957). *Proc. Second Intl. Cong. of Surface Activity,* **1,** 220 (London: Butterworths)
50. Hegde, M. G. and Slattery, J. C. (1971). *J. Colloid Interface Sci.,* **35,** 593

51. Osborne, M. F. M. (1968). *Koll. Z.*, **224**, 150
52. Goodrich, F. C. (1969). *Proc. Roy. Soc.*, **A310**, 359
53. Goodrich, F. C. and Chatterjee, A. K. (1970). *J. Colloid Interface Sci.*, **34**, 36
54. Goodrich, F. C., Allen, L. H. and Chatterjee, A. K. (1971). *Proc. Roy. Soc.*, **A320**, 537
55. Lucassen-Reynders, E. H. and Lucassen, J. (1969). *Advan. Colloid Interface Sci.*, **2**, 347
56. Mann, J. A. and Hansen, R. S. (1963). *J. Colloid. Sci.*, **18**, 757, 805
57. Lucassen, J. and Hansen, R. S. (1966). *J. Colloid Interface Sci.*, **22**, 32
58. Bendure, R. L. and Hansen, R. S. (1967). *J. Phys. Chem.*, **71**, 2889
59. Hansen, R. S., Lucassen, J., Bendure, R. L. and Bierwagen, G. P. (1968). *J. Colloid Interface Sci.*, **26**, 198
60. Garrett, W. D. and Zisman, W. A. (1970). *J. Phys. Chem.*, **74**, 1796
61. Battezzati, M. (1970). *J. Colloid Interface Sci.*, **33**, 24
62. Hegde, M. G. and Slattery, J. C. (1971). *J. Colloid Interface Sci.*, **35**, 183
63. Shuler, R. L. and Zisman, W. A. (1970). *J. Phys. Chem.*, **74**, 1523
64. Smith, T. (1967). *J. Opt. Soc. Amer.*, **57**, 1207; (1968), ibid., **58**, 1069
65. Trosper, T., Park, R. B. and Sauer, K. (1968). *Photochem. Photobiol.*, **7**, 451
66. Bernett, M. K., Halper, L. A., Jarvis, N. L. and Thomas, T. M. (1970). *I & EC Fundamentals*, **9**, 150
67. Sakata, E. K. and Berg, J. C. (1969). *I & E C Fundamentals*, **8**, 570
68. Ries, H. E., Jr. and Gabor, J. (1966). *Nature (London)*, **212**, 917
69. Heller, S. (1966). *Naturwissenschaften*, **53**, 429
70. Smith, T. (1967). *J. Colloid Interface Sci.*, **23**, 27
71. Motomura, K. and Matuura, R. (1969). *J. Colloid Interface Sci.*, **29**, 617
72. Cadenhead, D. A. and Demchak, R. J. (1968). *J. Chem. Phys.*, **49**, 1372
73. Motomura, K. (1967). *J. Colloid Interface Sci.*, **23**, 313
74. Robb, I. D. and Alexander, A. E. (1968). *J. Colloid Interface Sci.*, **28**, 1
75. Haydon, D. A. (1958). *J. Colloid Sci.*, **13**, 159
76. Lucassen-Reynders, E. H. (1966). *J. Phys. Chem.*, **70**, 1777
77. Vrij, A. (1965). *Kon. Vlaam. Acad. Wetensch., Lett. Shone Kunsten Belg., Kl. Wetensch., Collq. Grenslaagverschijnselen Vloeistoffilmen-Schuimen-Emulsies*, 13
78. Chattoraj, D. K. and Chatterjee, A. K. (1966). *J. Colloid Interface Sci.*, **21**, 159; Gershfeld, N. L. (1968), ibid., **28**, 240
79. Chattoraj, D. K. (1969). *J. Colloid Interface Sci.*, **29**, 407
80. Hachisu, S. (1970). *J. Colloid Interface Sci.*, **33**, 445
81. Payens, T. A. J. (1970). *J. Colloid Interface Sci.*, **33**, 480; Bell, G. M., Levine, S., Pethica, B. A. and Stephens, D., ibid., **33**, 482
82. Gershfeld, N. L. (1970). *J. Colloid Interface Sci.*, **32**, 167
83. Gershfeld, N. L. and Pagano, R. E. (1972). *J. Phys. Chem.*, in the press
84. Goddard, E. D., Smith, S. R. and Kao, O. (1966). *J. Colloid Interface Sci.*, **21**, 320; Goddard, E. D., Kao, O. and Kung, H. C. (1967) ibid., **24**, 297
85. Christodolou, A. P. and Rosano, H. L. (1968). *Advan. Chem. Series*, **84**, 210; Rosano, H. L., Christodolou, A. P. and Feinstein, M. E. (1969). *J. Colloid Interface Sci.*, **29**, 335
86. Patil, G. S., Matthews, R. H. and Cornwell, D. G. (1971). *J. Lipid Res.*, in the press
87. Shah, D. O. (1970). *J. Colloid Interface Sci.*, **32**, 570
88. Deamer, D. W., Meek, D. W. and Cornwell, D. G. (1967). *J. Lipid Res.*, **8**, 255
89. Goddard, E. D., Kao, O. and Kung, H. C. (1968). *J. Colloid Interface Sci.*, **27**, 616
90. Dreher, K. D. and Wilson, J. E. (1970). *J. Colloid Interface Sci.*, **32**, 248
91. Shah, D. O. and Schulman, J. H. (1965). *J. Lipid Res.*, **6**, 341
92. Hunt, E. C. (1969). *J. Colloid Interface Sci.*, **29**, 105
93. Müller, H., Friberg, S. and Hellsten, M. (1970). *J. Colloid Interface Sci.*, **32**, 132
94. Smith, T. and Serrins, R. (1967). *J. Colloid Interface Sci.*, **23**, 329; Smith, T., ibid., **25**, 443
95. Phillips, M. C. and Chapman, D. (1968). *Biochem. Biophys. Acta*, **163**, 301
96. Shah, D. O. and Schulman, J. H. (1967). *J. Lipid Res.*, **8**, 227
97. Standish, M. M. and Pethica, B. A. (1968). *Trans. Faraday Soc.*, **64**, 1113
98. Watkins, J. C. (1968). *Biochem. Biophys. Acta*, **152**, 293
99. Shah, D. O. and Schulman, J. H. (1967). *J. Lipid Res.*, **8**, 215
100. Papahadjopoulos, D. (1968). *Biochem. Biophys. Acta*, **163**, 240
101. Bergeron, J. A., Gaines, G. L., Jr. and Bellamy, W. D. (1967). *J. Colloid Interface Sci.*, **25**, 97
102. Aghion, J., Broyde, S. B. and Brody, S. S. (1969). *Biochemistry*, **8**, 3120

103. Ditmars, W. E., Jr. and Van Winkle, Q. (1968). *J. Phys. Chem.*, **72,** 39
104. Trosper, T. and Sauer, K. (1968). *Biochem. Biophys. Acta,* **162,** 97
105. Baaz, K. (1970). *Koll. Z.,* **236,** 154
106. Quintana, R. P., Garson, L. R. and Lasslo, A. (1969). *Can. J. Chem.,* **47,** 853; Quintana, R. P. and Owen, R. M. (1969). *J. Colloid Interface Sci.,* **29,** 692; Quintana, R. P. and Lasslo, A. (1970). ibid., **33,** 54
107. Cadenhead, D. A. and Phillips, M. C. (1967). *J. Colloid Interface Sci.,* **24,** 491
108. Cadenhead, D. A. and Demchak, R. J. (1971). *J. Colloid Interface Sci.,* **35,** 154
109. Chapman, D., Cherry, R. J., Finer, E. G., Hauser, H., Phillips, M. C., Shipley, G. G., and McMullen, A. I. (1969). *Nature (London),* **224,** 692
110. Smith, T. (1969). *J. Colloid Interface Sci..* **31,** 270
111. Cadenhead, D. A. and Demchak, R. J. (1967). *J. Colloid Interface Sci.,* **24,** 483; idem., (1969). *Biochem. Biophys. Acta,* **176,** 849
112. Cadenhead, D. A. and Osonka, J. E. (1970). *J. Colloid Interface Sci.,* **33,** 188
113. Healy, T. W. and Ralston, J. (1968). *Nature (London),* **220,** 1026
114. Loeb, G. I. (1965). *U.S. Naval Research Laboratory Report,* 6318
115. Malcolm, B. R. (1968). *Proc. Roy. Soc.,* **A305,** 363; (1970). *Amer. Chem. Soc. Polymer Preprints,* **11,** 1327; and references cited therein
116. Loeb, G. I. (a) (1968). *J. Colloid Interface Sci.,* **26,** 236; (b) (1969). ibid., **31,** 572; (c) (1970). *Amer. Chem. Soc. Polymer Preprints,* **11,** 1313
117. Jaffé, J., Ruysschaert, J. M. and Hecq, W. (1970). *Biochem. Biophys. Acta,* **207,** 11
118. Caspers, J., Ruysschaert, J. M. and Jaffé, J. (1970). *Amer. Chem. Soc. Polymer Preprints,* **11,** 1319
119. Tanford, C. (1968). *Advan. Protein Chem.,* **23,** 122
120. Blank, M., Lucassen, J. and van den Tempel, M. (1970). *J. Colloid Interface Sci.,* **33,** 94
121. Noll, W., Steinbach, H. and Sucker, C. (1970). *Koll. Z.,* **236,** 1; *Amer. Chem. Soc. Polymer Preprints,* **11,** 1341
122. Bernett, M. K. and Zisman, W. A. (1971). *Macromolecules,* **4,** 47
123. Jarvis, N. L. (1969). *J. Colloid Interface Sci.,* **29,** 649; idem., (1970). *Amer. Chem. Soc. Polymer Preprints,* **11,** 1334
124. Fox, H. W., Taylor, P. and Zisman, W. A. (1947). *Ind. Eng. Chem.,* **39,** 1401
125. Willis, R. F. (1971). *J. Colloid Interface Sci.,* **35,** 1
126. Gaines, G. L., Jr. (1970). *Amer. Chem. Soc. Polymer Preprints,* **11,** 1336
127. Sutherland, J. E. and Miller, M. L. (1969). *Polymer Lett.,* **7,** 871
128. Jaffé, J. and Ruysschaert, J.-M. (1968). *J. Polymer Sci., Part C,* **23,** 281
129. Labbauf, A. (1966). *J. Appl. Polymer Sci.,* **10,** 865; Labbauf, A. and Zack, J. R. (1971). *J. Colloid Interface Sci.,* **35,** 569
130. Hironaka, S. and Meguro, K. (1971). *J. Colloid Interface Sci.,* **35,** 367
131. Gabrielli, G., Pugelli, M. and Ferroni, E. (1970). *J. Colloid Interface Sci.,* **32,** 242, 657; **33,** 133; (1971). ibid.; **35,** 460; **36,** 401
132. Giles, C. H. and Agnihotri, V. G. (1967). *Chem. and Ind. (London),* 1874
133. Zatz, J. L. (1970). *J. Colloid Interface Sci.,* **33,** 465
134. Jaffé, J., Ruysschaert, J.-M. and Bricman, G. (1970). *J. Polymer Sci. Part A-2,* **8,** 817
135. Holliday, L. (1967). *Chem. Ind. (London),* 970
136. Beredjick, N. and Burlant, W. J. (1970). *J. Polymer Sci., Part A-1,* **8,** 2807
137. Nakahara, T., Motomura, K. and Matuura, R. (1966). *J. Polymer Sci., Part A-2,* **4,** 649; idem., (1967). *Bull. Chem. Soc. Japan,* **40,** 495
138. Ref. 1, pp. 270–1
139. Romeo, D. and Rosano, H. L. (1970). *J. Colloid Interface Sci.,* **33,** 84
140. Shah, D. O. and Dysleski, C. A. (1969). *J. Amer. Oil Chem. Soc.,* **46,** 645; Shah, D. O. (1970). *J. Colloid Interface Sci.,* **32,** 570
141. Crisp, D. J. (1949). *Surface Chemistry* (Supplement to *Research,* London), 23
142. Goodrich, F. C. (1957). *Proc. Second Intl. Congress of Surface Activity,* **1,** 85 (London: Butterworths)
143. Gaines, G. L., Jr. (1966). *J. Colloid Interface Sci.,* **21,** 315; also, Ref. 1, chap. 6
144. Shah, D. O. and Capps, R. W. (1968). *J. Colloid Interface Sci.,* **27,** 319
145. Pagano, R. E. and Gershfeld, N. L. (1972). *J. Phys. Chem.,* in the press
146. e.g. Defay, R., Prigogine, I., Bellemans, A. and Everett, D. H. (1966). *Surface Tension and Adsorption,* chap. 13 (New York: Wiley)
147. Gershfeld, N. L. and Pagano, R. E. (in press). *J. Phys. Chem.*

148. Kwong, C. N., Heikkila, R. E. and Cornwell, D. G. (1971). *J. Lipid Res.*, **12**, 31
149. Chapman, D., Owens, N. F., Phillips, M. C. and Walker, D. A. (1969). *Biochem. Biophys. Acta*, **183**, 458
150. Joos, P. (1969). *Bull. Soc. Chim. Belges*, **78**, 207; Phillips, M. C. and Joos, P. (1970). *Koll. Z.*, **238**, 499
151. Joos, P. (1971). *J. Colloid Interface Sci.*, **35**, 215
152. Spink, J. A. (1967). *J. Colloid Interface Sci.*, **23**, 9; *J. Electrochem. Soc.*, **114**, 646
153. Spink, J. A. (1967). *J. Colloid Interface Sci.*, **24**, 61
154. Petzny, W. J. and Quinn, J. A. (1969). *Science*, **166**, 751
155. Rose, G. D. and Quinn, J. A. (1967). *Science*. **159**, 636; *idem.*, (1968). *J. Colloid Interface Sci.*, **27**, 193
156. Miller, M. L., Sutherland, J. and Schmitt, T. (1968). *J. Colloid Interface Sci.*, **28**, 28
157. Larsson, K., Nordling, C., Siegbahn, K. and Stenhagen, E. (1966). *Acta Chem. Scand.*, **20**, 2880
158. Loeb, G. I. and Baier, R. E. (1968). *J. Colloid Interface Sci.*, **27**, 38
159. Takenaka, T., Nogami, K., Gotoh, H. and Gotoh, R. (1971). *J. Colloid Interface Sci.*, **35**, 395
160. Frans, R. P. and Davidson, F. D. (1965). *Rev. Sci. Inst.*, **36**, 230
161. Charles, M. W. (1969). *J. Colloid Interface Sci.*, **31**, 569; (1971). ibid., **35**, 167
162. Drexhage, K. H. and Kuhn, H. (1966). *Grundprobleme der Physik dunner Schichten*, Niedermayer, R. and Mayer, H. Eds, 339 (Göttingen: Vandenhoeck and Ruprecht); Bücher, H., Drexhage, K. H., Fleck, M., Kuhn, H., Möbius, D., Schäfer, F. P., Sondermann, J., Sperling, W., Tillman, P. and Wiegand, J. (1967). *Molec. Crystallogr.*, **2**, 199; Kuhn, H. (1967). *Naturwissenschaften*, **54**, 429; Kuhn, H. (1968). *Structural Chemistry and Molecular Biology*, Ed. by Rich, A. and Davidson, N. 566 (San Francisco: W. H. Freeman)
163. Drexhage, K. H. (1970). *Sci. Amer.*, **222**, No. 8, 108
164. Bücher, H., von Elsner, O., Möbius, D., Tillmann, P. and Wiegand, J. (1969). *Z. physik. Chem., N. F.* **65**, 152
165. Szentpály, L. v., Möbius, D. and Kuhn, H. (1970). *J. Chem. Phys.*, **52**, 4618
166. Bücher, H. and Kuhn, H. (1970). *Chem. Phys. Lett.*, **6**, 183; *Z. Naturforsch.*, **25b**, 1323
167. Möbius, D., Bücher, H., Kuhn, H. and Sondermann, J. (1969). *Ber. Bunsenges. Phy. Chem.*, **73**, 845
168. Deamer, D. W. and Branton, D. (1967). *Science*, **158**, 655; Adam, H. K. and Zull, J. E. (1970). *J. Colloid Interface Sci.*, **34**, 272
169. Kleinschmidt, A. K. (1968). *Methods in Enzymology*, Vol. XII, Nucleic Acids, Part B, Ed. by Grossman, L. and Moldave, K. 361 (New York: Academic Press)
170. Lang, D. and Coates, P. (1968). *J. Mol. Biol.*, **36**, 137
171. Trurnit, H. J. (1954). *Arch. Biochem. Biophys.*, **51**, 176
172. Muramatsu, M. and Ohno, T. (1971). *J. Colloid Interface Sci.*, **35**, 469
173. Charbonneau, J. E. and Kowkabany, G. N. (1970). *J. Colloid Interface Sci.*, **33**, 183
174. Jaffé, J., Bricman, G. and Berliner, C. (1967). *Polymer Lett.*, **5**, 153
175. Lagocki, J. W., Boyd, N. D., Law, J. H. and Kézdy, F. J. (1970). *J. Amer. Chem. Soc.*, **92**, 2923
176. Miller, I. R. and Ruysschaert, J.-M. (1971). *J. Colloid Interface Sci.*, **35**, 340
177. Shah, D. O. and Schulman, J. H. (1967). *J. Colloid Interface Sci.*, **25**, 107
178. Demel, R. A. and van Deenen, L. L. M. (1966). *Chem. Phys. Lipids*, **1**, 68
179. Demel, R. A., Crombag, F., van Deenen, L. L. M. and Kinsky, S. (1968). *Biochem. Biophys. Acta*, **150**, 1
180. Colacicco, G., Rapport, M. M. and Shapiro, D. (1967). *J. Colloid Interface Sci.*, **25**, 5
181. Dreher, K. H., Schulman, J. H. and Hofmann, A. F. (1967). *J. Colloid Interface Sci.*, **25**, 71
182. Snart, R. S. and Sanyal, N. N. (1968). *Biochem. J.*, **108**, 369
183. Pilpel, N. and Hunter, B. F. J. (1970). *J. Colloid Interface Sci.*, **33**, 615
184. Gaines, G. L., Jr. (1966). *Koll. Z.*, **210**, 150
185. Derjaguin, B. V., Fedoseyev, V. A. and Rosenzweig, L. A. (1966). *J. Colloid Interface Sci.*, **22**, 45
186. Snead, C. C. and Zung, J. T. (1968). *J. Colloid Interface Sci.*, **27**, 25
187. Derjaguin, B. V., Rosenzweig, L. A. and Fedoseyev, V. A. (1969). *Fiz. Aerodispersnykh Sist.*, 30; *Chem. Abs.*, **73**, 123825

188. Garrett, W. D. (1971). *J. Atmos. Sci.*, **28**, 816
189. Saylor, J. E. and Barnes, G. T. (1971). *J. Colloid Interface Sci.*, **35**, 143
190. Avetisyan, R. A. and Trapeznikov, A. A. (1968). *Koll. Zhn.*, **30**, 480 (*Coll. J. USSR*, **30**, 358)
191. Trapeznikov, A. A. and Avetisyan, R. A. (1968). *Koll. Zhn.*, **30**, 897 (*Coll. J. USSR*, **30**, 677)
192. Katti, S. S. and Patil, G. S. (1968). *J. Colloid Interface Sci.*, **28**, 227 Kuchhal, Y. K., Katti, S. S. and Biswas, A. B. (1969). ibid., **29**, 521; Patil, G. S. and Katti, S. S. (1969). ibid., **30**, 219; Katti, S. S. and Sansara, S. D. (1970). ibid., **32**, 361
193. MacRitchie, F. (1968). *Nature (London)*, **218**, 669; idem., (1969). *Science*, **163**, 929
194. Barnes, G. T. (1968). *Nature (London)*, **220**, 1025
195. Kappesser, R., Greif, R. and Cornet, I. (1969). *Science*, **166**, 403
196. Barnes, G. T., Quickenden, T. I. and Saylor, J. E. (1970). *J. Colloid Interface Sci.*, **33**, 236
197. Seimiya, T. and Sasaki, T. (1966). *J. Colloid Interface Sci.*, **21**, 229
198. Hickman, K. and White, I. (1971). *Science*, **172**, 718
199. Hawke, J. G. and White, I. (1966). *J. Phys. Chem.*, **70**, 3369. (1970). ibid., **74**, 2788
200. Hawke, J. G. and Wright, H. J. L. (1966). *Nature (London)*, **212**, 811
201. Overath, P., Schairer, H. U. and Stoffel, W. (1970). *Proc. Nat. Acad. Sci.*, **67**, 606
202. Hubbell, W. L. and McConnell, H. M. (1971). *J. Amer. Chem. Soc.*, **93**, 314
203. Blank, M., Goldstein, A. B. and Lee, B. B. (1969). *J. Colloid Interface Sci.*, **29**, 148; Blank, M. and Britten, J. S. (1970). ibid., **32**, 62; Blank, M. and Lee, B. B. (1971). ibid., **36**, 151
204. Galdston, M., Shah, D. O. and Shinowara, G. Y. (1969). *J. Colloid Interface Sci.*, **29**, 319; Galdston, M. and Shah, D. O. (1970). *Advan. Exptl. Medicine and Biol.*, **7**, 261
205. Gorter, E. and Grendel, F. (1925). *J. Exptl. Med.*, **41**, 439
206. Bar, R. S., Deamer, D. W. and Cornwell, D. G. (1966). *Science*, **153**, 1010; Cornwell, D. G., Heikkila, R. E., Bar, R. S. and Biagi, G. (1968). *J. Amer. Oil Chem. Soc.*, **45**, 297
207. Garrett, W. D. (1969). *Annal. der Meteorologie*, N. F., **4**, 25
208. Garrett, W. D. (1968). *J. Geophys. Res.*, **73**, 5145
209. Garrett, W. D. and Barger, W. R. (1970). *Envir. Sci. and Tech.*, **4**, 123
210. Robb, I. (1966). *Koll. Z.*, **209**, 162
211. Goddard, E. D. and Kung, H. C. (1969). *J. Colloid Interface Sci.*, **30**, 145

2
Membranes: Their Interfacial Chemistry and Biophysics

H. TI TIEN
Michigan State University

2.1	AN OVERALL VIEW OF MEMBRANE RESEARCH	26
2.2	EVOLUTION OF BIOLOGICAL MEMBRANE STRUCTURE	27
	2.2.1 *Historical summary*	27
	2.2.2 *Recent models*	28
	2.2.2.1 *Hydrophobic lipoprotein model*	29
	2.2.2.2 *Spherical micelle model*	30
	2.2.2.3 *Lipoprotein subunit model*	31
	2.2.2.4 *Lipid–protein association model*	31
	2.2.2.5 *Repeating lipoprotein subunit model*	31
2.3	CHEMISTRY OF MEMBRANES	33
	2.3.1 *Chemical composition*	33
	2.3.1.1 *Lipids*	33
	2.3.1.2 *Proteins*	34
	2.3.1.3 *Water*	35
	2.3.2 *Physical techniques*	35
	2.3.2.1 *Optical methods: o.r.d. and c.d.*	36
	2.3.2.2 *Differential scanning calorimetry (d.s.c.)*	36
	2.3.2.3 *Fluorescent probes*	37
	2.3.2.4 *Gel filtration*	37
	2.3.2.5 *Heat-burst calorimetry*	37
	2.3.2.6 *Nuclear magnetic resonance (n.m.r.)*	38
	2.3.2.7 *Non-aqueous dialysis*	38
	2.3.2.8 *Spin-labelling technique (e.s.r.)*	38
	2.3.2.9 *X-ray diffraction*	38
2.4	EXPERIMENTAL MODEL MEMBRANES	39
	2.4.1 *Bimolecular lipid membranes (BLM)*	40
	2.4.1.1 *General considerations and intrinsic properties*	40
	2.4.1.2 *Composition of BLM-forming solutions*	41
	2.4.1.3 *Structure of BLM and BLM modifiers*	41

             2.4.1.4 *Modifiers changing mechanical properties*   42
             2.4.1.5 *Modifiers altering electrical properties*   45
             2.4.1.6 *Modifiers conferring ion selectivity and
                       specificity*   47
             2.4.1.7 *Modifiers inducing electrical excitability*   49
             2.4.1.8 *Modifiers generating photoelectric effects*   50
      2.4.2 *Microvesicles (liposomes)*   52
             2.4.2.1 *Techniques of formation*   52
             2.4.2.2 *Structure of the microvesicles*   52
             2.4.2.3 *Osmotic properties and permeability to water*   53
             2.4.2.4 *Permeability to electrolytes*   53
             2.4.2.5 *The effects of modifiers*   54
             2.4.2.6 *Miscellaneous studies*   56
      2.4.3 *Monolayer and miscellaneous studies*   57
             2.4.3.1 *Monolayer films*   57
             2.4.3.2 *Miscellaneous studies*   58

2.5 BIOLOGICAL MEMBRANES   58
      2.5.1 *Plasma membranes*   58
      2.5.2 *Chloroplast membranes*   60
             2.5.2.1 *The primary processes*   61
             2.5.2.2 *Photophosphorylation and charge transport*   62
             2.5.2.3 *Oxygen evolution*   63
      2.5.3 *Mitochondrial membranes*   63
      2.5.4 *Nerve membranes*   66
      2.5.5 *Visual receptor membranes*   68

2.6 FRONTIERS OF MEMBRANE RESEARCH   70

2.1 AN OVERALL VIEW OF MEMBRANE RESEARCH

Membranes and their associated phenomena occur at interfaces. The formation of a membrane, its stability, adsorption, membrane potential, and a host of electrokinetic effects are intimately involved with interfaces. In living cells, the tremendous interfacial areas that exist between the membrane and its surroundings not only provide ample loci for carrying out activities vital to the living system, but also afford a clue for our understanding. Physically, an interface is characterised most uniquely by its interfacial free energy or tension, which is a result of the orientation of the constituent molecules. This and other interfacial properties of membranes can be understood to a large extent in terms of the laws of interfacial chemistry and physics that govern them. In fact, the current concept of the structure of biological membranes and their experimental models has been developed as a direct consequence of the applications of classical principles advanced by Langmuir, Adam, Harkins, McBain, and others.

  The purpose of this chapter is to review the current literature on membranes, both natural and artificial. More specifically, it focuses attention on

the chemistry and physics of membrane constituents, on theoretical and experimental models (bilayer lipid membrane, or BLM for short, and liposomes), and on natural membranes. In addition, relevant aspects of the interfacial chemistry of membrane phenomena are covered. In the case of natural membranes, emphasis is accented on the five basic types. These are:
  (i) plasma membranes of cell surfaces and intracellular organelles,
  (ii) thylakoid membranes of chloroplasts,
  (iii) cristae membranes of mitochondria,
  (iv) nerve membranes, and
  (v) visual receptor membranes of rods and cones.

Other topics dealing with physical techniques, structure and function, and certain aspects of membrane biochemistry have also been included.

It must be stated at the outset that, owing to space limitation, some of the listed topics are merely mentioned. Fortunately, a number of symposium volumes and specialised books detailing many aforementioned topics have been published in recent years[1-16]. For current advances in this vast and active field of membrane research, the *Biomembrane* section of *Biochimica et Biophysica Acta*, published regularly, should be consulted.

2.2 EVOLUTION OF BIOLOGICAL MEMBRANE STRUCTURE

2.2.1 Historical summary

In the 45 years following its suggestion by Gorter and Grendel, the bimolecular leaflet concept of the plasma membrane has been used as a starting point for numerous current models. The development of these models to a large extent has its origin in surface and colloid chemistry. A succinct account of the role played by surface chemistry in biology has been given by Lawrence[17]. The study of monomolecular layers (monolayers) at air/water interfaces introduced by Langmuir and others had provided a physical picture of the structure and orientation of lipid molecules at interfaces. The bimolecular lipid layer with absorbed protein layers was put into pictorial form by Davson and Danielli, who deduced their model from the results of interfacial tension measurements on cells. The bimolecular leaflet model for the structure of biological membranes may be simply stated as follows: The basic construct of all biological membranes consists of a bimolecular lipid leaflet (bilayer) with adsorbed non-lipid layers (mostly proteins). The lipid molecules are oriented with their polar groups facing outward, and their hydrocarbon chains away from interfaces forming the interior of the membrane. In the interior of the bilayer, hydrocarbon chains are held together by London–van der Waals forces and are in a liquid-like state. For details concerning the growth of the concept of the bimolecular leaflet model, the article by Davson[18] is of interest.

In the last decade, the bilayer leaflet model has been popularised by Robertson[19] who, based upon his extensive electron microscope observations, has come to the conclusion that all biological membranes, irrespective of their function or composition, are lipid bilayers or a modification thereof. Another line of evidence in support of the bimolecular leaflet model was the formation of a stable 'black' lipid membrane developed by Rudin and his co-workers (Tien and Wescott) in 1960. Later investigations carried out by many groups have established that black or bimolecular lipid membranes (BLM) are excellent experimental models for biological membranes (see Section 2.4.1). With the availability of BLM it was possible for the first time to investigate electrical properties and transport phenomena across an ultra-thin (<100 Å) lipid membrane separating two aqueous solutions.

The major historical landmarks in the development of biological membrane structure are shown chronologically in Table 2.1. The reader may also want to consult two of the most recent reviews on membrane structure written by Malhotra[20] and Hendler[21].

Table 2.1 Major landmarks in the development of biological membrane structure

	Year	
Soap bubbles (Boys, 1890)	1890	Permeability experiments (Overton 1890–1899)
Adsorption isotherm and soap films (Gibbs, 1906)	1900	
The monolayer technique (Langmuir, 1917)	1910	Lamellar micelles (McBain, 1913)
Electrical capacity measurements (Fricke, 1925)	1920	Bimolecular leaflet model (Gorter and Grendel, 1925)
Interfacial studies (Cole, 1932; Harvey and Danielli, 1934)	1930	Bimolecular lipoid layer (Davson and Danielli, 1935)
Polarisation microscopy of myelin (Schmitt et al., 1936)	1940	
	1950	Myelin figures (Frey-Wyssling, 1953) Myelin sheath (Geren, 1954) Unit membrane concept (Robertson, 1959)
BLM technique (Rudin, Tien and Wescott, 1960)	1960	Spherical micelle (Lucy, 1964) Lipoprotein subunit (Benson, 1966)
The microvesicles (Rendi 1964; Bangham et al., 1965)		Hydrophobic lipoprotein (Sjostrand, 1963–66; Lenard and Singer, 1966; Wallach and Zahler, 1966) Repeating lipoprotein subunit
	1970	(Green and Perdue, 1966)

2.2.2 Recent models

Investigation of biological membrane ultrastructure can be conveniently divided into three periods:
 (i) the classical period of Gorter–Grendel–Davson–Danielli,
 (ii) the unit-membrane period of Robertson, and
 (iii) the current period.
The current period that began roughly in 1963 up to the present time will be our main concern here.

Since much of the current work on membrane ultrastructure has been generated as an alternative to the unit membrane concept, which is derived directly from the classical model, it seems desirable for purposes of comparison to have such information in one place. Table 2.2 presents such a summary.

Table 2.2 A summary of some current models of biological membranes

Model	Evidence	Remarks
Bimolecular lipid leaflet (Gorter and Grendel, 1925)	Orientation of lipids at interfaces; monolayer studies with lipids extracted from red cells	Conceptually simple and elegant; energetically feasible; experimental bilayer lipid membranes (BLM) provide strongest support for the model
Lipoid layer with adsorbed proteins (Davson and Danielli, 1935)	Monolayer work of Gorter and Grendel; low surface tension values of marine eggs and a high surface tension at an oil–water interface which could be lowered by proteins; electrical capacitance of cells	The well-known model before 1960; a globular configuration was suggested for proteins; BLM studies indicate low surface tension can result from lipids alone
Unit membrane (Robertson, 1959)	Evidence cited above; x-ray studies on myelin; polarisation microscopy; electron microscopy on a variety of cell organelles	The most widely generalised model of biological membranes; protein configuration was assumed to be extended; lipid and protein interact electrostatically; a continuous bilayer lipid core presents difficulty in biological function
Spherical lipid micelle (Lucy and Glauert, 1964)	Behaviour of lipids in aqueous solution; x-ray diffraction studies; electron microscopy of lipid dispersions	It seems likely that spherical micelles of lipid can be in dynamic equilibrium with the bimolecular leaflet; this model in need of other experimental confirmation
Lipoprotein subunit (Benson, 1966)	Previous knowledge mentioned above and inferential evidence	An extreme view of lipid–protein interaction; inconsistent with spin-label experiments
Hydrophobic lipoprotein (Sjostrand, 1963–1966; Lenard and Singer, 1966; Wallach and Zahler, 1966)	Electron microscopy; i.r., o.r.d. and c.d. spectroscopy; Inferential evidence	These various models stress hydrophobic interactions between lipids and proteins; specific conformations are suggested for proteins; problems of isolating structural protein
Repeating lipoprotein (Green and Perdue, 1966)	Biochemical studies on mitochondria; electron microscopy of cell organelles and reconstituted systems, inferential evidence	This model emphasises the importance of protein particles; difficulties associated with the interpretation of electron micrographs

2.2.2.1 Hydrophobic lipoprotein model

Sjostrand[22], in a paper published in 1963, suggested that the globular particles of the membrane elements represent the lipids in the form of micelles. According to Sjostrand, the polar ends of the lipids are located at the surface of the micelles, which is covered by a layer of protein. The evidence for the hydrophobic lipoprotein model comes primarily from electron microscopy, although the work of Lucy and Glauert[23] on myelin figures produced from phospholipids and cholesterol has provided some indirect support (see next section).

2.2.2.2 Spherical micelle model

In this model globular micelles of lipid are in hexagonal close packing within the membrane (Figure 2.1). This model, somewhat similar to that suggested by Sjostrand, departs from the classical picture in that lipid molecules in a membrane are arranged as other than bimolecular leaflets. It has been

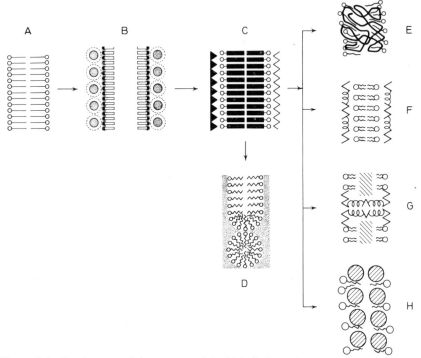

Figure 2.1 Development of the structure of the biological membrane; (A) Gorter–Grendel bimolecular leaflet model (1925), (B) Davson–Danielli protein–lipid leaflet model (1935), (C) unit membrane model (1959), (D) spherical micelle model (Sjostrand – 1963, Lucy – 1964), (E) lipoprotein model (1966), (F) and (G) Lenard and Singer model (1966) and (H) Green's protein subunit model (1966)

pointed out recently by Lucy[24] that the micellar structure is required for the fusion of two membranes; any situation which favours the micelles over the bimolecular leaflet should therefore tend to help membrane fusion. The work of Ohki and Aono[25] on phase transition between spherical and lamellar micelles, and that of Gulik-Krzywicki et al.[26] may be cited to support this model. The latter workers have reported that under certain conditions a mixture of protein and lipid may assume a highly ordered configuration.

2.2.2.3 Lipoprotein subunit model

This is by far the most elaborate, and hence, highly speculative of all current models in view of the meagre amount of data available on the chemical composition of membranes. The main thesis as suggested by Benson[27] is that lipids of membrane 'subunits' are bound by hydrophobic association of the hydrocarbon chains of the lipids, with complementary hydrophobic regions within the interior of the protein. The resulting two-dimensional lipoprotein aggregate, according to Benson, would possess the strongly negative charged groups of the phospholipids on its surface. The initial model was proposed for the chloroplast membrane.

2.2.2.4 Lipid–protein association model

This is the most recent arrival on the scene, and in a sense is a variation of the previous models including the model discussed in the following section. The evidence in support for this model has been gathered by its proponent, D. W. Deamer[28]. In this model, lipids are present as a bilayer, but each phospholipid has one chain interacting with a hydrophobic bonding site on membrane protein and the other chain directed into the non-polar membrane interior. Deamer has constructed a number of molecular models to substantiate his model. Whether this constitutes valid evidence is debatable, but the idea itself is certainly provocative.

2.2.2.5 Repeating lipoprotein subunit model

A considerable amount of energy has been expanded by Green and his colleagues[29] in suggesting an alternative model to the classical bimolecular leaflet where the lipid plays a primary role. In the repeating lipoprotein subunit model, on the other hand, the role of protein is strongly emphasised. Green and Perdue[30] have suggested that in mitochondrial membranes the so-called lipoprotein subunits are the basic building blocks. These subunits consist of basepieces and stalks. Evidence for the model comes mainly from electron microscopy and *in vitro* reconstitution. More recently Vanderkooi and Green[31] have proposed a somewhat different version, not unlike that of Benson's, where the hydrophobic areas are submerged in channels between the protein subunits of each repeating unit. Green and his colleagues believe that their model can explain many phenomena associated with the membranes, such as energy transduction and protein transport.

It seems appropriate to comment at this point on some of the major arguments raised against the bimolecular leaflet model by Green and his colleagues, and by others.

Korn[32], a long-time critic of the bimolecular leaflet model, has argued that neither the intracellular nor the plasma membranes are passive structural elements but, instead, are sites of important metabolic reactions. He has further stated that it is more reasonable to consider protein as the fundamental structural unit – it is the substance providing control of selectivity and specificity. Korn has suggested that protein might be laid down first, and lipids then added in an arrangement governed by the structure of the protein; the resultant lipoproteins interact to form the membrane. The unit membrane model seems to require that the lipid bilayer is formed first and this specifies the protein arrangement. The argument as to whether the red shift in o.r.d. and circular dichroism (see Section 2.3.2.1) should be interpreted as evidence of interactions between α-helical regions of adjacent protein molecules is unresolved at present. Both interactions require the protein helices to be buried in a hydrophobic region.

Another argument against the bimolecular leaflet model is based on the chemical composition. It has been found that the lipid to protein ratio in isolated membranes apparently varies widely. For example, the lipid contents vary all the way from 30% of the dry weight (in mitochondria) to about 80% in nerve myelin. The remaining weight is made up mostly of proteins (Section 2.3.1). The argument based on the ratio of lipid to protein constituents is a weak one, since at present the techniques used to isolate membranes from the rest of the cellular components are far from ideal. In many cases, meaningful chemical data simply do not exist. The problem of isolating membrane fractions of high purity from cells must first be solved. The fragmentary data now available should therefore be used with discretion.

The other frequently used argument against the bilayer leaflet model has been based on the uncertainty of electron microscopy. The interpretation of electron micrographs requires a great deal of skill and imagination. The authorities in the field do not seem to agree on what is seen in the electron microscope as artifacts and what is not. Thus, until better experimental techniques are available, the validity of results obtained from electron microscopy remains uncertain.

Perhaps one of the main shortcomings (or conversely, its strength) is the all-inclusive nature of the bilayer leaflet model which boldly claims that all biological membranes are fundamentally alike. A brief reflection on the diversity and functions of biological membranes, as they are understood today, would seem to indicate otherwise. This argument, although plausible, is not central to the point. The bilayer leaflet model sets the limit as to what is the most likely structure of the biological membrane. A proper interpretation perhaps is that the bilayer leaflet model should be viewed as a basic matrix upon which the diverse functions of biological membranes are built. The perturbation and modification of the basic bilayer structure by a host of lipid and non-lipid materials would appear to be more than adequate to accommodate all the functional diversities that are likely to arise in nature. On energetic considerations, the bilayer model provides not only an overall symmetry but also configurations that are natural for lipids and proteins. At the present time it is difficult to conceive of a more basic structure in two dimensions than the bilayer leaflet.

Therefore, it must be concluded that the bilayer leaflet model together with its modifications now constitutes a valid approach to the understanding of the structure and function of biological membranes. As a general framework, the bilayer leaflet model is not only useful theoretically, but also suggests positive and constructive experiments as are now being carried out with BLM and with lipid microvesicles (see Section 2.4).

2.3 CHEMISTRY OF MEMBRANES

2.3.1 Chemical composition

The main constituents of biological membranes are lipids and proteins. In most cells, membranous organelles constitute anywhere from 40 to 80% of the dry weight. O'Brien[33] has summarised the chemical composition of various types of membranes. The ubiquitous presence of water in membranes is often assumed but not explicitly stated. The role played by water, although obscure, must be crucial in maintaining the integrity of the membrane structure. Unfortunately, in giving the gross distribution of major constituents of biological membranes, water content has generally been omitted.

2.3.1.1 *Lipids*

The most unique property of lipids which sets them apart from proteins and other cellular constituents is their amphipathic or dual water–oil solubility. The principal types of lipids are phospholipids. In a paper entitled 'Phospholipids, liquid crystals and cell membranes,' Williams and Chapman[34] have recently reviewed the results of physical studies of these materials. Small[35] has given an interesting discussion of the behaviour of triglycerides and cholesterol and its esters. He has proposed a detailed classification of lipids based primarily upon their physical properties in bulk aqueous systems, and also considering their interfacial properties. The major classifications are non-polar and polar lipids. The second class is divided into various categories such as 'insoluble, non-swelling, amphiphilic lipids' (e.g. lecithins, sphingomyelins, monoglycerides); 'soluble amphiphiles capable of forming liquid crystals' (e.g. many classic detergents, lysolecithin); 'soluble amphiphiles not forming liquid crystals' (e.g. some bile salts, saponins). The interfacial properties and micelle-forming abilities of the various categories of polar lipids are also considered. Abramson[36] has considered the behaviour of phospholipids in water, in particular the effect of chain length and polar groups of single lipids and mixtures. The effect of hydrocarbon configuration and cholesterol interactions of choline phospholipids with sulphatide has been reported by Abramson and Katzman[37]. In a series of papers, Weinstein, Marsh, Glick and Warren[38] have reported their work on lipids of membranes isolated by the fluorescein mercuric method. They have suggested that lipid composition may be useful in classifying surface membranes. A summary of thermal studies of lipids and biological membranes has been published by Ladbrooke and Chapman[39]. Other excellent sources of information on lipids and biomembranes have been gathered by van Deenan[40], and by Rouser, Nelson, Fleischer and Simon[41].

2.3.1.2 *Proteins*

At the present time the membrane proteins have not been adequately characterised. One of the reasons for this lack of information is undoubtedly due to the experimental difficulties in isolating membrane proteins, since their solubility is poor in both aqueous and organic medium.

The current information on mitochondrial and chloroplast proteins has been reviewed by Criddle[42]. Zahler[43] has recently summarised the data on erythrocyte membrane proteins.

Basically three steps are involved in studying membrane proteins:
(i) isolation,
(ii) purification, and
(iii) characterisation.

In solubilising proteins two techniques are used: using the detergent sodium dodecyl sulphate (SDS), and using aqueous butanol or pentanol. A newer technique employing mixed ion-exchange resins has been recently reported by Mazia and Ruby[44]. Gel electrophoresis is one of the principal techniques used in the purification step. A recent paper on the use of this technique has been published by Lenaz et al.[45]. Characterisation is the third step of membrane protein research; a number of physical chemical techniques have been employed in recent years, including analytical ultracentrifugation[46], optical methods[47], amino acid analysis[48], and end group analysis[49]. The interested reader may want to consult the cited publications for further details.

Working with erythrocyte membranes, Blumenfeld[50] reported a new mild method for the isolation of membrane proteins of the human red cells. He reported that at least two classes of proteins are present: one contains all of the sialic acid and the other has strong affinity for the lipids of the membrane. In mitochondria, a structural protein (mol. wt. 23 000) has been characterised by Woodward and Munkres[51]. This protein constituted about 40% of the total protein found in the cytoplasm.

Opsin appears to be the only known protein in the visual receptor membrane. In another system Nachmansohn[52] has discussed at length certain proteins in excitable membranes. In influenza virus, envelope proteins have been reported by Tiffany and Blough[53]. It has been found that fatty acids of viruses with antigenically related envelope proteins show greater resemblance than those of an unrelated strain, which suggests that these proteins can influence the composition of membrane lipids at the site of viral release.

Physical chemical properties of a protein from red cell membranes has been reported recently by Marchesi and co-workers[54]. This protein is reported to be free of lipid and other non-protein material and has an apparent mol. wt. of 140 000. Whether this protein isolated by Marchesi et al. has any structural role in the membrane is uncertain. For relevant discussions on structural aspects of membrane proteins, the review by Kaplan and Criddle is of interest[55].

2.3.1.3 Water

As mentioned earlier, the role of water in membranes is not explicitly stated. This is again owing to our lack of information. A number of papers dealing with water in membranes have appeared recently. Drost-Hansen[56], in particular, has considered in detail the structural role of water in biological systems. He calls attention to the fact that water at or near interfaces frequently appears to undergo notable changes in properties and structure at a number of discrete temperature ranges (13–16 °C, 29–32 °C and 44–45 °C). He has made an interesting correlation between these temperature ranges and abrupt changes in biological systems.

Earlier, Hechter[57] proposed a membrane model that consists of lipid, protein and water. In this model, the protein phase is built of two layers of peptide–antipeptide units with each interlocked disc system being 6.9 Å thick, separated by two water layers in a hexagonal ice-like lattice, and with hydrogen bonded via second-neighbour relations to the carbonyl oxygens of the hydrophilic surface. Hechter has attempted to explain a number of vital membrane phenomena such as nerve conduction in terms of his model.

The water permeability of experimental membranes is considered in Section 2.4.

2.3.2 Physical techniques

In addition to model membrane systems, discussed in Section 2.4, sophisticated physical techniques such as e.s.r., n.m.r., d.s.c., etc. are being applied to a variety of biological membranes and their constituents. These physical

Table 2.3 Physical chemical techniques used in the investigation of biological membranes and their constituents

Method	*Remarks*	*References*
Bimolecular lipid membranes (BLM)	A widely used experimental model for biological membranes (see microvesicles)	89–95
Circular dichroism (c.d.)	Protein conformation (see o.r.d.)	61–63
Differential scanning calorimetry (d.s.c.)	Lipid and membrane phase transition studies	64–66
Fluorescent probes	Studies of conformational change	67–72
Gel filtration	Lipoprotein separation and interaction studies	73–74
Heat-burst calorimetry	Lipid–protein interaction studies	75–77
Microvesicles (liposomes)	A complementary model to BLM; useful for permeability studies, large interfacial area	187–192
Monolayers	Classical tool for studying lipids and proteins	13, 193, 220–233
Nuclear magnetic resonance (n.m.r.)	Lipid–protein interaction studies	78–79
Non-aqueous dialysis	Lipid–protein interaction	80–82
Optical rotatory dispersion (o.r.d.)	Protein conformation (see c.d.)	61–63
Spin-labelling e.s.r.	Lipid–protein interaction; the state of lipid	83–86
X-ray diffraction	Orientation and membrane thickness	87–88

techniques have yielded information in certain instances on the physical states of membranes. Previous reviews on the subject have been published by Chapman[58] and Finean[59]. The well-known methods such as electron microscopy will not be included in this section[60] (see Table 2.3).

2.3.2.1 Optical methods: o.r.d. and c.d.

The two most used optical methods are optical rotatory dispersion (o.r.d.) and its variation, circular dichroism (c.d.). Using these methods, Lenard and Singer[61], Urry and Krivacic[62] and Wallach and Zahler[63] have concluded that 50% or more of the membrane protein is present in an α-helical configuration. In addition, a red shift is also observed as compared with the standard spectrum of poly-L-lysine. It is interesting to note that infrared studies of erythrocyte and ascites tumour cells give no evidence of a β-configuration in these membranes.

2.3.2.2 Differential scanning calorimetry (d.s.c.)

Differential scanning calorimetry (d.s.c.) is a classic method, but application to biological systems is a recent event. The principle of d.s.c. is to heat a sample in a reference compound of known melting point and measure the difference in power input. If a transition takes place, more heat must be supplied to the sample than to the known compound. The heat of transition is proportional to the area of recording. The phase transitions observed in the d.s.c. can be correlated with changes in biological properties. For example, Steim et al.[64] have observed that the transition temperature of intact membranes of *M. laidlawii* occurs at the same temperature as in the extracted lipids dispersed in water. Using the d.s.c. technique, Ladbrooke et al.[65] have studied the effect of cholesterol on phase transitions of phospholipids. More recently Steim and co-workers[66] have provided evidence for the liquid crystalline state of lipids in biological membranes. The data indicate that 90% of the lipid in the membranes is in the bilayer conformation. It has been suggested that d.s.c. would also be suited to studying protein denaturation in membranes and even in intact cells.

2.3.2.3 Fluorescent probes

The use of dyestuffs in biology is well known. Recently some dyes have been used as 'fluorescent probes' owing to properties which depend on the state of the dielectric in which they are attached. The best known, ANS (8-anilino-1-napthalene sulphonic acid), has been employed by many workers in studies of changes of protein conformation and membranes. As a probe of conformational change, ANS is known to exhibit a strong fluorescence at 520 nm in non-polar solvents, but it fluoresces poorly in aqueous solution. Azzi, Chance, Radda and Lee[67] have shown that ANS fluorescence is enhanced some 25–35 times on binding to mitochondrial membranes. Freedman and Radda[68], and Rubalcava, de Munoz and Gitler[69] have used ANS on red blood cells, and Vanderkooi and Martonosi[70] have used it on mitochondrial membranes. All these investigators reported the effects of salts on ANS binding. Similar findings have been reported recently by Gomperts, Lantelme and Stock[71] on microsomal membranes and lecithin micelles.

In artificial membranes, the recent paper of Smekal et al.[72] is of interest (see Section 2.4.1).

2.3.2.4 Gel filtration

The gel filtration technique has been used by Fleischer et al.[73] and by Scanu et al.[74]. The latter investigator used Sephadex G-200 to separate the products of the high density lipoprotein (HDL). Fleischer et al. used the method to investigate the interaction of γ-lipoprotein with a phospholipid. This method is of value in determining the stoichiometry of binding between lipids and protein.

2.3.2.5 Heat-burst calorimetry

The technique has been used in biochemistry since the work of Kitzinger and Benzinger[75]. It has been used by Lovrien and Anderson[76] to study the interaction between sodium dodecyl sulphate and γ-lactoglobulin, and by Klopfenstein[77] to determine the enthalpy change on binding of lysolecithin by bovine serum albumin. The technique is of potential value in providing quantitative thermodynamic data useful in the understanding of the nature of binding forces between the lipid and protein.

2.3.2.6 Nuclear magnetic resonance (n.m.r.)

Interactions between lipid and protein can be investigated by comparing the spectrum with those of the lipid and protein alone. Chapman and associates[78] have reported n.m.r. spectra from phospholipids, bile salts, serum lipoproteins and erythrocyte ghosts. Serum lipoprotein has also been studied by Steim[79]. The conclusions drawn from these studies are that interaction between protein and lipid results in the restriction of the mobility of the hydrocarbon chains, which presumably were in a liquid state. The nature of interaction between lipid and protein is believed to be hydrophobic. N.M.R. spectroscopy of red cell membrane ghosts has also been reported by Chapman et al.[78]. They suggested that the fatty acid chains in the membrane may be relatively immobile.

2.3.2.7 Non-aqueous dialysis

The technique involves a partition of lipid in a non-aqueous solvent separated by a dialysing membrane immersed in an aqueous protein solution. Ji, Hess and Benson[80] studied the products of chloroplast lamellae by this technique. Interactions between fatty acids and serum albumin have been investigated by Spector et al.[81] and by Arvidsson and Belfrage[82].

2.3.2.8 Spin-labelling technique (e.s.r.)

The basis of spin-labelling technique is that a tagged molecule (spin label) on a macromolecule such as a protein can be measured with an electron spin resonance (e.s.r.) spectrometer. The spin labels thus far used are nitroxide free radicals such as di-ti-butyl nitroxide and 2,2,6,6-tetramethylpiperidine-1-oxyl. Except in detecting instrumentation, the use of a spin label is very much similar to introducing a dye into a biological material. The advantages of spin labels are sensitivity to the local environment and ability to measure very rapid molecular motion[83]. To date, the spin-label technique has been used on rabbit nerves, excitable membranes, bovine erythrocyte ghosts, and model systems. The conclusion drawn from these experiments is that biological membranes possess hydrophobic regions consistent with the current bilayer leaflet model. Very recently Hsia, Schneider and Smith[84] reported e.s.r. spectra of spin-labelled cholesterol and phospholipids. Distinct changes in the spectra were interpreted in terms of molecular orientation. Whether the technique can provide other information concerning the state of aggregation and the molecular structure of biological membranes remains to be seen. The technique is unique in many ways and undoubtedly will be further exploited in the study of biological and model membranes[85, 86].

2.3.2.9 X-ray diffraction

Conventional x-ray diffraction techniques and small-angle x-ray scattering continue to be used on dry membrane preparations (see Section 2.3.1) and model lipoprotein systems. Shipley, Leslie and Chapman[87] have obtained scattering curves for some of the octane-soluble complexes of cytochrome C with phospholipids, and suggested that the two complexes observed are not simple monomers but are aggregated. Recently, Engleman has reported studies using x-ray diffraction on biological membranes[88].

2.4 EXPERIMENTAL MODEL MEMBRANES

The fundamental life processes include energy transduction, active transport, signal transmission, and replication. In broad terms, all of these vital processes are determined by the structure and function of the cell membranes. Because of complex structural and environmental factors associated with biological membranes, it is realised that approaches using experimental model systems are essential to an understanding of the fundamental life processes in physical and chemical terms. Therefore, a vast number of investigations using different techniques have been carried out on model systems (for early references, see Ref. 14, p. 361). It is obvious that the chemical composition and thickness of any adequate model should be similar to those of biological membranes. A good experimental model should therefore possess a thickness of about 100 Å. This has been the least met criterion in all the models used before 1960. Biological membranes generally exist as 'liquid' structures bounded on either side by an aqueous phase; thus there are *two* co-existing liquid–liquid interfaces associated with each membranous structure. Furthermore, many biological membranes have been shown to function in a vectorial or directional manner, e.g. glucose accumulation in red blood cells by carrier-facilitated or active transport. These characteristics impose corresponding requirements on any model membrane system: the model (i) should be capable of separating dissimilar aqueous phases, (ii) should enable the formation of structural and chemical asymmetry, and (iii) should be simple operationally so that unidirectional, vectorial functions, such as transport, can be investigated.

As has been discussed in Section 2.2.1, the majority of biological membranes thus far examined under the electron microscope has a thickness on the order of a bimolecular leaflet (~ 100 Å). Therefore, it has been evident for some time that if the bimolecular leaflet were indeed the major structural component of biological membranes, knowledge concerning the properties and the formation of such a structure *in vitro* would be of considerable significance, both experimentally and theoretically. It was also readily apparent that a detailed physical chemical description of biological membranes would be best approached by studies of simpler well-defined models.

The resorting to model systems has led to the discovery of two important methods of forming experimental lipid membranes of bimolecular thickness, whose physical and chemical properties resemble closely those of cell membranes. The first is the BLM system (BLM denotes bimolecular, black, or bilayer lipid membrane(s)) which has been reviewed several times in recent years[89–92]. The most recent reviews have been written by Goldup, Ohki and Danielli[93] and by Tien[94]. In this section only the highlights of previous work and the work published in the last 2 years will be described.

The second experimental model membrane system is known as the liposome or microvesicle, which is quite complementary to the BLM system. This second model together with recent monolayer and other studies will also be reviewed.

2.4.1 Bimolecular lipid membranes (BLM)

2.4.1.1 *General considerations and intrinsic properties*

The formation of a bilayer lipid membrane (BLM) in aqueous solution is, conceptually, extremely simple. It essentially involves creating two co-existing solution/membrane interfaces or a biface. In general, thinning of the lipid solution takes place spontaneously and under favourable circumstances leading to a BLM. BLMs are generally formed by one of the three basic methods. The original method calls for a small brush and a BLM is generated literally by painting the BLM-forming solution across a small hole in a polyethylene or Teflon cup (5–10 ml capacity) immersed in a salt solution. The basic methods of BLM formation can be described by referring to Figure 2.2. The cell assemblies usually consist of two chambers. The wall of the inner chamber made of Teflon or polyethylene contains a small hole (1–2 mm diameter) in which the BLM is formed. The second method, useful for studying the formation characteristics and thickness determination, avoids the transfer of the BLM-forming solution through air. This can be done by passing a Teflon loop or frame through an oil/water interface (or vice versa). The third method of BLM formation, much less used than the other two methods, involves 'blowing a bubble'[95]. With this last method, a spherical BLM of a

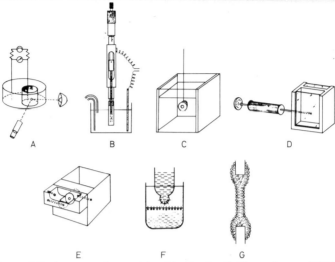

Figure 2.2 Apparatus for studying bimolecular lipid membranes (BLM); (A) from Mueller, Rudin, Tien and Wescott (1964), (B) from Hanai and Haydon (1966), (C) from Huang and Thompson (1966), (D) from Vreeman (1966), (E) from Andreoli, Bangham and Tosteson (1967), (F) from Tsofina, Liberman and Babakov (1966), (G) diagram of a supported BLM separating two aqueous solutions
(By courtesy of Dr. A. D. Bangham)

Table 2.4 Comparison of properties of unmodified BLM and a layer of liquid hydrocarbon of equivalent thickness

Property	Liquid hydrocarbon (Extrapolated)	Unmodified BLM (experimental)	Reference
Thickness (Å)	100	40–130	95, 99, 113
Resistance (Ω cm^2)	10^8	10^8–10^9	93, 94
Capacitance (μF cm^{-2})	~ 1.0	0.3–1.3	93, 95
Breakdown voltage (V cm^{-1})	10^5–10^6	10^5–10^6	1, 92
Dielectric constant	2–5	2–5	1, 93
Refractive index	1.4–1.6	1.4–1.6	99
Water Permeability (μm s^{-1})	~ 35	8–24	95, 115, 116
Interfacial tension (dyn cm^{-1})	~ 50	0.2–6.0	96, 106, 109
Potential difference per tenfold concentration of KCl (mV)	~ 0	~ 0	94, 95
Electrical excitability	None	None	95, 157
Photoelectric effects	None	None	174

centimetre or more may be formed under optimal conditions[96, 97]. For a detailed description of various BLM techniques, the reader is referred to the literature and to a chapter in a book on techniques of surface chemistry and physics[98].

The intrinsic properties of an unmodified BLM generated from either lecithin or oxidised cholesterol in an alkane solvent are strikingly similar to those expected of a layer of liquid hydrocarbon of equivalent thickness. A comparison of the properties of the unmodified BLM with the extrapolated properties of a 100 Å layer of liquid hydrocarbon is summarised in Table 2.4.

2.4.1.2 Composition of BLM-forming solutions

In the manner described above, BLMs can be formed from a variety of materials. These include brain lipids, proteolipids, phospholipids (e.g. lecithin), *E. coli* lipids, chloroplast extracts, surfactants, and oxidised cholesterol[93, 94]. Depending upon the BLM-forming solution used, a BLM can be generated in a wide range of salt concentrations, pH values and temperatures. The fact that a BLM can be formed from a variety of materials and under quite different conditions suggests that the phenomenon is of general occurrence in nature. It should be mentioned that the presence of at least one liquid hydrocarbon has been found necessary in the formation of a stable BLM.

2.4.1.3 Structure of BLM and BLM modifiers

It has been assumed that BLM adopts the 'neat' or smectic mesomorphic form. The polar groups of the lipids face outward in contact with the aqueous solution. The hydrocarbon chains in the BLM are assumed to be in the liquid state[99]. In the fashion depicted, the lipid molecules can thereby minimise their free energy by aggregating their hydrocarbon chain through van der Waals' interactions, and the polar portion of the molecules can maximise their coordination energy with each other and with the aqueous solutions. The experimental facts in support of this picture come from the following lines of evidence: (a) the liquid nature of the BLM can be demonstrated by the fact that a BLM may be probed with a fine object (thin wire or hair) which can be moved within the membrane and then withdrawn without rupturing the membrane; (b) the 'hydrocarbon-like' interior is deduced from the materials used; (c) water permeability studies have shown that to a first approximation, the BLM may be considered as a continuous liquid hydrocarbon layer less than 100 Å thick. The permeability coefficient for water of such a layer would be expected to be about 35 $\mu m\ s^{-1}$ as estimated from the solubility and diffusion data; (d) the d.c. resistance of the BLM is about $10^8\ \Omega\ cm^2$. This corresponds to a bulk resistivity of about $10^{14}\ \Omega\ cm$. The resistivity of most wet liquid hydrocarbons is of the same order of magnitude; and (e) the dielectric breakdown strength of the BLM is usually in the range of 10^5–10^6 $V\ cm^{-1}$, which again corresponds closely to that of a long-chain liquid hydrocarbon[94]. Another line of indirect evidence is provided by a spin-label experiment using nitroxides (see Section 2.3.2.8). In view of these data, it seems reasonable to consider that the interior of the BLM is essentially an ultra-thin, continuous layer of liquid hydrocarbon saturated with water.

It is evident from the data presented in Table 2.4 that unmodified BLM appear to be poor models for biological membranes. For instance, it is well known that biological membranes are ion-selective. In the case of nerve membrane, electrical 'excitability' is one of the most unique features.

In an attempt to modify the intrinsic properties of the BLM, literally several hundreds of compounds have been evaluated. Among these the the following groups of materials have been tried: common proteins, enzymes, surfactants, fermentation products, vitamins, tissue extracts (e.g. retina), antibiotics, uncoupling agents of phosphorylation, and a variety of simple organic and inorganic compounds. The very first modifier, termed 'excitability inducing material' (or EIM), not only dramatically reduced the BLM resistance but induced electrical 'excitability' as well[95]. Beginning with EIM, a number of modifying agents (or modifiers) have been discovered, which, when present in the BLM, impart new properties that are of biological interest. To facilitate the discussion, the BLM modifiers may be conveniently grouped into five categories:

(i) those altering the passive electrical properties,
(ii) those changing the mechanical properties,
(iii) those conferring ion selectivity,
(iv) those inducing electrical excitability and
(v) those generating photoelectric effects.

The effects of these modifiers on the intrinsic properties of BLM will be considered under separate headings.

2.4.1.4 Modifiers changing mechanical properties

Under this heading the stability, thickness, bifacial tension, and permeability of the BLM together with the effects of modifiers on these properties will be considered.

The mechanical stability of most BLMs is usually enhanced by the addition of cholesterol and its derivatives. This fact may account for the extensively studied BLMs formed from oxidised cholesterol. The presence of cholesterol also increases the capacitance of lecithin BLM to about $0.6\ \mu F\ cm^{-2}$. Explanation for this change may be owing either to the thickness reduction or to the dielectric constant increase. Finkelstein and Cass[100] have found that the water permeability coefficient, P_0, is markedly dependent on the molar ratio of cholesterol to phospholipid in the BLM-forming solution. For example, P_0 is about 42 $\mu m\ s^{-1}$ for a brain phospholipid BLM (molar ratio phospholipid/cholesterol = 0:1), whereas P_0 is only about 8 $\mu m\ s^{-1}$ when the ratio is 1:8. Similar effect has been noted when ergosterol was used instead. This decrease in permeability to water is attributed to an increase in the viscosity of the hydrocarbon region of the BLM, resulting in a decrease of the diffusion coefficient of water within this phase. It is interesting to note that the presence of cholesterol causes a decrease of the apparent area occupied by lecithin molecules in monolayer systems[101]. Therefore, the condensing effect of cholesterol is likely to reduce the available 'free' volume for permeating water molecules.

The mechanical properties of BLM are affected by a number of surfactants, polyene antibiotics, and proteins. The early investigators[95] have shown that surfactants such as digitonin and sodium oleate tend to break BLMs formed from brain lipids. HDTAB (hexadecyltrimethylammonium bromide) at concentrations higher than 4.5×10^{-4} M effectively prevents the formation of stable BLM[96]. Van Zutphen et al.[102] have reported that polyene antibiotics (filipin) interact preferentially with cholesterol-containing BLM and appreciably reduce their mechanical stability. The effect of filipin on BLM stability has also been examined by Finkelstein and Cass[100]. These workers used BLM made from ox-brain lipids plus DL-α-tocopherol and cholesterol. In addition Finkelstein and Cass have also found that both nystatin and amphotericin B rupture their BLM at high concentrations. Papahadjopoulos and Ohki[103] have examined certain phospholipid BLM. They have found that these BLM are more stable and possess higher electrical resistance when formed in Na^+ ions alone. The BLM instability becomes more evident when they are formed under conditions of asymmetric distribution of Ca^{2+}. Ohki and Papahadjopoulos[104] suggest that the instability of their BLM is due to the difference in interfacial energy across the biface.

Although a bimolecular lipid membrane separating two aqueous solutions may be considered as two monomolecular layers with the hydrocarbon chains joined together, the physical properties are quite different, and they are not simply additive. A BLM is unique in that it possesses two identical interfaces, insofar as the dielectric constants of the two bulk phases are concerned. As a first approximation, a BLM has two aqueous-solution/membrane interfaces, or a *biface*[105]. The bifacial tension (rather, the interfacial tension) of BLM for a variety of BLM systems in general is less than 6 dyn cm^{-1}. Recently, Moran and Ilani[106] measured the bifacial tension of a cholesterol–lecithin–methyloleate BLM and its water/bulk lipid solution interfacial tension, and concluded that the difference between this free energy of formation is more than 1 erg cm^{-2}. A value of 3.6 kcal mol^{-1} had been reported earlier[105]. In an ingenious device, Wobschall[107] has studied the bifacial tension, elasticity, breaking strength, and thinning rate of a surfactant–cholesterol BLM. Elastic constants of 150–300 dyn cm^{-1} and breaking tensions of 3–10 dyn above the bifacial tension value (5–10 dyn cm^{-1}) are obtained. In addition, Wobschall has discussed the effect of aperture geometry on BLM stability and thinning rates. Good[108], in a theoretical study, has analysed the formation of BLMs. Consistent with the experimental findings, Good has shown that the free energy of formation of BLMs is generally negative. Other thermodynamic approaches using Gibb's adsorption equation have been reported[105]. Haydon and Taylor[109] have estimated the London–van der Waals forces from contact angle measurements.

Quite recently Ter Minassian-Saraga and Wietzerbin have reported the action of HDTAB on lecithin–cholesterol BLM. At a concentration greater than 5×10^{-7} M, the lifetime of the membrane is appreciably reduced[110].

The presence of modifiers such as proteins can have a profound effect on the BLM stability and bifacial tension. The early workers[95] have observed that BLM formed from lipids alone can have appreciable interfacial tension, and in this respect resemble drum-taut air soap films, which shatter violently when broken. A bulge formed by hydrostatic pressure upon the BLM will flatten out again upon release of the pressure. In contrast, BLM formed from total lipids containing proteolipids can show very little interfacial tension. Once bulged, this type of BLM does not always return to its planar configuration after the pressure is equalised. Instead the membrane can become floppy, moving back and forth with the convection currents in the bathing solutions. The oxidised cholesterol BLM exhibit similar characteristics[94]. Evidently, the presence of protein is not an absolute requirement for these properties. It should be mentioned that the presence of proteins in the aqueous solution usually makes the formation of stable BLM exceedingly difficult, if not impossible. This, however, does not necessarily imply that protein-modified BLM are inherently less stable. It seems more likely that proteins and their related compounds are also highly surface-active materials which can modify the BLM support to such an extent that adhesion is difficult for the lipid solution.

The total protein derived from red blood cells has been tried on lecithin BLM by Maddy, Huang and Thompson[111]. The principal effects are a reduction of the bifacial tension of the membrane and an increase in optical reflectance. The latter effect has also been observed using oxidised cholesterol BLM[112]. The following proteins were tested: alcohol dehydrogenase, insulin, γ-globulin, chymotrypsin, trypsin and ribonuclease. In the case of alcohol dehydrogenase, the thickness of the membrane increased from 50 Å to about 100 Å. The other proteins increased the membrane thickness to about 80 Å. In most of these BLM–protein interaction experiments, the pH of the bathing solution had a marked effect on membrane reflectivity.

The reflectivity of the BLM has been used by many investigators interested in the membrane thickness. Most recently, Cherry and Chapman[113] have reported that the reflectivity of the BLM depends only on the average refractive index and total thickness of the membrane. However, other workers[96, 114] have found that a triple layer structure is a more realistic model for the BLM. In general, the thickness is about 10% higher when interpreting BLM reflectance data in terms of the triple-layered optical model[105].

Permeability to water in BLM has been measured by two methods: the osmotic method and the tagged water method[115]. In the tagged water method the flow of isotopically tagged water through the BLM is determined in the absence of an osmotic gradient. The permeability coefficient is P_d. In the osmotic method a concentration gradient is created, and the net flow of water across the BLM is measured. The permeability coefficient, P_0, is then calculated. The apparent difference between P_d and P_0 found by earlier workers is owing to inadequate stirring which gave rise to stagnant layers at the interfaces[100].

Regarding the mechanism of water permeation through the BLM, two specific mechanisms have been proposed: (i) the solubility-diffusion mechanism in which the BLM is assumed to be an ultra-thin layer of liquid hydrocarbon having bulk properties[116], and (ii) the energy-barrier mechanism[117]. In the latter scheme, it is envisaged that the water molecules get across the barrier (solution/BLM/solution) by 'barging' their way through. This means the creation of 'holes' along the pathway as the molecules permeate across the system. In terms of the absolute reaction rate theory[118], only those water molecules striking the interfacial region with sufficient energy will get across the BLM at the biface, implying that the permeating water molecules are in an activated state[117]. Pechhold[119] has suggested that paraffins and polyethylene may contain structural defects owing to the conformational changes in the hydrocarbon chains. The resulting change in free energy is estimated to be about 7 kcal mol^{-1}. In view of Pechhold's work, it is entirely reasonable to assume that the structural defects may exist as well in the hydrocarbon phase within a BLM.

Permeability of BLM to other compounds has also been reported. Lippe[120] has found that the lipid composition of the BLM is important in the regulation of the permeability to urea and thiourea. The effects of amphotericin B and other polyene antibiotics have been reported by two groups of workers. Andreoli and colleagues[121] have shown that P_0 is influenced by the antibiotics only in the cholesterol-containing BLMs. Holz and Finkelstein[122] suggest that polyene antibiotics create aqueous channels in the BLM, the effective radius of these channels being about 4 Å. The similarity of results obtained for the human red cells and for BLM is discussed in the Holz–Finkelstein paper.

Another paper by Dennis and Stead[123] discusses molecular aspects of polyene-dependent and sterol-dependent pore formation in BLM.

The mechanical stability of the BLM could also be affected by local anaesthetics as reported by Ohki[124]. These compounds in the presence of Ca^{2+} also lowered BLM resistance, the order of effectiveness being nupercaine > tetracaine > cocaine > procaine.

Jung[125] has employed spherical BLM of total lipid extracts of human red cell ghosts in a permeability study. The observed permeabilities of the membranes agree quantitatively with what is predicted by an analysis of non-specific movements of non-electrolytes across the cell membranes.

2.4.1.5 Modifiers altering electrical properties

The intrinsic electrical properties of a BLM can be modified by a variety of materials. These include simple inorganic and organic ions, surfactants, polypeptides and proteins. In addition, physical parameters such as temperature and applied electrical field can also alter the electrical properties of BLM. Among simple electrolytes, KI has been found by Lauger et al.[126] to lower lecithin BLM resistance by a factor of 10^3 or more. In HDTAB–cholesterol BLM it has been found[94] that the order of effectiveness in reducing resistance follows a lyotropic anion sequence: $I^- > Br^- > SO_4^{2-} > Cl^- > F^-$. Cations such as Fe^{3+} at aqueous concentrations below 10^{-5} can also decrease the resistance of lecithin BLM[127]. Rosenberg and Jendrasik[128] have also investigated iodine-modified BLM, and have suggested that a charge-transfer complex might be formed in the membrane, which could participate in the conduction process. Contrary to the above suggestion, Finkelstein and Cass[100] favoured an ionic conduction mechanism involving the polyiodides such as I_3^- and I_5^-. Recent experiments using oxidised cholesterol BLM provide additional interesting observations. In the presence of I^--I_2, the oxidised cholesterol BLM behaved like an ideal iodide electrode[94].

The effect of surfactants on the electrical properties of BLM reported by the earlier workers[94,95] has also been investigated by Hashimoto[129]. In addition, Hashimoto found that the resistance is nearly independent of the temperature (7–45 °C) and pH. Based upon his findings, Hashimoto suggested that the structural change of the membrane may be caused by both electrostatic and hydrophobic interactions of ionic surfactants with the membrane, and that the electrostatic interaction is necessary for the change in the BLM structure.

The effect of lipid composition on the electrical conductivity has been studied by Noguchi and Koga[130]. They have found that the BLM conductivity is dependent on lipid composition but depends little on the concentration and kind of electrolyte used. However, the membrane resistance was found to be pH dependent.

In a different type of study, Mehard, Lyons and Kumamoto[131] have employed BLM in a test for the mechanism of ethylene action. Although these investigators did not observe any effects due to ethylene, it is of interest to note that chloroform gas markedly altered the conductivity of the BLM in a reversible manner. The idea of using BLM as a model for the sensory detector is worth pursuing[132]. In this connection the study of Gutknecht and Tosteson[133] is of interest. They reported that in the presence of aliphatic alcohols the BLM resistance is appreciably lowered, but not the osmotic permeability.

It should be mentioned that both the lipid composition and the solvent used have a pronounced effect on the electrical properties of BLM. For example, lecithin BLM made from n-decane solvent had a resistance some two orders of magnitude higher than those formed from chloroform–methanol–tetradecane mixture. It is of interest also to compare the value of sorbitan tristearate BLM reported by Bradley and Dighe[134]. The BLM was found to be ohmic up to 300 mV and the average resistance was $5 \times 10^6 \, \Omega \, cm^2$. The lipid solvent was chloroform. Thus, it seems probable that the use of polar lipid solvents such as chloroform or alcohol may be an important factor responsible for the reduction of BLM resistance. The effect of lipid composition on the BLM resistance has been reported by van Zutphen and van Deenan[135], and by Seufert, Beauchesne and Belanger[136]. The latter group formed BLM from α-tocopherol and tetradecane, and reported that these BLM are considerably less resistant to mechanical shock than lecithin BLM. It is worth mentioning that α-tocopherol or vitamin E has long been known as a biological antioxidant. Pryor[137], in a stimulating article, has summarised the effects of vitamin E on biological membranes in relation to ageing and pathology, in which free radicals are believed to play a crucial role.

Leslie and Chapman[138] have found that incorporating either β-carotene or all-*trans*-retinal into lecithin BLM does not make the electrical conductivity unduly different from that of the unmodified BLM. The effect of the polar phospholipids on the ion permeability of BLM has been reported by Hopfer, Lehninger and Lennarz[139]. Their data indicate that the polar head groups of the lipids play an important role; BLM of positively charged lipids are anion selective, whereas BLM of negatively charged lipids are sensitive to cations, as would be expected. A paper by Neumcke[140], describing ion transport across BLM with charged surfaces, is of interest in this connection.

The effects of electric fields across a BLM can also alter its basic properties, as has been reported by Babakov, Ermishkin and Liberman[141] and more recently by Rosen and Sutton[142] and by White[143]. Using oxidised cholesterol BLM, White has found that the BLM capacitance is related to voltage by a simple equation, and the variations in membrane capacitance are interpreted as resulting from thickness changes. Walz, Bamberg and Lauger[144] have considered non-linear electrical effects in a series of papers. Gillespi[145], in a theoretical paper, has solved the Poisson–Boltzmann equation numerically for conditions approximating the surface of a lipid membrane and also compared the variation of potential, electrical field, range of potential, and adsorbed charge with electrolyte concentration and type with some published data.

An interesting technique has been devised by Del Castillo et al.[146] to detect antigen–antibody reactions using BLM. These investigators prepared BLM from the bovine brain lipid extract in 0.1 N NaCl. In immunological experiments, albumins were used as the antigen and immune serums as the antibody. A large number of enzymes with appropriate substrates were also tested (e.g. trypsin + BAEE, urease + urea, chymotrysin + ovalbumin).

Another interesting finding using BLM has been reported recently by Jain, Strickholm and Cordes[147]. These investigators formed BLM from a mixture of oxidised cholesterol and a synthetic surfactant. ATP is added to one side of the BLM, which is followed by the addition of a preparation of membrane-bound ATPase (Na-K dependent fraction). After these additions, a decrease in the membrane resistance is observed. Jain et al. observed a positive current flow across the BLM. This is interpreted as being due to the transport of Na^+ which is analogous to that found in the natural membrane. The current flow in their reconstituted system is said to be dependent on the presence of Na^+ in the bathing solution, and the magnitude of current depends on ionic strength. Further, the presence of ouabain, a compound known to block Na^+ transport in red cells, inhibits the current flow across the ATPase-modified BLM. It seems probable that, by introducing ATPase into the bathing solution, 'molecular sodium pumps' could have been inserted into the BLM, which may account for the observation. If so, this implies that ATPase (or BLM-modified ATPase) is actually the 'sodium pump' which has been described by numerous investigators in the literature[148].

Redwood and co-workers[149] have reported the physical properties of a BLM formed from 1,2-diphytanoyl-3-SN-phosphatidyl choline. This BLM is said to possess good mechanical stability. Most recently, Parisi, Rivas and De Robertis[150] have examined the effect of acetylcholine on the BLM conductivity. They reported that a rapid and transient resistance change could be observed only in the presence of a special proteolipid from the electric organ of *Electrophorus*.

2.4.1.6 Modifiers conferring ion selectivity and specificity

The intrinsic properties of lecithin or oxidised cholesterol BLM show little ion selectivity (apart from I^- noted above), which is consistent with the expected properties of the ultrathin hydrocarbon layer of the BLM (Table 2.4). These poor ion-selective properties of BLM, however, can be modified by a number of neutral macrocyclic molecules. The cyclic peptide valinomycin[151, 152], for instance, has been found to confer a selective order among the alkali metal cations as follows: $Rb^+ > K^+ > Cs^+ > Na^+ > Li^+$. The effect of valinomycin was earlier observed by Bangham, Standish and Watkins[153] and by Chappell and Crofts[154] in microvesicles (see Section 2.4.2). The observation of similar effect in BLM by Mueller and Rudin[151] and by Gotlib et al.[155] has been extended by other investigators. Andreoli et al.[156] have shown that the action of valinomycin is independent of the lipid composition. It should be mentioned that the mechanism by which valinomycin produces its effect on BLM and certain biological membranes is not clearly understood. At present, at least two different mechanisms have been proposed. Mueller and Rudin[157] favour the creation of specific pores in the BLM, whereas Lardy et al.[158] and Pressman[159] suggest that valinomycin may act as a molecular carrier. Recently, Eisenman et al.[160] have suggested a number of possible mechanisms to account for the observed effects.

In several recent papers the effect of the so-called uncouplers of phosphorylation on BLM has been extensively investigated. Bielawski and coworkers[161] have found that 2,4-dinitrophenol (DNP) when added in small amounts to the aqueous solution causes a marked reduction in BLM resistance. Other compounds related to DNP which produce similar effects on BLM have also been reported[162]. The effects of uncouplers of oxidative phosphorylation have been examined by several workers. Liberman and associates[163] treated their BLM with tetrachloro-2-trifluoromethylbenzimidazole (TTFB), carbonylcyanide-p-trifluoromethoxyphenylhydrazone (FCCP), carbonylcyanide-m-chlorophenylhydrazone (CCCP), and other compounds such as tetraphenylboron (TϕB). They have found pronounced effects on the BLM conductivity which increases linearly, or with the square of the concentration of additives. Current–voltage curves of BLM treated with some of these compounds exhibit a region with negative resistance. It is suggested that current–voltage characteristics with negative resistance can be related to diffusion processes not only in the BLM itself but at bifacial regions as well. A recent paper by Rosenberg and Pant[164] reports some electrical properties of the oxidised cholesterol BLM modified by picric acid and iodine. Owing to a limited amount of available data, Rosenberg and Pant have concluded that the nature of the charge carriers across the membrane, whether electronic or ionic, cannot be determined.

The effect of uncouplers on BLM has also been reported by Ting, Wilson and Chance[165]. Although they have also observed a resistance decrease by a number of compounds, the conclusion they reached is that no simple correlation exists between the relative effectiveness of the uncouplers on the conductance of BLM. Since literally hundreds of agents are known at the present time, a theory is needed to explain the observed phenomena. Bruner[166] has in fact attempted to establish a correlation between the pK value and the membrane potential arising from ion concentration gradients. He suggests that charge transport is by direct transfer of either protons or anions of the uncouplers. A fixed charge density at the membrane surface is required in Bruner's theory. Recently, iodine has been incorporated directly into the BLM-forming solutions (oxidised cholesterol or chloroplast extract)[167]. The BLM modified in this fashion have been found to respond ideally to iodide. The behaviour of I_2-containing BLM is reminiscent of a metallic electrode reversible to its ion (e.g. Ag electrode and Ag^+). As will be considered later in Section 2.4.1.8, BLM generated from chloroplast extract exhibit photoelectric effects. The photoelectric phenomena in BLM provide strong evidence that electronic charge carriers are involved in the membrane. This type of BLM should therefore be capable of functioning as a *redox* electrode similar to the platinum electrode. To test this idea, a chloroplast BLM (Chl-BLM) was formed in the usual manner separating two buffered solutions. To one side of the BLM a known quantity of Fe^{2+} was added, producing a few mV across the membrane. The Fe^{2+} is then titrated with $KMnO_4$. A titration curve of classical form was obtained with a rapid rise at the equivalent concentration[94]. This piece of evidence together with the iodine-containing BLM suggests that a BLM (after suitable modification) can behave like a redox electrode. In other words, a modified BLM can be considered either as a reversible electrode (i.e. a non-metal and its corresponding anions) or as an inert electrode analogous to a Pt electrode, which merely serves as a conductor for facilitating electron transfer[167].

2.4.1.7 Modifiers inducing electrical excitability

The most well known but still incompletely understood modifier, called EIM (Excitability Inducing Material), can induce the so-called electrical excitability in experimental BLM[95]. An EIM-modified black lipid membrane whose resistance has been lowered by 3 to 4 orders of magnitude will respond to constant current pulses. The membrane resistance is voltage dependent, and at a definite threshold the potential increases regeneratively by a factor of 5–10 fold[168]. In addition to EIM, a number of chemically defined compounds can also induce electrical excitability in BLM. The most outstanding example is alamethicin, a cyclopeptide antibiotic. This compound, in the hands of Mueller and Rudin[157], developed a cationic conductivity in BLM made of mixed brain lipids, egg lecithin, or oxidised cholesterol. The most unusual finding is that, when alamethicin is brought together with a minute quantity of protamine, and in the presence of an ionic gradient, it develops characteristics of negative resistance. Further, these specially treated BLM also exhibit delayed rectification, bistable changes of membrane potential, and single or rhythmic firing[157]. In a recent paper, Ehrenstein, Lecar and Nossal[169] suggest the existence of ion-containing channels capable of undergoing transitions between two states of different conductance in the EIM-modified BLM. The change in conductivity of BLM under applied electrical field may be a result of combinations of several factors. Some of these are: (i) an increase in the number of charge carriers, (ii) an increase in charge carrier mobility, (iii) the formation of pathways by trapped emulsified droplets and (iv) the formation of bridges by induced dipoles in strong fields[94]. Any one of these factors may contribute to the eventual dielectric breakdown of the BLM. Partial breakdown or electrical transients (excitability) in the BLM mentioned above may be similarly explained in terms of these factors. It is conceivable that a periodic build-up and release of space charges at the biface can give rise to complex electrokinetic phenomena. In connection with these interesting observations, mention must be made of the findings obtained by

Monnier[170] and Shashoua[171]. These investigators used membranes prepared according to their own methods, which, although two to three orders of magnitude larger in thickness, also exhibit rhythmic firing under applied voltages. For further discussion of electrical excitability associated with ultra-thin lipid membranes, the papers of Rudin and Mueller[157] and Monnier[170] should be consulted.

Electrical excitability can also be initiated by mechanical means. Ochs and Burton[172], and Ochs[173] have observed that a BLM can function as a mechanical electrical transducer. That is, when the BLM is subjected to mechanical vibration, a capacitance change is induced in the membrane, which follows the wave-form of the vibration. After an extensive study, Ochs has concluded that the primary effect is due to a variation in BLM area; the magnitude of the electrical response is inversely proportional to the thickness of the membrane. Furthermore, it is said that the response is significant only in a very thin membrane.

2.4.1.8 Modifiers generating photoelectric effects

Recent work has shown that BLM containing light-sensitive pigments such as chlorophylls[174] exhibit interesting photoelectric effects. In the form of a thin membrane (1000 Å to 0.1 mm thick) such effects have also been observed but are much less pronounced. When a chloroplast BLM is illuminated by light of various wavelengths, a photoaction spectrum can be obtained which is practically identical to the absorption spectrum[175]. The observation of the photo-e.m.f. (open-circuit voltage induced by light) together with the photo-electric action spectrum of the Chl–BLM provides strong evidence for the separation of *electronic* charges (electrons and holes) in the membrane. It is suggested that the presence of electrons and holes across the BLM may initiate redox reactions. It seems probable that these reactions will be demonstrated in suitable modified BLM to mimic certain aspects of biological processes such as the photophosphorylation in the plant chloroplasts.

Recently, a different series of experiments has been initiated with the aim of constituting BLM-containing carotenoid pigments and rhodopsin. The lipid solutions used for BLM formation consisted of various carotenoids (e.g. all-*trans*-retinal, 9-*cis*-retinal, all-*trans*-retinol, and β-carotene), phospholipids, and oxidised cholesterol, dissolved in liquid alkanes. The dark d.c. resistance of the membrane was ohmic ranging from 10^6–$10^7 \Omega$ cm^2, which was about two to three orders of magnitude lower than that of carotenoid-free BLM. All carotenoid BLM were found to be photoactive. In addition, the photoresponses of these carotenoid BLM have been found to be complex as the experimental conditions are altered. Depending upon the carotenoid pigments used and external factors (e.g. the presence of modifiers, temperature, pH gradient, and applied voltage), the voltage/time curves can vary from a simple monophasic response to a typical biphasic wave-form not unlike those found in vertebrate retina. The finding, from the standpoint of the structure of the outer sac membranes of visual organelles, is said to be an ideal system for investigating not only the initial energy transduction mechanism from photons to electrons and holes but also the triggering mechanism for ionic permeability across the membrane[176, 177].

The light-induced effects in photoactive BLM have also been reported by several other investigators. Hesketh[178] has devised a method to exclude thermal effects from the photoelectric measurements. In the presence of I_2 and I^-, Mauzerall and Finkelstein[179] have observed light-induced changes in the photoconductivity of BLM formed from brain lipids, α-tocopherol and cholesterol, and suggested that the photoactive material should be located in the membrane in order to be more efficient in light conversion. Using α-tocopherol and sphingomyelin instead of brain lipids, Kay and Chan[180] have also studied the photoresponses in the presence of I^-, and hypothesised that changes in the BLM induced by u.v. radiation are an important part of the mechanism leading to the photoelectric effect in their system. Other inorganic ions such as ferric cyanide and ferric chloride have been investigated by Pant and Rosenberg[181] using oxidised cholesterol BLM. They have reported an increase in the conductivity by a factor of 100–200 when the ions are excited by light in the u.v. region. From their data Pant and Rosenberg propose a model system which is said to suggest a new mechanism of the visual receptor process in biological photoreceptors.

In two short papers on chlorophyll-containing BLM, Alamuti and Lauger[182], and Trissl and Lauger[183] in the first paper report a fluorescence method by which the chlorophyll concentration in the BLM can be estimated. In the second paper they have observed very pronounced photoeffects in the presence of N,N,N',N'-tetramethyl-p-phenylenediamine. Trissl and Lauger suggest a similar redox reaction, as described earlier, occurring across the membrane.

Cherry, Hsu and Chapman[184] have recently measured the absorption spectrum of chlorophyll in a BLM. The similarity between the absorption spectrum obtained by their method and the action spectrum reported earlier[175] is quite striking. The original papers may be consulted for experimental details.

In addition to chlorophylls and retinals, a number of dyestuffs, when incorporated into BLMs which are otherwise not photoactive, can photosensitise the membrane. In a short note Ullrich and Kuhn[185] reported the photoresponses of a brain-lipid BLM in the presence of cyanine dyes. A systematic study of photosensitisation by a variety of dyes has been reported[167]. These include methyl orange, rhodamine B, methyl red, methylene blue, thionine, and methyl viologen. Most recently, photopotentials of a chloroplast BLM elicited by light flashes have been reported[186]. It has been found that the appearance of photopotentials across the membrane requires less than 1 μs, and the wave-forms of the photo response can be altered by (i) a chemical gradient, ΔC, (ii) a hydrogen ion gradient, ΔpH, and (iii) an electrical potential gradient, ΔV. The effects of the combinations of these gradients across the BLM have also been examined. The observed electrical transients elicited by light flashes are explained in terms of a simple model in which the choroplast BLM is considered as an ultra-thin liquid crystal of poor conductance separating two highly conducting aqueous solutions. When efficient electron acceptors (e.g. Fe^{3+}) are present at the solution/membrane interfacial region, the rapid formation of an electrical field across the BLM is interpreted as due to the ejection of photoelectrons from the membrane to the electron acceptors, which presumably originate from the excited chlorophylls[174]. The fast appearance of the electrical field with no detectable latency across the membrane in the microsecond range suggests strongly the production and separation of electronic charge carriers. The location of the electron acceptors which determines the polarity of the photopotential being always negative with respect to the acceptor-free side, is said to be in accord with this explanation[167].

2.4.2 Microvesicles (liposomes)

The second widely used model system consists of fragmented myelinic figures formed by sonicating (or mechanical agitating) a suspension of hydrated amphipathic lipids. These cylindrical and spherical vesicles of minute size ranging from several hundred Ångstroms to fractions of a millimetre have recently been studied by a number of investigators. These vesicles or microvesicles (also known as liposomes, spherules, or smectic mesophase) are of special interest, since their limiting structure is also of bimolecular thickness; they offer a much larger surface area than planar BLM described in the previous section. A comprehensive review of microvesicles covering the literature up to 1967 has been prepared by Bangham[187]. The use of these microvesicles as models for biological membranes has been summarised in a concise review by Sessa and Weissmann[188]. Details contained in the previous reviews are not repeated here. Instead, this section will give a brief description of recent contributions.

2.4.2.1 Techniques of formation

Detailed procedure for the preparation of microvesicles has been given by several investigators[189–192].

In brief, dried phospholipids are allowed to swell in an electrolyte solution with gentle agitation at room temperature. Mechanical agitation facilitates fragmentation of the micelles, producing microvesicles of varying sizes in the range of 5–50 μm in diameter. Instead of using mechanical agitation, ultrasonication can be used, which produces microvesicles of a much smaller size (~ 500 Å). The procedure is as follows: lyophilised phospholipids are suspended in a 0.1 M NaCl buffer with tris at pH 8.5. The suspension at a 3% concentration of phospholipids is ultrasonicated under nitrogen for 160 min at 2 °C. The resulting suspension is then isolated by molecular sieve chromatography on large-pore agarose gels. This method is said to yield a homogeneous fraction of microvesicles which are bounded by a single bilayer[191].

Similar to BLMs, microvesicles have been formed from a wide variety of synthetic and natural lipids. In contrast to BLMs, it should be mentioned that hydrocarbon or organic solvents are not required for the formation of the microvesicles.

2.4.2.2 Structure of the microvesicles

Unlike fatty acid soaps which undergo phase transitions in water, forming spherical micelles and dispersed molecules in most cases, phospholipids and the like form continuous or 'closed' bilayer sheets separating two aqueous solutions. The structural characteristics of microvesicles generated from various phospholipids have been investigated by optical techniques, electron microscopy, and x-ray diffraction[187, 193]. The phospholipid microvesicles in the light microscope exhibit birefringence. The size of birefringence depends upon the surface charge of microvesicles and the ionic strength of the aqueous phase. X-ray diffraction of microvesicles shows the presence of multilamellar structures with repeating distances varying from 54–75 Å, depending upon the lipid used[190]. The electron micrographs of microvesicles stained negatively with ammonium molybdate show concentric lamellar layers, each approximately 50 Å thick. In the case of microvesicles prepared from bovine brain lipids fixed with OsO_4, electron micrographs indicate that the boundaries consist of a single or a few bilayer leaflets having a peak-to-peak distance of 45 ± 5 Å per bilayer. The size distribution of the microvesicles in a given preparation has been determined by electron microscopy and turbidity measurement using light-scattering techniques[194].

2.4.2.3 Osmotic properties and permeability to water

Microvesicles are about 100 million times more permeable to water than to electrolyte, and are therefore osmotically active[187]. Using photometric and volumetric methods Rendi[195, 196] has shown that the water present in the microvesicles is extruded by the addition of osmotically active compounds. The loss of water obeys the Boyle–Van't Hoff law. Bangham, de Gier and Greville[197] have found that microvesicles made from charged phospholipids behave as almost perfect osmometers when alkali metal salts, glucose, or mannitol are used as solutes. The water permeability coefficients for lecithin microvesicles are about 44 $\mu m\ s^{-1}$ at 25 °C and 70 $\mu m\ s^{-1}$ at 37 °C. From these data the actuation energy for water permeation is estimated to be 8.2 kcal mol^{-1}. Reeves and Dowben[198] suggest that their data can be explained in terms of a solubility mechanism; however, further studies are necessary to establish its validity (see Section 2.4.1.4). It is interesting to note that microvesicles containing cholesterol show lower water permeability, which has also been observed in the BLM system[100, 117].

Other solutes such as erythriotol, malonamide, urea, glycerol, propionamide, ammonium acetate, ethylene glycol, methylurea, and ethylurea are most permeable[197]. The rapid volume changes combined with the measured total external surface area are used to calculate water permeability coefficients. Values in the range of 0.8–16 $\mu m\ s^{-1}$ have been obtained by Bangham, de Gier and Greville[197].

2.4.2.4 Permeability to eletrolytes

Microvesicles containing positively charged lipids such as long-chain secondary and quaternary amines are very impermeable to univalent cations (e.g. Na^+ and K^+), but are permeable to anions. Bangham, Standish and Watkins[189] have found that the diffusion rate for anions follows the sequence

$I^- > Cl^- > F^- > NO_3^- > SO_4^{2-} > HOP_4^{2-}$. The diffusion rate for cations is largely controlled by the size and magnitude of the surface charge. However, in contrast to anions, no significant differences are found for the cation series Li^+, Na^+, K^+, Rb^+ and choline. The effect of divalent cations has also been reported[187].

Owing to the minute size of microvesicles, any direct measurements of electrical properties have thus far been precluded. Nevertheless Papahadjopoulos and Watkins[190] have estimated a resistance value of $10^7 \Omega$ cm^2 using their Cl^- flux data. This estimated value, although one to two orders lower than that of BLM, is consistent with the postulated lipid interior and the osmotic properties discussed in the preceding sections (2.4.1.1 to 2.4.1.3).

Recently the electrical capacitance of a suspension of microvesicles has been measured by Schwan, Takashima and co-workers[199]. These investigators used a frequency range from 1 kHz–100 MHz and have concluded that (i) at low frequencies the microvesicles undergo a counterion-induced relaxation (movement of counter-ions tangential to the membrane surface), and (ii) a Maxwell–Wagner dispersion results at high frequencies (above 1 MHz). The dispersion indicates a dielectric constant value of 10 or higher for the microvesicular membrane. It is further stated that the water inside the microvesicles is of the 'free' type and is entirely comparable to that outside. This conclusion at first sight is at variance with the electrical properties of the red cell reported earlier by Pauly and Schwan[200]; it is pointed out however that the microvesicles contain no proteins inside.

In another interesting paper the effects of essential fatty acids on ion permeability of lecithin microvesicles were reported by Moore, Richardson and DeLuca[201]. At 37 °C the rate of Na^+ efflux was greater for the microvesicles prepared from lecithins deficient in essential fatty acids. The results imply that dietary induced variation in fatty acid composition of membrane lipids might play an important role in membrane ion permeability.

2.4.2.5 The effects of modifiers

A variety of modifying agents such as surfactants, steroids, antibiotics, certain proteins, and anaesthetics can significantly increase the permeability of microvesicles. For instance, the effects of a larger number of steroids and other lytic agents, studied by Bangham, Standish and Weissmann[202] can be summarised in two kinds: those agents that increase the permeation and those that retard the permeation. The first group includes diethylstibestrol, deoxycorticosterone, testersterone, progesterone, filipin, nystatin, amphotericin B, and others[203]. Compounds that reduce the permeability or stabilise the microvesicles are cortisone, cortisol, chloroquine, and 17-β-estradiol.

The effect of a very well known surfactant, Triton X-100 (a mixture of p-t-octyl poly(phenoxyethoxy) ethanols, Union Carbide Chemical Co.), has been studied by Weissmann, Sessa and Weissmann[203]. These workers found that Triton X-100 accelerates anion release from microvesicles and causes substantial ultrastructural changes. Other modifiers that exhibit pronounced effects are bacterial toxins. Weissmann, Keiser and Bernheimer[204] have found that these haemolysins rupture both microvesicles and certain natural membranes. It is suggested that since bacterial toxins possess lipid-like structure (amphipathic), insertion of such molecules into lamellar membranes could produce weak spots and eventually lead to structural failure.

Interactions of microvesicles (liposomes) with proteins have been reported only in a few papers. Papahadjopoulos and Miller[205] and Papahadjopoulos and Watkins[190] have examined the effects of proteins on the properties of microvesicles. They reported that the presence of cytochrome C and to a lesser extent bovine serum albumin (BSA) tends to produce finer microvesicles (200–1000 Å). The presence of proteins on the vesicles appeared to increase salt capture in the case of phosphatidylcholine. However, salt capture was decreased in the case of acidic phospholipids such as phosphatidylserine in the presence of cytochrome C. This phenomenon is explained in terms of complex formation, thereby diminishing the available ionic groups which presumably are involved in the salt capture process. Additional work on phospholipid microvesicle protein interactions has been reported by Reeves and Dowben[192], and Sweet and Zull[206]. It is reported that BSA when adsorbed upon the exterior boundary of these structures enhances glucose diffusion from the interior through the bilayer-like walls. A pH-dependent conformational change in the albumin is assumed, but the properties of the system including its detailed structure are not defined.

In the presence of local anaesthetics such as n-butanol, n-octanol, diethyl ether and chloroform, Bangham, Standish and Miller[207] have found that the K^+ permeability of microvesicles is increased. More recently, Johnson and Bangham[208] reported the effect of temperature on the K^+ permeability with and without the above-mentioned compounds. The conclusions reached by these workers can be summarised as follows:

(i) The main barrier to cation permeation is located at the solution–membrane interface,
(ii) near this interface the structure (of water?) is highly ordered, and
(iii) the effect of local anaesthetics increased the disorder at the interfacial region.

It is worth noting that very similar suggestions have been made earlier concerning water permeability in the BLM systems[117]. (see Section 2.4.1.4).

The effect of five other local anaesthetics on microvesicles has been reported by Papahadjopoulos[209]. He measured the effects of dibucaine, tetracaine, cocaine, lidocaine and procaine on the efflux of $^{22}Na^+$ across microvesicles formed from a phosphatidylserine–phosphatidic acid mixture. The most interesting observation is that in the presence of Ca^{2+}, the calcium-induced Na^{2+} efflux is completely inhibited by these compounds, which is said to be comparable to the effect observed in nerves[210]. In a related system (smectic mesophase) Murthy and Rippie[211] have studied hydrolysis reactions of procaine and its quaternary derivatives.

In a preliminary note, McElhaney et al.[212] have reported, in addition to the fatty acid composition, the effect of cholesterol on the permeability of microvesicles and *M. laidllawii* B cells. Daemen and Bonting[213] have reported that addition of retinaldehyde to phosphatidylethanolanine microvesicles increases cation permeability. It is interesting to note that bilayer lipid membranes containing carotenoid pigments, besides being sensitive to light, also exhibit increased ion conductance[174]. Their findings are of value in the understanding of the mechanism of visual excitation.

2.4.2.6 Miscellaneous studies

Similar to the BLM system, microvesicles have been used as a model for the mitochondrial and chloroplast membranes. Chapman and Fast[214] have carried out photochemical studies with microvesicles containing chlorophyll in the presence of cytochrome C in the aqueous solution. When the chlorophyll microvesicles were illuminated by light at 435 nm fluorescence was observed at 673 nm with a red shift of some 10 nm, as compared with the spectrum of chlorophyll in ether. Evidence was also presented by Chapman and Fast for light-initiated reduction of cytochrome C.

Interactions between cytochrome C and microvesicles prepared from a mixture of lecithin and cardiolipin have been reported recently by Kimelberg and Lee[215]. They reported certain similarities and differences between their observations those reported by Das and Crane[216] and by Reich and Wainio[217] on lipid cytochrome complexes. The reduced rate of reaction of cytochrome C with ascorbate and dithionite is attributed to the permeability characteristic of the membrane. The presence of either TMPD (tetramethyl p-phenylene diamine) or PMS (phenazine methosulphate) which greatly increased the rate of reductions is cited as evidence in support of their explanations.

More recently microvesicles formed from chloroplast extracts have been used in the author's laboratory for studies related to photosynthetic and redox reactions. For instance, evolution of oxygen from chlorophyll-containing microvesicles, as detected by a Clark-type electrode, has been provisionally observed in our preliminary experiments[218].

A third model membrane system, closely related to the microvesicles, has been used to study the effects of permeation and transport properties of cells. Stable vesicles much larger in diameter (1–100 μm) than the microvesicles described above, have been prepared from nylon, collodion, or heparin–collodion and utilised as artifical cells. Chang and his associates[219] have found that these vesicles, known as microcapsules, can mimic many simpler properties of cells. Attempts have been made to administer a variety of microcapsules for the recirculation of peritoneal fluid, and of dialysis fluid of artificial kidneys. The use of semipermeable microcapsules in the treatment of certain enzyme-deficiency diseases or disorders due to the accumulation of toxins is suggested[219].

Before ending this section on microvesicles, it may be useful to point out the relative merit of the microvesicles and BLM systems. First, the microvesicles (liposomes) provide large surface areas and therefore are well suited for the study of transport properties of ultra-thin membranes (~ 100 Å). Second, the uniformity of size of microvesicles, as demonstrated recently by Huang[191], offers an opportunity for investigating lipid–protein interaction using the well-developed physical methods such as light scattering and ultracentrifugation. Other techniques such as electron microscopy, x-ray diffraction, n.m.r., and e.s.r. using spin labels (see Section 2.3.2.8) may also be applied. The main drawback of microvesicles as an experimental model for the biological membrane (see Section 2.5) is that, owing to their minute size, the powerful electrical methods cannot be applied at the present time. Perhaps one of the best approaches is to carry out the studies concurrently using both BLM and microvesicles[193]. These two model systems are not mutually exclusive; on the contrary, they are complementary to each other, since both types of membranes are derived from the amphipathic lipid.

2.4.3 Monolayer and miscellaneous studies

2.4.3.1 *Monolayer films*

Research in this field of monolayers has continued at a steady pace. Literature relevant to the present consideration has been reviewed by Gaines[220], Lakshminarayanaiah[13], James and Augenstein[221], and by Tien and James[193].

Although there exists a considerable body of information about lipid monolayers and some information about insoluble protein monolayers when present separately, the physical state of an interface at which lipids and proteins are simultaneously present is not well defined. Dawson[222] has given a detailed summary of phospholipase reactions at the phospholipid–water interface. In these systems the zeta potential of the interface along with the charge on the enzyme are highly important. Here, of course, the enzyme must be adsorbed at the interface in such a way that its conformation does not change sufficiently to destroy the enzymic activity. As has been pointed out by James and Augenstein[221], proteins frequently undergo marked conformational changes at interfaces sufficient to cause decreased solubility, loss of biological activity and denaturation. However, where the energy of the interface is low or the area available per protein molecule is small, adsorption may not be followed by denaturation. Lucy[223] has suggested that unfolding at interfaces may represent simply a loss of tertiary protein structure, hence specific interactions between protein and lipids should be considered, perhaps, in light of the possible stability of α-helices at lipid–water interfaces. By means of indications, it has been found that the α-helix may be the stable configuration in polypeptide monolayers at air–water and perhaps at oil–water interfaces[224].

The possibility of demonstrating complex formation between lipid and protein, lipid and lipid, or other combinations at interfaces is still somewhat in dispute. Vilallonga *et al.*[225] have reported studies of a mixed film formed by spreading protein into a film of cholesterol and found a linear relationship between the surface pressure of the mixed monolayer and the mole fraction of amino acid residues in the film. From results obtained in radiotracer studies, Matsubara[226] has refuted earlier reports of complex formation in mixed films of lipids and sodium alkyl sulphate. The detergent is found to adsorb and exert its film pressure independent of the lipid film.

Studies of the interaction of protein injected beneath a lipid monolayer were once thought to indicate that a layer of spread protein was formed beneath the monolayer, followed often by a second layer of unspread, globular protein molecules beneath the first. Recent studies by Colacicco and Rapport[227] have tended to refute the idea that the hydrophobic moiety of protein penetrates between lipid hydrocarbon chains. Some evidence cited is the following: pronase or trypsin injected into the subphase beneath other protein in equilibrium with a lipid monolayer produced no fall of surface pressure, although protein hydrolysis in the subphase was extensive. Further, Colacicco[228] has discussed the problem of specific lipid–protein interactions in monolayers, and concluded that proteins which are found associated with lipids in nature seem to have unique surface activity and form the mixed films spontaneously when injected.

Other properties of the monolayer which can be studied are the surface potential, surface viscosity, and material transport. Some practical applications have been reported by Blank[229], Miller[230], and by Cadenhead and Phillips[231]. Most recently, Fromherz[232] has reported electron microscope studies of lipid–protein (ferrifin) monolayers. Kleuser and Bucher[233] have studied electrochromic effects in chlorophyll monolayers.

It should be pointed out that the use of monolayers as models for biological membranes is much less satisfactory, since a lipid monolayer, with or without sorbed proteins, cannot be identified as one half of a lipid bilayer. From energetic considerations the molecular interaction and organisation may be quite different in the two systems.

2.4.3.2 *Miscellaneous studies*

Many kinds of other model systems have been investigated. These include micelles[234], liquid crystals[235], lipid-in-bulk phases[236, 237] and ion-exchange membranes[238–240]. For the interfacial chemistry of these systems, the literature has been reviewed by Ottewill[241]. The literature and the subject matter dealing especially with transport phenomena have been published by Lakshminarayanaiah[13].

2.5 BIOLOGICAL MEMBRANES

Modern biochemistry and molecular biology have conclusively demonstrated the preponderance of lamellar membranous structures and the intimate role of these structures associated with the basic energy transducing processes occurring in mitochondria and chloroplasts. The specific permeability properties of living cells and conduction of nerve impulse, found earlier and greatly reinforced recently, are also attributed to lamellar membranes.

To facilitate this review, biological membranes are classified in terms of their functions. Five basic types of biological membranes can be easily recognised. These are the plasma membrane of cells, the thylakoid membrane of chloroplasts, the cristae membrane of mitochondria, the nerve membrane of axon, and the visual receptor membrane. Each of these membranes is considered under a separate heading in this section (see Figure 2.3).

2.5.1 Plasma membranes

Among the five basic types of biological membranes listed above, the plasma membrane is by far the most frequently studied living membrane. Although it has not been conclusively demonstrated, it seems probable that all the other biological membranes may have evolved from the plasma membrane. The most familiar example of a plasma membrane is that of erythrocytes or red blood cells. Insofar as is known, the function of the plasma membrane, in addition to serving as a phase boundary between the cytoplasm and its environment, is to create and maintain the interior content of the cell by the active transport of ions and nutrients, and to eliminate the waste products. The plasma membrane is also known to be antigenic. Depending upon the species, the gross composition of the erythrocyte plasma membrane is about 60–80% protein and 40–20% lipid[242, 243]. The major lipids are phospholipids and cholesterol. Phosphatidylcholine (PC), phosphatidylethanolamine (PE), phosphatidylserine (PS) together with sphingomyelin (SM) account for about 50% of the phospholipids. Cholesterol content for different species varies from 22 to 32% of the total weight of the lipids[40]. Other lipids are glycolipids and gangliosides. Remarkable differences exist in the fatty acid compositions of the plasma membranes. In general, there are about 40% polyunsaturated C_{18}–C_{22} fatty acids[33].

The protein of erythrocyte plasma membranes has been very difficult to isolate and study. Maddy[244] has extracted proteins from erythrocyte ghosts using aqueous butanol. The proteins showed one major and three minor peaks by electrophoresis. By gel filtration the molecular weight was estimated to be between 2–7×10^5. Morgan and Hanahan[245] showed that the protein can be separated from the lipid by sonication of erythrocyte ghosts in 15% butanol.

Passow[246], after reviewing the literature on the passive ion permeability of the red cell membrane, concluded that many features of ion permeability as a function of pH and concentration can be explained by the presence of dissociable fixed charges in the membrane. The pK value of the membrane is estimated to be about 9, but whether this is due to lipid or protein NH_2 groups is uncertain. Passow has also pointed out that at pH values below 6.8, anion permeability of red cell membranes passes through a maximum, and the fixed charge concept at the present time is inadequate to explain the decrease of anion flux across the membrane with decreasing pH of the medium. Further speculations about alternatives to the fixed charge hypothesis of membrane permeation are anticipated.

At a recent symposium on 'Red Cell Membranes', Weinstein[247] summarised the ultrastructure of the membrane, Hanahan[243] reported lipid–protein interactions, whereas a detailed account of the lipid composition was given by Sweeley and Dawson[242]. Other aspects of the red cell membrane were also covered by leading investigators in the field[11]. The surface chemistry of red cell membranes has been extensively reviewed by Tenforde[248].

In recent years despite the generally obtained unit membrane pattern in electron micrographs of the red cell membrane or in red cell ghosts, a dispute has arisen about the detailed membrane structure. For example, Wallach and Zahler[63] have concluded that erythrocyte ghosts do not contain large amounts of protein in the extended β-configuration, based upon infrared and fluorescence spectroscopy. Optical rotatory dispersion (o.r.d.) measurements suggest that portions of the membrane protein exist in regions of high refractive index supporting the suggestions of hydrophobic interactions.

A model is proposed in which hydrophilic peptide portions lie at the membrane surface connected by hydrophobic helical rods penetrating the apolar core of the membrane normal to its surface. The interior of these rods could be polar, providing for hydrophilic 'pores' through the membrane[249]. Other authors agree that protein extracted from the plasma membrane is not in the β-configuration and consider, instead, that it is a typical globular protein with low α-helical content. Spin-labelling experiments indicate that erythrocytes, their ghosts, and mitochondrial membranes among others do not contain liquid-like, hydrophobic regions of low viscosity capable of binding the TEMPO spin label[251]. These regions are presumably phospholipid bilayers similar to those found in artificially prepared phospholipid vesicles (see Section 2.3.2.8).

Low-frequency impedance studies on beef erythrocyte suspensions have failed to show any dispersion below 1 kHz which casts doubt on the possibility of lipids being present in a continuous bimolecular layer in these red cell membranes[61]. Human red cell membranes upon treatment with the enzyme phospholipase C lose about 70% of their total phosphorus by hydrolysis. However, the membrane remains intact under phase microscopy and this protein content as measured by circular dichroism is virtually unchanged. The model proposed is stabilised by hydrophobic interactions and locates polar and ionic heads of lipids on the outer surface of the membrane accessible to enzyme. Non-polar interactions between hydrocarbon chains of lipid and protein are also suggested on the basis of n.m.r. studies of erythrocyte membrane fragments after sonic disruption[252]. Absorption peaks of methylene protons of hydrocarbon residues of proteins are unresolved until after treatment with urea or trifluoroacetic acid.

Additional difficulties arise from interpretation of electron micrographs of plasma membranes. Recent work has been summarised by Chapman and Wallach[253]. Among the findings which must either be made to fit the membrane model or be explained away as artifacts are more than twofold variations in thickness, and granular or 'pebbly' substructure revealed principally by shadowing techniques. Recent small-angle x-ray diffraction studies by Moretz et al.[254] have indicated that standard electron microscopy fixatives, OsO_4 and $KMnO_4$, did not prevent rearrangement of the structure of the myelin membrane of frog sciatic nerve during subsequent acetone and alcohol dehydration. The observed changes in periodicity and intensity are said to be associated with a considerable extraction of cholesterol and a lesser extraction of polar lipids.

2.5.2 Chloroplast membranes

The importance of membranes in photosynthesis has been increasingly recognised in recent years. The membranes, in particular the thylakoid membrane, are implicated in the process of quantum conversion, phosphorylation, and O_2 evolution. Recent reviews on the photochemical systems by Boardman[255], on the membrane structure by Branton[256], and on ion movements and energy conservations by Walker[257] are of interest.

Chloroplast is considered the basic photosynthetic unit of green plants and algae. The internal membraneous organelles are enclosed within a limiting plasma membrane termed the chloroplast envelope. One of the obvious functions of the chloroplast envelope membrane is to entrap all the necessary chemicals and enzymes in close proximity for the CO_2 reduction. The internal organelles are comprised of flattened membrane sacs or thylakoids. These are frequently called thylakoid membranes by Menke[258]. From the images obtained under the electron microscope, the structure of the thylakoid membrane has been variously interpreted. According to Park and his colleagues[259, 260] the membrane consists of an ultra-thin lamella with attached globular units termed quantasome, which measure $160 \times 150 \times 100$ Å. Similar interpretations have been given by Branton and Park[261] from results obtained on freeze-etched membranes. Weier and Benson[262] have interpreted their images in a somewhat similar fashion, and conjured up a highly intricate picture. Using the freeze-etching technique pioneered by Moor, Muehlethaler, Waldner and Frey-Wyssling[263], Muehlethaler[264] has suggested a picture for the thylakoid membrane, which is based upon the bimolecular leaflet model. The salient feature in all these models lies in their oriented lipid core onto which other important cellular constituents such as proteins may interact through either ionic or van der Waals attractions or both. Muehlethaler's picture is of special interest in view of the BLM experiments described in Section 2.4.1.8.

The gross composition of thylakoid membranes consists of 45% protein and 55% lipids. The major lipids are digalactosyldiglyceride and monogalactosyldiglyceride[265]. Phospholipids and phosphatidyl glycerol make up 75% of the total lipids. The remaining 25% are chlorophylls, carotenoids and other pigments. Sterols and glycerides are very minor constituents[266]. The structural proteins have been obtained by Criddle and Park[267]. Only one major peak was seen in the analytical ultracentrifuge. The molecular weight was 23 000. Other studies on the protein component have been carried out by Lockshin and Burris[268] and by Thornber, Stewart, Hatton and Bailey[269]. The latter workers used sodium dodecyl sulphate in their preparations. Further details may be found in a recent review on structural proteins of chloroplasts by Criddle[42]. A recent paper by Ji, Hess, and Benson[80] discussed the nature of interactions between lipids and proteins in chloroplast lamellar membranes.

The membrane aspects of photosynthesis may be divided into three major categories:
  (i) the primary processes,
  (ii) photophosphorylation and ion transport and
  (iii) oxygen evolution
The exact role of the membrane of chloroplasts in some of these processes is by no means clear at the present time.

2.5.2.1 *The primary processes*

The involvement of membranes in the primary photophysical and photochemical events is implicit in the various formulations of the mechanisms of photosynthesis. The scheme that two photosystems are necessary in driving electrons from water to a final state capable of CO_2 reduction is still in vogue. A good review of the literature is that of Boardman[255] cited earlier.

The primary event of charge separation following the absorption of light is undoubtedly the most crucial step. In Van Niel's scheme it is represented by a redox reaction: $H_2O + hv \rightarrow (H) + (OH)$, which in modern terminology may be translated into 'electrons' and 'holes' associated with their respective molecular species. Recent evidence has been provided by Witt, Rumberg and Junge[270], using an electrochromic method which is based on the measurement of transient absorption difference changes between 200 to 800 nm. The speed of the absorption change at 515 nm is extremely fast ($\tau \simeq 10^{-8}$ s), and has been interpreted in terms of charge separation across the membrane. Other studies using absorption difference spectroscopy and related spectrophotometric methods in photosynthesis research have been reviewed by Fork and Amesz[271].

2.5.2.2 Photophosphorylation and charge transport

In the two photosystems of 'Z' scheme mentioned in the preceding section, it is generally accepted that Photosystem I and Photosystem II together provide reduced nicotinamide adenine dinucleotide phosphate (NADPH) and adenosine triphosphate (ATP) for the CO_2 reduction. However, the details of these systems remain obscure and are the focal point of much current activity. Avron and Neumann[272] have summarised the effort of many investigators on possible mechanisms for photophosphorylation. Avron has suggested that the light-induced pH gradient is not the primary condition for phosphorylation but its presence may help to drive the negatively charged phosphorylation agents across the thylakoid membranes. In contrast, in the chemiosmotic hypothesis (COH) of Mitchell[273] a pH gradient and/or an electrical field of sufficient magnitude is a prerequisite for phosphorylation. In Mitchell's formulation the membranes must be relatively impermeable to ions. The driving force for phosphorylation, according to Mitchell, is due to membrane potentials which could come about as a result of pH gradient produced in turn by the electron transport chain located in or on the membrane. The results obtained by Jagendorf, Neumann and Hind[274, 275] with chloroplast suspensions subjected to pH 'shock' are generally cited as evidence in support of the chemiosmotic hypothesis. It should be stated that the Mitchell mechanism centred around ion-impermeable membranes as his main thesis has many critics (see Section 2.5.3).

In connection with the above discussion on charge transport Barber and Kraan[276] have observed that with pre-illuminated chloroplast suspensions upon addition of alkali metal salts the phenomenon requires the presence of a membrane, as predicted by the chemiosmotic hypothesis. The salt-induced light emission was reported earlier by Miles and Jagendorf[277]. The latter investigators have also reported an anion specificity at low salt concentration (about 0.3 M) but little difference in the ability of Na^+ and K^+ to induce luminesence. These interesting observations as pointed out by Barber and Kraan[276] lend support to the idea that the thylakoid membranes have a low passive permeability to H^+. The efflux of protons as charge carriers driven by an electrical gradient can result in an increase in the rate of recombination of electrons and holes in the membrane.

Other papers dealing with electrical potentials across the thylakoid and chloroplast envelope membranes have been published by Vredenberg[278], and by Luttge and Pallaghy[279].

2.5.2.3 Oxygen evolution

That isolated chloroplast retained the ability to produce oxygen in the light was shown by Hill in 1937; however, the mechanism of O_2 evolution is still a mystery. To what extent the thylakoid membranes are involved will be highly speculative at the present time. Nevertheless, the requirement of manganese has been shown by Cheniae and Martin[280]. Current investigations are concerned with quantum requirement, the effects of chemicals, and physical parameters. Kok and Cheniae[281] have published a review of the literature on the kinetics and intermediates of the O_2 evolution. Kok, Forbush and McGloin[282] in a recent paper reported oxygen evolution in flashing light. They have observed that the amount of O_2 evolved follows an oscillation with a period of four flashes, the yield being highest at the third flash. Barbieri and co-workers[283], using a highly sensitive oxygen electrode, have obtained higher yields with four flashes. The co-operation of four reaction sites occurring successively on the same photoactive centre is indicated. W. P. Williams[284] has recently discussed the concept of an 'oxygen unit', conceived earlier by Izawa and Good[285], and suggested its usefulness in resolving some controversial points in the field of photosynthesis research.

From a biophysical point of view, the use of whole chloroplasts to understand the aforementioned processes at the molecular level does not seem satisfying. The translocation of ions such as protons in a complex system such as a chloroplast is obviously very difficult. The interactions between various fluxes (ion and water movements, electron and hole transport, etc.) and their conjugate forces (osmotic pressure and electrical potential, etc.) are far too complex (if not totally unknown) to be amenable to a simple analysis. It must be concluded that, as has been remarked by Gaffron[286], there is an obvious need to study entirely artificial function membranes.

2.5.3 Mitochondrial membranes

Besides the chloroplast, the mitochondrion is the second most important physiological unit that can be easily recognised for its specific function (Figures 2.3). Like chloroplasts, mitochondria can be studied in isolation. The sequences of events taking place in the mitochondrion have been described in considerable detail by biochemists. The overall process consists of three separate but closely related steps:
  (i) the production of ATP during the oxidation of the coenzymes,
  (ii) the rearrangement and oxidation of carbohydrates to form reduced coenzymes,

(iii) the oxidation of these coenzymes by molecular oxygen.

However, the details of these steps are by no means known with certainty. There are no definite answers, for instance, to questions like: exactly how is energy transduced and trapped in a chemical bond of ATP and what is the membranous nature of this energy transducer

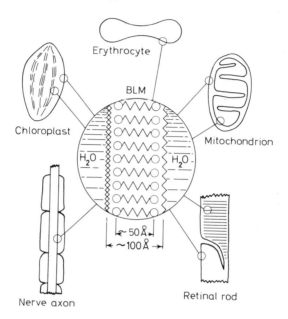

Figure 2.3 Schematic illustration of the five basic types of biological membranes as they are generally visualized under the electron microscope and their molecular interpretation according to the bimolecular leaflet model (BLM)
(From Tien[94], by courtesy of Marcel Dekker, Inc.)

Nevertheless, two types of mitochondrial membranes are generally recognised: the outer membrane and the inner (cristae) membrane. The outer membrane contains about 55% protein and 45% lipids. Of the total lipids a small portion is cholesterol[33]. The inner membrane contains all the cytochromes and other enzymes, and less lipid than the outer membrane. Green and associates[29, 30, 287] have obtained four protein fractions. Racker[288] has isolated the so-called F-fraction, an oligomycin-sensitive ATPase, and shown it to be a sphere of diameter 85 Å. Extraction of 90% or more of the mitochondrial lipid with acetone–water mixtures does not materially alter the gross appearance of the OsO_4-fixed specimen in electron micrographs. Green and Fleischer[287] have suggested that protein may be primary in determining mitochondria membrane structure. Respiratory activity is lost following such de-lipidation but regained when phospholipid is restored, even though the added phospholipid may be greatly different in composition from the original. Fleischer et al.[73] have suggested that ionic, hydrogen and hydrophobic bonding are involved in anchoring the lipid within the protein particulate.

Crane and Hall[289] have re-interpreted electron micrographs of stained mitochondrial sections in terms of a binary membrane consisting of a double layer of lipoprotein globules. The mitochondrial membrane appears as a five-layered structure—three dark lines separated by two unstained regions. The proposal had its origin in fragmentation studies of cristae. It is said that fracture through the central polar regions of the membrane is consistent with the lipoprotein bilayer membrane but not with the unit membrane consisting of a single layer of lipoprotein particles as described by Green and Perdue[30].

Prezbindowski et al.[290] have found the mitochondrial cristae membranes can be split by detergent treatment into two fractions each of which shows a globular structure. The thicker membrane is said to be re-formed when the two fractions are recombined. From a consideration of amino acid composition of soluble and membranous lipoproteins, Hatch and Bruce[291] conclude that the marked differences in the types of lipoproteins are related to the different environments to which they are exposed. Other aspects of the molecular structure of mitochondrial membranes have been thoroughly reviewed by MacLennan[292].

In a series of papers, Grinius, Jasaitis, Kadziauskas, Liberman, Skulachev, Topali, Tsofina, Vladimirova, Bakeeva, Kuliene, Levitsky, Severina, Samuilov and Isaev[293] have discussed energy transduction in membranes of mitochondria, submitochondrial particles and bacterial chromatophores. Liberman, Skulachev[294] and their colleagues have concluded that penetrating ions move in the electric field oriented across the membrane as a result of electron and H^+ transfer or ATP hydrolysis.

Turning now to the major problem of mitochondrial research, the mechanism of ATP synthesis, we see a huge volume of literature has been generated in recent years. Only the controversy will be briefly mentioned here. Stated simply, there are two major points of contention. The first point is the contention between the chemical intermediate hypothesis (CIH) and the chemiosmotic hypothesis (COH), and the second point, closely related to the chemiosmotic hypothesis, concerns the membrane conformation during energy transduction.

The bone of contention between the CIH and the COH has been succinctly presented by Greville[295]. The highlights are as follows. In the CIH the primary process is linked to the redox reactions in the formation of a high-energy intermediate. This is represented by an accumulation of $X \sim I$, the unknown chemical intermediate. Further, in the CIH the chemical mechanism of oxidative phosphorylation in principle does not require the presence of any membrane structure. In contrast to the above stipulations, in the COH the redox and ATPase reactions are linked through an H^+ movement across the mitochondrial (cristae) membrane, the driving force being either a pH gradient or a membrane potential or both[273]. Finally, it is an essential requirement of the COH not only that the presence of a membrane is necessary, but also that it must be relatively impermeable to ions. The strength of COH lies in its various predictions which may be tested experimentally. However, at the present time, direct tests with the mitochondrial membranes are difficult. Experimental evaluation of the Mitchell hypothesis using model membrane systems should be a useful approach (see Section 2.4.1.5).

The second controversial point is on membrane conformation changes during energy transduction. Green and his colleagues[296] suggest that changes in the shapes of membrane constituents are essential during stages of ATP formation. As a result of electron transport, membrane components change from the so-called 'non-energised' to 'energised' conformation. The free energy stored in the process is somehow later utilised to make ATP. Green and his colleagues[296] have supported their thesis with information derived from electron microscopy, light scattering, and measurement of proton jumps. Papers critical of the above scheme have appeared recently. In particular, Packer and his colleagues[297] have stated, after defining two types of changes (configurational and conformational) that changes in configuration during energy conservation and transduction are secondary events which can be produced by conformational changes, H^+ movements and osmotic mechanisms. According to Packer *et al.* configurational changes involve alterations in the shape or folding of membranes, whereas conformational changes reflect changes in structure at the molecular level. Packer *et al.* concluded that their data obtained using similar methods used also by Green and co-workers indicated the configurational changes can occur without any accompanying conformational changes and *vice versa*. The latter changes are crucial in Green's conformational thesis. Packer *et al.* believe that the current evidence is insufficient to decide between the hypothesis of Mitchell in which the primary driving force for ATP production is the energy stored in a pH gradient across the membrane, and the hypothesis of Green in which the energy is stored in the membrane itself.

Studies by Sordahl *et al.*[298], Stoner and Sirak[299], and by Weber and Blair[300] on mitochondria provide negative support for the conformational hypothesis. These investigators either failed to observe configurational changes under 'energising' conditions or interpreted their electron micrographs as artifacts of fixation. These conflicting views and results will, no doubt, stimulate further experimental effort by all interested investigators.

As mentioned earlier, the attractiveness of Mitchell's hypothesis stems from its simplicity and suggests further experimentation. The hypothesis is claimed to explain a number of phenomena associated with an electron-transport system, membrane potential, charge separation, photo- and oxidative-phosphorylation and ion exchange. Concerning the antagonists' view of the hypothesis, it should be mentioned that the work done by Dilley[301], Cockrell *et al.*[302], Chance *et al.*[303], Thore *et al.*[304], and by McCarty[305] are of interest. Briefly, these authors' experiments show that proton gradients are not necessary for phosphorylation. They do not, however, exclude the possibility that the 'proton motive force' has been reduced to one of its hypothetical components as postulated by Mitchell.

2.5.4 Nerve membranes

The most unique property of this type of membrane is its electrical excitability or action potential. The classical work on this subject has been reviewed by Hodgkin[306], and more recently by Noble[307] and by Hille[308].

Unlike other membranes the nerve membrane, as typified by myelin, contains the smallest percentage of protein. For example, human myelin consists of 80% lipid and only 20% protein. The fatty acids of the lipids are the mono-unsaturated oleic and saturated stearic acid together with some polyunsaturated C_{28}–C_{24} acids[33]. From lipid analyses useful generalisations can be made. Myelin and probably the plasma or cell-surface membranes are characterised by high cholesterol and sphingolipid content together with a predominance of saturated and mono-unsaturated fatty acids. In contrast, mitochondria, endoplasmic reticulum and perhaps cytoplasmic membranes as a group possess generally low levels of cholesterol and sphingolipids and contain a relatively high proportion of polyunsaturated fatty acids. Although myelin is a rather atypical membrane system, low in protein and metabolically rather static, studies of spinal cord *in vitro* do show higher metabolic activity for the protein portion[309].

Myelin seems to be the exceptional structure which provides the best experimental support for the bilayer lipid model (see Section 2.2.1). Studies of birefringence of the myelin sheath of peripheral nerve led W. J. Schmidt[310] to postulate a structure of alternating protein and lipid layers encircling the nerve axon. The lipid chains are considered to be oriented radially, and successive lipid layers are separated by thin layers of protein. The process of myelin formation has been seen to arise from a kind of spiral wrapping up of the axon by an extension of the surface membrane of a Schwann cell[59]. The repeat period of myelin of different origins varies from 100–120 Å in the central nervous system, to about 160 Å in optic nerve and white matter of brain and spinal cord, to nearly 180 Å in mammalian peripheral nerve. Each myelin layer, of course, encompasses two cell membranes of opposite orientation. Thus the thickness of myelin is in adequate agreement with the concept that it includes two lipid bilayers alternating with two non-lipid regions (perhaps largely protein in the configuration) of thickness perhaps 30–45 Å each. Finean[59] has reviewed the physical aspects of nerve membrane structure.

Recently, Worthington and Blaurock[311] have reported additional x-ray diffraction studies on myelin and suggested that the results are consistent with the classical bimolecular leaflet model. The overall myelin thickness according to Worthington and Blaurock is 77 Å.

The classical work of Hodgkin, Huxley and Katz has suggested a scheme that the sum of the electrical currents of Na^+ and K^+ can be used to account for the observed resting potential and the transmembrane conductance during the action potential. Tasaki and his colleagues[312, 313] (Watanabe and Lerman) however, have recently shown in their experiments that neither Na^+ or K^+ is an absolute requirement for electrical excitability. They have demonstrated that, even when internal K^+ is replaced by Na^+, the axon membrane is still excitable. In view of their data, Tasaki *et al.* have proposed a general mechanism based on a monovalent–divalent ion-exchange reaction, in which the outer region of the membrane contains divalent ions such as Ca^{2+} in the resting state. Upon stimulation by an outward directed current, univalent ions such as K^+ or Na^+ derived from the internal solution are exchanged for divalent ions. During this ion-exchange reaction it is suggested that a change in membrane conformation occurs, accompanied by a change in membrane permeability. The work of Tasaki is exciting and is being watched by interested investigators in the field. It must be pointed out, however, that the suggested ion-exchange mechanism, even if shown to be correct, would not necessarily invalidate the classical work but would show that the explanation advanced by the earlier workers was a special case.

A great deal of other work on axons and action potentials has been reported. The subject of ionic channels in nerve membranes has been recently reviewed by Hille[308], as already mentioned. In an earlier paper, Goldman[314] discussed the structure of axon membrane and its behaviour. Brinley and Mullins[315], and Foster, Gilbert and Shaw[316] reported ion transport experiments using perfused and dialysed axons. A paper by Baker, Hodgkin and Ridgway[317] mentioned calcium permeability changes during the action potential of giant axons of *Loligo*. Nachmansohn[52] discussed the effect of acetylcholine and proteins in excitable membranes. Changes in fluorescence in squid axon during activity have been reported by Conti and Tasaki[318], and by Cohen, Hille and Keynes[319].

In the Soviet Union, Berestovsky, Liberman, Lunevsky and Frank[320] have carried out research by optical methods on structural changes in nerve during excitation.

2.5.5 Visual receptor membranes

Sjostrand[22] has suggested that the inner layers of the retina or double-membrane sacs of the outer segments of the rods and cones might have their origin in the plasma membrane, as do other types of biological membranes. As mentioned earlier (Section 2.3.1), the gross composition of the rod outer segments is about 40–50% protein, 20–40% lipids and 4–10% retinenes[321]. The visual pigment, rhodopsin, accounts for about 35% of the dry weight, the rest being mainly lipids[322]. The protein moiety of the visual pigment, known as opsin, is less stable when isolated from rhodopsin, which has a molecular weight between 40 000–60 000. Of the retinenes, 11-*cis* plays the most crucial role in visual excitation, as strongly emphasised by Wald[323].

Phospholipids, especially phosphatidyl ethanolamine, are implicated in the visual receptor membranes[324]. Whether 11-*cis*-retinal, the principal chronophore, is linked directly to opsin or bound by a Schiff base linkage to the phospholipid is uncertain at the present time[325]. The suggestion that the photolytic transfer of the retinal from lipid to protein may be involved in visual excitation is intriguing and needs to be tested experimentally. Perhaps the use of carotenoid BLM for such studies may prove helpful (see Section 2.4.1.8).

The structure of the outer segments appears to be a highly ordered array of repeating units and a thickness of about 55 Å per membrane is indicated[326]. Similar to the four types of biological membranes already described, the lamellar membrane of retinal rods is also interpreted in terms of a Gorter–Grendel bilayer model. Gras and Worthington[327] have proposed a model to account for their data obtained from x-ray diffraction studies. They suggest that the membrane surface facing the intra-sac space is 40 Å thick, the lipid region is 16 Å thick, and the surface facing the cytoplasm is 18.5 Å thick. Blaurock and Wilkins[328] have reported similar observations.

Biochemical aspects of the visual process have been investigated. For example, complexes between phospholipid and retinals have been reported by Daemen and Bonting[329], and Adams, Jennings and Sharpless[330], confirming an earlier observation of Poincelot et al.[325]. It should be mentioned that these views are at variance with those of Bownds[331], and of Akhtar, Blosse and Dewhurst[332].

In a theoretical paper, Pullman, Langlet and Berthod[333] have calculated the electron density of the fine retinal isomers both in their ground and excited states, using the semi-empirical SCF–PP method.

A paper by Poincelot and Abrahamson[334] examines the phospholipid composition and extra stability of bovine rod outer segments and rhodopsin micelles. In particular, they point out that lipids play a significant role in the visual process, not only through binding the pigments, but presumably in modulation of the membrane potential and permeability properties. In this connection the earlier paper of McConnell, Rafferty and Dilley[335] is of interest. These workers have observed that fragments of bovine retinal outer segments containing intact photoreceptor sacs accumulate hydrogen ions upon illumination. Most significantly, the bleached samples were not able to do so. This finding is seen to be in accord with results reported earlier by Falk and Fatt[336], who, after investigating suspensions of retinal outer segments with flash light, suggested that the uptake of protons by rhodopsin was responsible for the observed potential change in their system.

Turning to more physiological aspects of the visual process, the early receptor potential (ERP), first reported by Brown and Murakami[337], has been the focal point of much current research[338]. The ERP is of interest for it is believed to be the primary event which can be elicited from the eye. The ERP at room temperature consists of two components: the initial corneo-positive phase and the second corneo-negative phase, which have been termed, respectively, as R1 and R2 by Cone[339]. The following characteristics of ERP have been reported. Pak[340] has shown that the ERP is not sensitive to the ionic environment. Similar observations have also been reported by Brindley and Gardner-Medwin[341]. Cone[342] reported that the ERP is not affected by anoxia. At the present time the ERP has not been observed in aqueous suspensions of rhodopsin or any other pigments, or with the aqueous suspension of rods and cones. This fact strongly suggests that the generation of ERP requires the presence of an organised pigment arrangement in membrane structure as has been indicated in the work of Hagins and McGaugh[343], Cone and Brown[344], Wald[323], and by Goldstein and Berson[345]. It is interesting to note that wave-forms similar to that of ERP can be observed in non-ocular pigmented epithelium and the leaves of green plants[346]. In spite of the minor differences in response time and sensitivity toward temperature variations, all photoresponses observed in these pigmented tissues exhibit essentially the same features of the ERP, and hence, indicate the common mechanisms underlying its generation. At present even with the sophisticated intracellular recording techniques advanced by Toyoda, Nosaki and Tomita[347], it has not been possible to place recording electrodes on either side of the visual receptor membrane, and to explore the basic nature of ERP owing to the minute size of the photoreceptor membrane.

As has been realised by many investigators, the mechanism of visual excitation cannot be understood without a fuller understanding of the structure and function of the membranes[323]. For instance, in order to investigate generations of the ERP at the molecular level it is necessary to examine changes in the membrane properties upon illumination. The study of a well defined model system is therefore of interest (see Section 2.4.1.8).

2.6 FRONTIERS OF MEMBRANE RESEARCH

Having summarised current work in biological and artificial membranes and their closely related fields, it seems appropriate to speculate and remark on the trends and frontiers of membrane research so that all those in any way concerned with the biological membranes may anticipate the future developments. Obviously there are many risks involved in any speculations, especially in this active and rapidly expanding field of membrane research. Nevertheless, it is thought useful to draw attention to what are some outstanding problems in membrane research. A partial list of these outstanding problems includes light conversion in photosynthesis and vision, energy migration, active transport, ion selectivity and specificity, generation and conduction of nervous impulse, redox reactions and electron transfer, oxidative and photophosphorylation, DNA replication, protein synthesis, immunological and pharmacological reactions, lipid–protein interaction, and membrane structure.

Experimentally, although at the present time the monolayer technique is much less used, it still constitutes a powerful tool for obtaining information concerning the manner in which lipid and protein molecules orient and interact at interfaces. In the bulk phase, liposomes (or microvesicles) appear to be a better system for studying lipid–protein interaction than micelles. However, the fact that micelles, and globular micelles in particular, may play an important role in biological membranes, should not be overlooked.

In view of the importance attached to the lipid bilayer, both theoretically and experimentally, it is relevant to point out that the formation of BLM (bilayer lipid membranes) separating two aqueous solutions permits for the first time direct characterisation of the postulated bimolecular leaflet model of the biological membranes in physical chemical terms. At present the modified BLM systems constitute realistic experimental models for a variety of biological membranes[94]. It remains to be seen whether many outstanding problems in membrane biophysics and biochemistry can be elucidated with the use of experimental BLMs and their closely related system, liposomes.

In sum, it is anticipated that, in order to find the molecular basis of life processes, even greater attention will be focused in the coming years on biological membranes. In areas of membrane biophysics and biochemistry, the task of correlating structure and function will be increasingly undertaken in the present decade by investigators with diverse backgrounds.

References

1. Blank, M. (1968). Ed., *Surface Chemistry of Biological Systems*, (New York: Plenum Press)
2. Branton, D. and Park, R. B. (1968). Eds., *Biological Membrane Structure*, (Boston: Little, Brown & Co.)

3. Bronner, F. and Kleinzeller, A. (1970). Eds., *Current Topics in Membranes and Transport,* (New York: Academic Press)
4. Brown, H. D. (1971). Ed., *Chemistry of the Cell Interface,* (New York: Academic Press)
5. Burton, R. M. (1967). Ed., *Proceedings of Symposium on Lipid Monolayer and Bilayer Models, and Cellular Membranes,* (The American Oil Chemists' Society). Also published in *J. Amer. Oil Chemists' Soc.,* **45,** 107, 201, 297 (1968)
6. Chinard, F. P. (1968). Ed., *Biological Interfaces: Flows and Exchanges.* (Boston: Little, Brown & Co.)
7. Danielli, J. F., Riddiford, A. C. and Rosenberg, M. D. (1970). Eds., *Recent Progress in Surface Science,* Vol. III. (New York: Academic Press)
8. Dowben, R. M. (1969). Ed., *Biological Membranes.* (Boston: Little, Brown & Co.)
9. Goddard, E. D. (1968). Ed., *Molecular Association in Biological and Related Systems,* Advan. Chem. Series, No. 84, (Washington, D.C.: American Chemical Society)
10. Hair, M. L. (1971). Ed., *The Chemistry of Bio-Surfaces,* (New York: Marcel Dekker, Inc.)
11. Jamieson, G. A. and Greenwalt, T. J. (1969). Eds, *Red Cell Membranes,* (Philadelphia: J. B. Lippincott Co.)
12. Kotyk, A. and Janacek, K. (1970). *Cell Membrane Transport,* (New York: Plenum Press)
13. Lakshminarayanaiah, N. (1969). *Transport Phenomena in Membranes,* (New York: Academic Press)
14. Tien, H. T. (1971). *Surface and Colloid Science,* Vol. 4, p. 361 (Matijević, E., Ed.). (New York: John Wiley & Sons, Inc.)
15. Rogers, H. J. and Perkins, H. R. (1968). *Cell Walls and Membranes,* (London: E. & F. N. Spon Ltd.)
16. Schmitt, F. O., Melnechuk, T., Quarton, G. and Adelman, G. (1970). Eds, *Neurosciences Research Symposium Summaries,* Vol. 4 (Cambridge, Massachusetts: MIT Press)
17. Lawrence, A. S. C. (1958). *Surface Phenomena in Chemistry and Biology,* 9, (Danielli, J. F., Pankhurst, K. G. A. and Riddiford, A. C., Eds.). (London: Pergamon Press)
18. Davson, H. (1962). *Circulation,* **26,** 1022
19. Robertson, J. D. (1967). *Protoplasma,* **63,** 218
20. Malhotra, S. K. (1970). *Progress in Biophysics and Molecular Biology,* Vol. 20, 67, (Butler, J. A. V. and Noble, D., Eds.). (New York: Pergamon Press)
21. Hendler, R. W. (1971). *Physiol. Rev.,* **51,** 66
22. Sjostrand, F. S. (1963). *J. Ultrastruct. Res.,* **9,** 340
23. Lucy, J. A. and Glauert, A. M. (1968). *Symposium of International Society for Cell Biology,* Vol. 6, 19. (New York: Academic Press)
24. Lucy, J. A. (1970). *Nature (London),* **227,** 815
25. Ohki, S. and Aono, O. (1970). *J. Coll. Interface Sci.,* **32,** 270
26. Gulik-Krzywicki, T., Shechter, E. and Luzzati, V. (1969). *Nature,* **223,** 1116
27. Benson, A. A. (1966). *J. Amer. Oil Chemists' Soc.,* **43,** 265
28. Deamer, D. W. (1970). *Bioenergetics,* **1,** 237
29. Green, D. E., Allman, D. W., Bachmann, E., Baum, H., Kopaczyk, K., Korman, E. F., Lipton, S., MacLennan, D. H., McConnell, D. G., Perdue, J. F., Rieske, J. S. and Tzagoloff, A. (1967). *Arch. Biochem. Biophys.,* **119,** 312
30. Green, D. E. and Perdue, J. (1966). *Proc. Nat. Acad. Sci. U.S.,* **55,** 1295
31. Vanderkooi, G. and Green, D. E. (1970). *Proc. Nat. Acad. Sci. U.S.,* **66,** 615
32. Korn, E. D. (1969). *Ann. Rev. Biochem.,* **38,** 263
33. O'Brien, J. S. (1967). *J. Theoret. Biol.,* **15,** 307
34. Williams, R. M. and Chapman, D. (1970). *Progress in the Chemistry of Fats and Other Lipids,* 1, (Holman, R. T., Ed.). (New York: Pergamon Press)
35. Small, D. M. (1968). *J. Amer. Oil Chemists' Soc.,* **45,** 108
36. Abramson, M. B. (1970). *Advan. Expt. Med. Biol.,* **7,** 37
37. Abramson, M. B. and Katzman, R. (1970). *Advan. Expt. Med. Biol.,* **7,** 85
38. Weinstein, D. B., Marsh, J. B., Glick, M. C. and Warren, L. (1969). *J. Biol. Chem.,* **244,** 4103
39. Ladbrooke, B. D. and Chapman, D. (1969). *Chem. Phys. Lipids,* **3,** 304
40. van Deenen, L. L. M. (1965). *Progr. Chem. Fats,* **8,** 1
41. Rouser, G., Nelson, G. J., Fleischer, S. and Simon, G. (1968). *Biological Membranes,* 5, (Chapman, D., Ed.). (New York: Academic Press)
42. Criddle, R. S. (1969). *Ann. Rev.-Plant Physiol.,* **20,** 239

43. Zahler, P. (1969). *Experientia,* **25,** 449
44. Mazia, D. and Ruby, A. (1968). *Proc. Nat. Acad. Sci. U.S.,* **61,** 1005
45. Lenaz, G., Haard, N. F., Lauwers, A., Allman, D. W. and Green, D. E. (1968). *Arch. Biochem. Biophys.,* **126,** 746
46. Schachman, H. (1968). *Ultracentrifugation in Biochemistry,* (New York: Academic Press)
47. Urry, D. W. and Ji, T. H. (1968). *Arch. Biochem. Biophys.,* **128,** 802
48. Rosenberg, S. A. and Guidotti, G. (1969). *J. Biol. Chem.,* **244,** 5118
49. Gros, C. and Labouesse, B. (1969). *European J. Biochem.,* **7,** 463
50. Blumenfeld, O. (1968). *Biochem. Biophys. Res. Commun.,* **30,** 200
51. Woodward, D. and Munkres, K. (1967) in *Proc. Rutgers Symposium Organizational Biosynthesis,* (New York: Academic Press)
52. Nachmansohn, D. (1970). *Science,* **168,** 1059
53. Tiffany, J. M. and Blough, H. A. (1969). *Science,* **163,** 573
54. Marchesi, S. L., Steers, E., Marchesi, V. T., and Tillack, T. W. (1970). *Biochemistry,* **9,** 50
55. Kaplan, D. M. and Criddle, R. S. (1971). *Physiol. Rev.,* **51,** 249
56. Drost-Hansen, W. (1971), in *Chemistry of the Cell Interface,* (Brown, H. D., Ed.). (New York: Academic Press)
57. Hechter, O. (1965). *Fed. Proc.,* **24,** 91
58. Chapman, D. (1969). *Lipids,* **4,** 251
59. Finean, J. B. (1969). *Quart. Rev. Biophys.,* **2,** 1
60. Wehrli, T., Muehlethaler, K. and Moor, H. (1970). *Expt. Cell. Res.,* **59,** 336
61. Lenard, J. and Singer, S. J. (1966). *Proc. Nat. Acad. Sci. U.S.,* **56,** 1828
62. Urry, D. W. and Krivacic, J. (1970). *Proc. Nat. Acad. Sci. U.S.,* **65,** 845
63. Wallach, D. F. H. and Zahler, P. (1966). *Proc. Nat. Acad. Sci. U.S.,* **56,** 1552
64. Steim, J. M., Edner, O. J. and Bargoot, F. G. (1968). *Science,* **162,** 909
65. Ladbrooke, B. D., Williams, R. M. and Chapman, D. (1968). *Biochem. Biophys. Acta,* **150,** 333
66. Steim, J. M., Tourtellotte, M. E., Reinert, J. C., McElhaney, R. N. and Radar, R. L. (1969). *Proc. Nat. Acad. Sci. U.S.,* **63,** 104
67. Azzi, A., Chance, A., Radda, G. K. and Lee, C. P. (1969). *Proc. Nat. Acad. Sci. U.S.,* **62,** 612
68. Freedman, R. B. and Radda, G. K. (1969). *Fed. Europ. Biochem. Soc. Lett.,* **3,** 150
69. Rubalcava, B., de Munoz, M. and Gitler, C. (1969). *Biochem.,* **8,** 2742
70. Vanderkooi, J. and Martonosi, A. (1969). *Arch. Biochem. Biophys.,* **133,** 153
71. Gomperts, B., Lantelme, F. and Stock, R. (1970). *J. Memb. Biol.,* **3,** 241
72. Smekal, E., Ting, H. P., Augenstein, L. G. and Tien, H. T. (1970). *Science,* **168,** 1108
73. Fleischer, B., Sekuzu, I. and Fleischer, S. (1967). *Biochem. Biophys. Acta,* **147,** 552
74. Scanu, A., Pollard, H., Hirz, R. and Kothary, K. (1969). *Proc. Nat. Acad. Sci. U.S.,* **62,** 171
75. Kitzinger, C. and Benzinger, T. H. (1960) in *Methods of Biochemical Analysis,* (Glick, D., Ed.). (New York: Interscience)
76. Lovrien, R. and Anderson, W. (1969). *Arch. Biochem. Biophys.,* **131,** 139
77. Klopfenstein, W. E. (1969). *Biochem. Biophys. Acta,* **181,** 323
78. Chapman, D., Kamat, V. B., DeGier, J. and Penkett, S. A. (1968). *J. Mol. Biol.,* **31,** 101
79. Steim, J. M. (1968). *Advan. Chem. Ser.,* **84,** 259
80. Ji, T. H., Hess, J. L. and Benson, A. A. (1968). *Biochim. Biophys. Acta,* **150,** 676
81. Spector, A. A., John, K. and Fletcher, J. E. (1969). *J. Lipid Res.,* **10,** 56
82. Arvidsson, E. O. and Belfrage, P. (1969). *Acta Chem. Scand.,* **23,** 232
83. Griffith, O. H. and Waggoner, A. S. (1969). *Accounts Chem. Res.,* **2,** 17
84. Hsia, J. C., Schneider, H. and Smith, I.C.P. (1970). *Biochem. Biophys. Acta,* **202,** 399
85. McConnell, H. M. and McFarland, B. G. (1970). *Quart. Rev. Biophys.,* **3,** 91
86. Seelig, J. (1970). *J. Amer. Chem. Soc.,* **92,** 3881
87. Shipley, G. G., Leslie, R. B. and Chapman, D. (1969). *Biochim. Biophys. Acta,* **173,** 1
88. Engleman, D. M. (1970). *J. Molec. Biol.,* **47,** 115
89. Kajiyama, M. (1968). *Biophysics (Japan),* **8,** 1
90. Rothfield, L. and Finkelstein, A. (1968). *Ann. Rev. Biochem.,* **37,** 463
91. Castleden, J. A. (1969). *J. Pharm. Sci.,* **58,** 149
92. Henn, F. A. and Thompson, T. E. (1969). *Ann. Rev. Biochem.,* **38,** 241
93. Goldup, A., Ohki, S. and Danielli, J. F. (1970). *Recent Progress in Surface Science,* Vol. 3, 193. (New York: Academic Press)

94. Tien, H. T. (1971). *The Chemistry of Bio-Surfaces*, (Hair, M. L., Ed.). (New York: Marcel Dekker, Inc.)
95. Mueller, P., Rudin, D. O., Tien, H. T. and Wescott, W. C. (1964). *Recent Progress in Surface Science*, Vol. 1, 379, (New York: Academic Press)
96. Tien, H. T. (1967). *J. Phys. Chem.*, **71**, 3395
97. Pagano, R. and Thompson, T. E. (1968). *J. Molec. Biol.*, **38**, 41
98. Tien, H. T. and Howard, R. E. (1971). *Techniques of Surface Chemistry and Physics*, Vol. 1, (Good, R. J., Stromberg, R. R. and Patrick, R. L., Eds). (New York: Marcel Dekker, Inc.)
99. Tien, H. T. (1967). *J. Theoret. Biol.*, **16**, 97
100. Finkelstein, A. and Cass, A. (1968). *J. Gen. Physiol.*, **52**, 145s
101. Dervichian, D. G. (1968). *Advan. Chem. Ser.*, **84**, 78
102. Van Zutphen, H., van Deenen, L. L. M. and Kinsky, S. C. (1966). *Biochem. Biophys. Res. Commun.*, **22**, 393
103. Papahadjopoulos, D. and Ohki, S. (1969). *Science*, **164**, 1075
104. Ohki, S. and Papahadjopoulos, D. (1970). *Surface Chemistry of Biological Systems*, 155 (New York: Plenum Press)
105. Tien, H. T. (1968). *J. Gen. Physiol.*, **52**, 125s
106. Moran, A. and Ilani, A. (1970). *Chem. Phys. Lipids*, **4**, 169
107. Wobschall, D. (1971). *J. Coll. Interface Sci.*, **36**, 385
108. Good, R. J. (1969). *J. Coll. Interface Sci.*, **31**, 540
109. Haydon, D. A. and Taylor, J. L. (1968). *Nature (London)*, **217**, 739
110. Ter Minassian-Saraga, L. and Wietzerbin, J. (1970). *Biochem. Biophys. Res. Commun.*, **41**, 1231
111. Maddy, A. H., Huang, A. and Thompson, T. E. (1966). *Fed. Proc.*, **25**, 933
112. Smekal, E. and Tien, H. T. (1969). Unpublished results
113. Cherry, R. J. and Chapman, D. (1969). *J. Theoret. Biol.*, **24**, 137
114. Simons, R. (1970). *Biochem. Biophys. Acta*, **203**, 209
115. Huang, C. and Thompson, T. E. (1966). *J. Molec. Biol.*, **15**, 539
116. Hanai, T. and Haydon, D. A. (1966). *J. Theoret. Biol.*, **11**, 370
117. Tien, H. T. and Ting, H. P. (1968). *J. Coll. Interface Sci.*, **27**, 702
118. Zwolinski, B. J., Eyring, H. and Reese, C. E. (1949). *J. Phys. Chem.*, **53**, 1426
119. Pechhold, H. (1968). *Kolloid Z.*, **228**, 1
120. Lippe, C. (1969). *J. Molec. Biol.*, **39**, 669
121. Andreoli, T. E., Dennis, V. W. and Weigl, A. M. (1969). *J. Gen. Physiol.*, **53**, 133
122. Holz, R. and Finkelstein, A. (1970). *J. Gen. Physiol.*, **56**, 125
123. Dennis, V. W. and Stead, N. W. (1970). *J. Gen. Physiol.*, **55**, 375
124. Ohki, S. (1970). *Biochim. Biophys. Acta*, **219**, 18
125. Jung, C. Y. (1971). *J. Memb. Biol.*, **5**, 200
126. Lauger, P., Lesslauer, W., Marti, E. and Richter, J. (1967). *Biochim. Biophys. Acta*, **135**, 20
127. Miyamoto, V. K. and Thompson, T. E. (1967). *J. Coll. Interface Sci.*, **25**, 16
128. Rosenberg, B. and Jendrasik, G. L. (1968). *Chem. Phys. Lipids*, **2**, 47
129. Hashimoto, M. (1968). *Biochem. Soc. Japan*, **41**, 2823
130. Noguchi, S. and Koga, S. (1969). *J. Gen. Appl. Microbiol.*, **15**, 41
131. Mehard, C. W., Lyons, J. M. and Kumamoto, J. (1970). *J. Memb. Biol.*, **3**, 173
132. Davies, J. T. and Taylor, F. H. (1959). *Biol. Bull.*, **117**, 222
133. Gutknecht, T. J. and Tosteson, D. C. (1970). *J. Gen. Physiol.*, **55**, 359
134. Bradley, J. and Dighe, A. M. (1969). *J. Coll. Interface Sci.*, **29**, 157
135. Van Zutphen, H. and van Deenen, L. L. M. (1967). *Chem. Phys. Lipids*, **1**, 389
136. Seufert, W. D., Beauchesne, G. and Belanger, M. (1970). *Biochim. Biophys. Acta*, **211**, 356
137. Pryor, W. A. (1971). *Chem. Eng. News*, **49**, 34
138. Leslie, R. B. and Chapman, D. (1967). *Chem. Phys. Lipids*, **1**, 143
139. Hopfer, U., Lehninger, A. L. and Lennarz, W. J. (1970). *J. Memb. Biol.*, **2**, 41
140. Neumcke, B. (1970). *Biophysik*, **6**, 231
141. Babakov, A. V., Ermishkin, L. N. and Liberman, E. A. (1966). *Nature*, **210**, 933
142. Rosen, D. and Sutton, A. M. (1968). *Biochim. Biophys. Acta*, **163**, 226
143. White, S. (1970). *Biophys. J.*, **10**, 1127
144. Walz, D., Bamberg, E. and Lauger, P. (1969). *Biophys. J.*, **9**, 1150
145. Gillespi, C. J. (1970). *Biochim. Biophys. Acta*, **203**, 47

146. Del Castillo, J., Rodriguez, A., Remero, C. A. and Sanchez, V. (1966). *Science,* **153,** 185
147. Jain, M. K., Strickholm, A. and Cordes, E. H. (1969). *Nature (London),* **222,** 871
148. Skou, J. C. (1965). *Physiol. Rev.,* **45,** 596
149. Redwood, W. R., Pfieffer, F. R., Weisbach, J. A. and Thompson, T. E. (1971). *Biochim. Biophys. Acta,* **223,** 1
150. Parisi, M., Rivas, E. and De Robertis, E. (1971). *Science,* **172,** 56
151. Mueller, P. and Rudin, D. O. (1967). *Biochem. Biophys. Res. Commun.,* **26,** 398
152. Lev. A. A. and Buzhinsky, E. P. (1967). *Z. Evolyu. Biokhim. Fiziol.,* **9,** 102
153. Bangham, A. D., Standish, M. M. and Watkins, J. C. (1965). *J. Mol. Biol.,* **13,** 238
154. Chappell, J. B. and Crofts, A. R. (1966). *Biochim. Biophys. Acta Library,* 1, (Tagar, J. M., Papa, S., Quagliariello, E. and Slater, E. C., Eds). (Amsterdam: Elsevier)
155. Gotlib, V. A., Buzhinsky, E. P. and Lev, A. A. (1968). *Biophysic. Rev.,* **13,** 675
156. Andreoli, T. E., Bangham, J. A. and Tosteson, D. C. (1967). *J. Gen. Physiol.,* **50,** 1729
157. Mueller, P. and Rudin, D. O. (1968). *J. Theoret. Biol.,* **18,** 222
158. Lardy, H. A., Graven, S. N. and Estrada-O, S. (1967). *Fed. Proc.,* **26,** 1355
159. Pressman, B. C. (1968). *Fed. Proc.,* **27,** 1283
160. Eisenman, G., Ciani, S. M. and Szabo, G. (1968). *Fed. Proc.,* **27,** 1289
161. Bielawski, J., Thompson, T. E. and Lehninger, A. L. (1966). *Biochem. Biophys. Res. Commun.,* **24,** 948
162. Sotnikov, P. S. and Melnik, E. I. (1968). *Biofizika,* **13,** 185
163. Liberman, E. A., Topaly, V. P. and Silberst, A. Y. (1970). *Biochem. Biophys. Acta,* **196,** 221
164. Rosenberg, B. and Pant, H. C. (1970). *Chem. Phys. Lipids,* **4,** 203
165. Ting, H. P., Wilson, D. F. and Chance, B. (1970). *Arch. Biochem. Biophys.,* **141,** 141
166. Bruner, L. J. (1970). *Biophysik,* **6,** 241
167. Tien, H. T. and Verma, S. P. (1970). *Nature (London),* **227,** 1232
168. Mueller, P., Rudin, D. O., Tien, H. T. and Wescott, W. C. (1962). *Nature (London),* **194,** 979
169. Ehrenstein, G., Lecar, H. and Nossal, R. (1970). *J. Gen. Physiol.,* **55,** 119
170. Monnier, A. M. (1968). *J. Gen. Physiol.,* **51,** (pt. 2) 26s
171. Shashoua, V. E. (1967). *Nature (London),* **215,** 846
172. Ochs, A. L. and Burton, R. M. (1968). *Abst. Biophys. Soc. 12th Ann. Meeting,* A-27
173. Ochs, A. L. (1969). *A Lipid Membrane Vibration Response,* (Ph.D. Thesis). (St. Louis, Mo.: Washington University)
174. Tien, H. T. (1968). *J. Phys. Chem.,* **72,** 4512
175. Van, N. T. and Tien, H. T. (1970). *J. Phys. Chem.,* **74,** 3559
176. Tien, H. T. and Kobamoto, N. (1969). *Nature (London),* **224,** 1107
177. Kobamoto, N. and Tien, H. T. (1971). *Biochem. Biophys. Acta,* **241,** 129
178. Hesketh, T. R. (1969). *Nature (London),* **224,** 1026
179. Mauzerall, D. and Finkelstein, A. (1969). *Nature (London),* **224,** 690
180. Kay, R. E. and Chan, H. (1969). *Radiat. Res.,* **40,** 177
181. Pant, H. C. and Rosenberg, B. (1971). *Photochem. Photobiol.,* **14,** 1
182. Alamuti, N. and Lauger, P. (1970). *Biochem. Biophys. Acta,* **211,** 362
183. Trissl, H. W. and Lauger, P. (1970). *Z. Naturforsch. B. B.,* **25,** 1059
184. Cherry, R. J., Hsu, K. and Chapman, D. (1971). *Biochem. Biophys. Res. Commun.,* **43,** 351
185. Ullrich, H. M. and Kuhn, H. (1969). *Z. Naturforsch. B. B.,* **24,** 1342
186. Hueber, J. S. and Tien, H. T. (1972). *Biochem. Biophys. Acta,* in press
187. Bangham, A. D. (1968). *Progress in Biophysics and Molecular Biology,* Vol. 18, 29 (Butler, J. A. V. and Noble, D., Eds). (New York: Pergamon Press)
188. Sessa, G. and Weissmann, G. (1968). *J. Lipid Res.,* **9,** 310
189. Bangham, A. D., Standish, M. M. and Watkins, J. C. (1965). *J. Molec. Biol.,* **13,** 238
190. Papahadjopoulos, D. and Watkins, J. C. (1967). *Biochim. Biophys. Acta,* **135,** 639
191. Huang, C. (1969). *Biochemistry,* **8,** 344
192. Reeves, J. P. and Dowben, R. M. (1969). *J. Cell. Physiol.,* **73,** 49
193. Tien, H. T. and James, L. K., Jr. (1971). *Chemistry of the Cell Interface.* (Brown, H. D., Ed.). (New York: Academic Press)
194. Miyamoto, V. K. and Stoeckenius, W. (1971). *J. Memb. Biol.,* **4,** 252

195. Rendi, R. (1964). *Biochim. Biophys. Acta*, **84**, 694
196. Rendi, R. (1967). *Biochim. Biophys. Acta*, **135**, 333
197. Bangham, A. D., de Gier, J. and Greville, G. D. (1967). *Chem. Phys. Lipids*, **1**, 225
198. Reeves, J. P. and Dowben, R. M. (1970). *J. Memb. Biol.*, **3**, 123
199. Schwan, H. P., Takashima, S., Miyamoto, V. K. and Stoeckenius, W. (1970). *Biophys. J.*, **10**, 1102
200. Pauly, H. and Schwan, H. P. (1966). *Biophys. J.*, **6**, 621
201. Moore, J. L., Richardson, T. and DeLuca, H. F. (1969). *Chem. Phys. Lipids*, **3**, 39
202. Bangham, A. D., Standish, M. M. and Weissmann, G. (1965). *J. Molec. Biol.*, **13**, 253
203. Weissman, G., Sessa, G. and Weissmann, S. (1966). *Biochem. Pharmacol.*, **15**, 1537
204. Weissmann, G., Keiser, H. and Bernheimer, A. W. (1965). *J. Exptl. Med.*, **118**, 205
205. Papahadjopoulos, D. and Miller, N. (1967). *Biochim. Biophys. Acta*, **135**, 624
206. Sweet, C. and Zull, J. E. (1969). *Biochim. Biophys. Acta*, **173**, 94
207. Bangham, A. D., Standish, M. M. and Miller, N. G. A. (1965). *Nature*, **208**, 1295
208. Johnson, S. M. and Bangham, A. D. (1969). *Biochim. Biophys. Acta*, **193**, 92
209. Papahadjopoulos, D. (1970). *Biochim. Biophys. Acta*, **211**, 467
210. Narahashi, T., Frazier, D. T. and Yamada, M. (1970). *J. Pharmacol. Expt. Therap.*, **171**, 32
211. Murthy, K. S. and Rippie, E. G. (1970). *J. Pharm. Sci.*, **59**, 459
212. McElhaney, R. N., de Gier, J. and van Deenen, L. L. M. (1970). *Biochim. Biophys. Acta*, **219**, 245
213. Daemen, F. J. M. and Bonting, S. L. (1969). *Biochim. Biophys. Acta*, **183**, 90
214. Chapman, D. and Fast, P. G. (1968). *Science*, **160**, 188
215. Kimelberg, H. K. and Lee, C. P. (1969). *Biochem. Biophys. Res. Commun.*, **34**, 784
216. Das, M. L. and Crane, F. L. (1964). *Biochem.*, **3**, 696
217. Reich, M. and Wainio, W. W. (1961). *J. Biol. Chem.*, **236**, 3058
218. Sahu, S. and Tien, H. T. (1971). Unpublished studies
219. Chang, T. M. S., MacIntosh, F. C. and Mason, S. G. (1966). *Can. J. Physiol. and Pharm.*, **44**, 115
220. Gaines, G. L. (1965). *Insoluble Monolayers at Liquid–Gas Interfaces*. (New York: Interscience)
221. James, L. K., Jr. and Augenstein, L. G. (1966). *Advances in Enzymology*, Vol. 28, 1, (Nord, F. F., Ed.). (New York: Interscience)
222. Dawson, R. M. C. (1968). *Biological Membranes*, (Chapman, D., Ed.). (New York: Academic Press)
223. Lucy, J. A. (1968). *Biological Membranes*, (Chapman, D., Ed.), 233. (New York: Academic Press)
224. Malcolm, B. R. (1968). *Proc. Roy. Soc. (London)*, **A305**, 363
225. Vilallonga, F., Altschul, R. and Fernandez, M. S. (1967). *Biochim. Biophys. Acta*, **135**, 406
226. Matsubara, A. (1965). *Bull. Chem. Soc. Japan*, **38**, 1254
227. Colacicco, G. and Rapport, M. M. (1968). *Advan. Chem. Series*, **84**, 157
228. Colacicco, G. (1969). *J. Coll. Interface Sci.*, **29**, 345
229. Blank, M. (1968). *Biological Interfaces: Flows and Exchanges*, (Boston: Little, Brown & Co.)
230. Miller, I. R. (1968). *J. Gen. Physiol.*, **52**, 209s
231. Cadenhead, D. A. and Phillips, M. C. (1968). *Advan. Chem. Series*, **84**, 131
232. Fromherz, H. (1971). *Nature (London)*, **231**, 268
233. Kleuser, D. and Bucher, H. (1969). *Z. Naturforsch. B. B.*, **24**, 1371
234. Hamori, E. and Michaels, A. M. (1971). *Biochim. Biophys. Acta*, **231**, 496
235. Knapp, F. F. and Nicholas, H. J. (1970). *Molec. Cryst.*, **6**, 319
236. Khaiat, A. and Miller, I. R. (1969). *Biochim. Biophys. Acta*, **183**, 309
237. Bikhazi, A. B. and Higuchi, W. I. (1971). *Biochim. Biophys. Acta*, **233**, 676
238. Sollner, K. (1969). *J. Macromolec. Sci. Chem.*, **3**, 1
239. Green, M. E. and Yafuso, M. (1968). *J. Phys. Chem.*, **72**, 4072
240. Furukawa, T., Uematsu, Y., Asakawa, K. and Wada, Y. (1968). *J. Appl. Polymer. Sci.*, **12**, 2675
241. Ottewill, R. H. (1969). *Ann. Rep. Progr. Chem.*, Sect A, **66**, 183
242. Sweeley, C. C. and Dawson, G. (1969). *Red Cell Membranes*, 172, (Jamieson, G. A. and Greenwalt, T. J., Eds). (Philadelphia: Lippincott Co.)

243. Hanahan, D. J. (1969). *Red Cell Membranes*, 83, (Jamieson, G. A. and Greenwalt, T. J., Eds). (Philadelphia: Lippincott Co.)
244. Maddy, A. H. (1966). *Int. Rev. Cytol.*, **20**, 1
245. Morgan, T. E. and Hanahan, D. J. (1966). *Biochemistry*, **5**, 1050
246. Passow, H. (1969). *Progress in Biophysics and Molecular Biology*, Vol. 19, 425, (Butler, J. A. V. and Noble, D., Eds). (London: Pergamon Press)
247. Weinstein, R. S. (1969). *Red Cell Membranes*, 36, (Jamieson, G. A. and Greenwalt, T. J., Eds). (Philadelphia: Lippincott Co.)
248. Tenforde, T. (1970). *Advan. Biol. Med. Physics*, **13**, 43
249. Wallach, D. F. H. and Gordon, A. (1968). *Fed. Proc.*, **27**, 1263
250. Maddy, A. H. and Malcolm, B. R. (1966). *Science*, **153**, 213
251. Hubbell, W. L. and McConnell, H. M. (1968). *Proc. Nat. Acad. Sci. U.S.*, **61**, 12
252. Chapman, D. and Kamat, V. B. (1968). *Regulatory Functions of Biological Membranes*, 99, (Jarnefely, J., Ed.). (Amsterdam: Elsevier)
253. Chapman, D. and Wallach, D. F. H. (1968). *Biological Membranes*, (Chapman, D., Ed.). (New York: Academic Press)
254. Moretz, R. C., Akers, C. K. and Parsons, D. F. (1969). *Biochim. Biophys. Acta*, **193**, 1
255. Boardman, N. K. (1968). *Advan. Enzymol.*, **30**, 1
256. Branton, D. (1969). *Ann. Rev. Plant Physiol.*, **20**, 209
257. Walker, D. A. and Crofts, A. R. (1970). *Ann. Rev. Biochem.*, **39**, 389
258. Menke, W. (1967). *Brookhaven Sym. Biol.*, **19**, 328
259. Park, R. B. and Pon, N. G. (1963). *J. Molec. Biol.*, **6**, 105
260. Park, R. B. and Biggins, J. (1964). *Science*, **144**, 1009
261. Branton, D. and Park, R. B. (1968). *Selected Papers on Biological Membrane Structure*, (Boston: Little, Brown & Co.)
262. Weier, T. E. and Benson, A. A. (1966). *Amer. J. Biol.*, **54**, 389
263. Moor, H., Muehlethaler, K., Waldner, H. and Frey-Wyssling, A. (1961). *J. Biophys. Biochem. Cytol.*, **10**, 1
264. Muehlethaler, K. (1966). *Biochemistry of Chloroplasts*, 49, (Goodwin, T. W., Ed.). (New York: Academic Press)
265. Ongun, A., Thompson, W. W. and Mudd, J. B. (1968). *J. Lipid Res.*, **9**, 409
266. Lichtenthaler, H. K. and Park, R. B. (1963). *Nature (London)*, **198**, 1070
267. Criddle, R. S. and Park, L. (1964). *Biochem. Biophys. Res. Commun.*, **17**, 74
268. Lockshin, A. and Burris, R. H. (1966). *Proc. Nat. Acad. Sci. U.S.*, **56**, 1564
269. Thornber, J. P., Stewart, J. C., Hatton, M. W. C. and Bailey, J. L. (1967). *Biochem.*, **6**, 2006
270. Witt, H. T., Rumberg, B. and Junge, W. (1968). *Mosbach Colloquim*, 262, (Berlin: Springer-Verlag)
271. Fork, D. C. and Amesz, J. (1970). *Photophysiology*, Vol. 5, 97, (Giese, A. C., Ed.). (New York: Academic Press)
272. Avron, M. and Neumann, J. (1968). *Ann. Rev. Plant Physiol.*, **19**, 137
273. Mitchell, P. (1966). *Biol. Rev.*, **41**, 445
274. Jagendorf, A. T. and Hind, G. (1965). *Biochem. Biophys. Res. Commun.*, **18**, 702
275. Jagendorf, A. T. and Neumann, J. S. (1965). *J. Biol. Chem.*, **240**, 3210
276. Barber, J. and Kraan, G. P. B. (1970). *Biochim. Biophys. Acta*, **197**, 49
277. Miles, C. D. and Jagendorf, A. T. (1969). *Arch Biochem. Biophys.*, **129**, 711
278. Vredenberg, W. J. (1969). *Biochem. Biophys. Res. Commun.*, **37**, 785
279. Luttge, U. and Pallaghy, C. K. (1969). *Z. Pflanz.*, **61**, 58
280. Cheniae, G. M. and Martin, I. F. (1967). *Biochem. Biophys. Res. Commun.*, **28**, 89
281. Kok, B. and Cheniae, G. M. (1966). *Current Topics in Bioenergetics*, Vol. 1, 1 (Sanadi, D. R., Ed.). (New York: Academic Press)
282. Kok, B., Forbush, B. and McGloin, G. (1970). *Photochem. Photobiol.*, **11**, 457
283. Barbieri, G., DeLosme, R. and Joliot, P. (1970). *Photochem. Photobiol.*, **12**, 197
284. Williams, W. P. (1970). *Nature (London)*, **225**, 1214
285. Izawa, S. and Good, N. E. (1965). *Biochim. Biophys. Acta*, **102**, 20
286. Gaffron, H. (1968). *Comparative Biochemistry and Biophysics of Photosynthesis*, 431, (Shibata, K., Takamiya, A., Jagendorf, A. T. and Fuller, R. C., Eds). (Tokyo: University of Tokyo Press)
287. Green, D. E. and Fleischer, S. (1964). *Metabolism and Physiological Significance of Lipids*, (Dawson, C. R. M. C. and Rhodes, D. N., Eds). (New York: Wiley)

288. Racker, E. (1967). *Fed. Proc.*, **26**, 1335
289. Crane, F. L. and Hall, J. D. (1969). *Biochem. Biophys. Res. Commun.*, **36**, 174
290. Prezbindowski, K. S., Ruzicka, F. J., Sun, F. F. and Crane, F. L. (1968). *Biochem. Biophys. Res. Commun.*, **31**, 164
291. Hatch, F. T. and Bruce, A. L. (1968). *Nature (London)*, **218**, 1166
292. MacLennan, D. H. (1970). *Current Topics in Membranes and Transport*, 177–232, (Bronner, F. and Kleinzeller, A., Eds). (New York: Academic Press)
293. Grinius, L. L., Jasaitis, A. A., Kadziauskas, Yu. P., Liberman, E. A., Skulachev, V. P., Topali, V. P., Tsofina, L. M., Vladimirova, M. A., Bakeeva, L. E., Kuliene, V. V., Levitsky, D. O., Severina, I. I., Isaev, P. I. and Samuilov, V. D. (1970). *Biochim. Biophys. Acta*, **216**, 1, 13, 22
294. Liberman, E. A. and Skulachev, V. P. (1970). *Biochim. Biophys. Acta*, **216**, 30
295. Greville, G. D. (1969). *Current Topics in Bioenergetics*, Vol. 3, (Sanadi, D. R., Ed.). (New York: Academic Press)
296. Harris, R. A., Williams, C. H., Caldwell, M., Green, D. E. and Valdivia, E. (1969). *Science*, **165**, 700
297. Packer, L., Murakami, S., Wrigglesworth, J. M., House, D. R. and Donovan, M. P. (1970). *Bioenergetics Bull.*, **2**, 1; see also (1971). *Bioenergetics*, **1**, 33
298. Sordahl, L. A., Blailock, Z. R., Kraft, G. H. and Schwartz, A. (1969). *Arch. Biochem. Biophys.*, **132**, 404
299. Stoner, C. D. and Sirak, H. D. (1969). *Biochem. Biophys. Res. Commun.*, **35**, 59
300. Weber, N. E. and Blair, P. V. (1969). *Biochem. Biophys. Res. Commun.*, **36**, 987
301. Dilley, R. A. and Vernon, L. P. (1965). *Arch. Biochem. Biophys.*, **111**, 365
302. Cockrell, R. S., Harris, E. J. and Pressman, B. C. (1967). *Nature*, **215**, 1487
303. Chance, B., Lee, C. P. and Mèla, L. (1967). *Fed. Proc.*, **26**, 1341
304. Thore, A., Keister, D. L. and San Pietro, A. (1968). *Biochemistry*, **7**, 3499
305. McCarty, R. E. (1968). *Biochem. Biophys. Res. Commun.*, **32**, 27
306. Hodgkin, A. L. (1964). *The Conduction of the Nervous Impulse.* (Springfield, Ill.: Thomas)
307. Noble, D. (1966). *Physiol. Rev.*, **46**, 1
308. Hille, B. (1970). *Progr. Biophys. Mol. Biol.*, **21**, 1
309. Finean, J. B. (1966). *Progr. Biophys. Mol. Biol.*, **16**, 143
310. Schmidt, W. J. (1936). *Z. Zellforsch Mikroskop. Anat.*, **23**, 657
311. Worthington, C. R. and Blaurock, A. E. (1968). *Nature (London)*, **218**, 87
312. Tasaki, I., Lerman, L. and Wantanabe, A. (1969). *Amer. J. Physiol.*, **216**, 130
313. Tasaki, I. (1968). *Nerve Excitation.* (Springfield, Ill.: Thomas)
314. Goldman, D. E. (1967). *Ber. Bunsenges. Phys. Chem.*, **71**, 799
315. Brinley, F. J. and Mullins, L. J. (1968). *J. Gen. Physiol.*, **52**, 181
316. Foster, R. F., Gilbert, D. S. and Shaw, T. L. (1969). *Biochem. Biophys. Acta*, **183**, 401
317. Baker, P. F., Hodgkin, A. L. and Rodgway, E. B. (1970). *J. Physiol.*, **208**, 80
318. Conti, F. and Tasaki, I. (1970). *Science*, **169**, 1322
319. Cohen, L. B., Hille, B. and Keynes, R. D. (1970). *J. Physiol.*, **211**, 495
320. Berestovsky, G. N., Liberman, E. A., Lunevsky, V. Z. and Frank, G. M. (1970). *Biofizika*, **15**, 62
321. Eichberg, J. and Hess, H. H. (1967). *Experientia*, **23**, 993
322. Krinsky, N. I. (1958). *Arch. Ophthal.*, **60**, 688
323. Wald, G. (1968). *Science*, **162**, 238
324. Abrahamson, E. W. and Ostroy, S. E. (1967). *Progr. Biophys. Mol. Biol.*, **17**, 179
325. Poincelot, R. P., Miller, P. G., Kimbel, R. L. and Abrahamson, E. W. (1969). *Nature (London)*, **221**, 256
326. Blasie, J. K., Dewey, M. M., Blaurock, A. E. and Worthington, C. R. (1965). *J. Molec. Biol.*, **14**, 143
327. Gras, W. J. and Worthington, C. R. (1969). *Proc. Nat. Acad. Sci. U.S.*, **63**, 233
328. Blaurock, A. E. and Wilkins, M. H. F. (1969). *Nature (London)*, **223**, 906
329. Daemen, F. J. M. and Bonting, S. L. (1969). *Nature (London)*, **222**, 879
330. Adams, R. G., Jennings, W. H. and Sharpless, N. E. (1970). *Nature (London)*, **226**, 270
331. Bownds, D. (1967). *Nature (London)*, **216**, 1178
332. Akhtar, M., Blosse, P. T. and Dewhurst, P. B. (1967). *Chem. Commun.*, **13**, 631
333. Pullman, B., Langlet, J. and Berthod, H. (1969). *J. Theoret. Biol.*, **23**, 492
334. Poincelot, R. P. and Abrahamson, E. W. (1970). *Biochemistry*, **9**, 1820

335. McConnell, D. G., Rafferty, C. N. and Dilley, R. A. (1968). *J. Biol. Chem.*, **243**, 5820
336. Falk, G. and Fatt, P. (1966). *J. Physiol.*, **183**, 211
337. Brown, K. T. and Murakami, M. (1964). *Nature (London)*, **201**, 626
338. Brown, K. T. (1968). *Vision Res.*, **8**, 633
339. Cone, R. A. (1964). *Nature (London)*, **204**, 736
340. Pak, W. L. (1965). *Cold Spring Harbor Sym. Quant. Biol.*, **30**, 493
341. Brindley, G. S. and Gardner-Medwin, A. R. (1966). *J. Physiol.*, **182**, 185
342. Cone, R. A. (1965). *Cold Spring Harbor Sym. Quant. Biol.*, **30**, 483
343. Hagins, W. A. and McGaugh, R. E. (1967 & 1968). *Science*, **157**, 813 (1967); **159**, 213 (1968)
344. Cone, R. A. and Brown, P. K. (1967). *Science*, **156**, 536
345. Goldstein, E. B. and Berson, E. L. (1969). *Nature (London)*, **222**, 1272
346. Arden, G. B., Bridge, C. D. B., Ikeda, H. and Shiegel, I. M. (1966). *Nature (London)*, **212**, 1235
347. Toyoda, J., Nosaki, H. and Tomita, T. (1969). *Vision Res.*, **9**, 453

3
Recent Advances in the Study of Solid Surfaces

W. P. ELLIS*
University of California, Los Alamos Scientific Laboratory, New Mexico

3.1	INTRODUCTION	79
	3.1.1 *Resurgence of surface studies*	79
	3.1.2 *The low-energy electron surface probe*	80
3.2	ELASTICALLY SCATTERED LOW-ENERGY ELECTRONS	82
	3.2.1 *Difficulty of intensity analysis*	82
	3.2.2 *Reciprocal lattice considerations*	82
	3.2.3 *Dynamical approaches*	84
	3.2.4 *Optical transforms*	86
	3.2.4.1 *Assignment of the two-dimensional space group*	86
	3.2.4.2 *The low-energy electron microscope*	87
3.3	INELASTICALLY SCATTERED LOW-ENERGY ELECTRONS	88
	3.3.1 *Auger electron spectroscopy (AES)*	88
	3.3.2 *X-ray appearance potential spectroscopy*	92
	3.3.3 *Ionisation spectroscopy*	93
3.4	PHOTOELECTRON SPECTROSCOPY (PES)	94

3.1 INTRODUCTION

3.1.1 Resurgence of surface studies

Within the preceding decade there has been a remarkable resurgence of studies directed toward an understanding of the atomic properties of clean solid surfaces and the reactions that occur there. In a large measure this renewed interest has resulted from the recent availability of all metal, ultra-

---

*Work performed under the auspices of the U.S. Atomic Energy Commission.

high-vacuum equipment, low-energy electron diffraction systems (LEED), and inelastic secondary-electron spectrometers. The literature on these and related subjects has grown prodigiously in the last few years, almost explosively in fact, prompting even the emergence of several new journals. It is of course quite beyond the capacity of this article to include all of the individual studies, and some significant areas, for example, field-ion microscopy, calculations of surface energies, flash desorption, or scanning electron microscopy, are not discussed at all.

Instead, a more modest approach is taken in the following pages in which a few selected examples were chosen to demonstrate current thinking about some persistent problems of a broad nature in experimental surface analysis, and the more recent progress in solving those problems. In particular the emphasis will be directed toward the low-energy electron surface probe in (i) the elastically scattered mode, i.e. LEED, and (ii) the inelastically scattered spectral mode, i.e. secondary-electron spectroscopy.

3.1.2 The low-energy electron surface probe

A half-dozen years ago, in the late 1960s with the widespread introduction of commercial LEED units, there was a budding optimism in the field of surface physics that at last a relatively simple tool was available which would solve the major structural problems[1,2]. The method promised a renaissance in the area of surface crystallography with applications to the study of epitaxy, adsorption, corrosion and oxidation, catalysis and so on. Indeed, significant advances have been made[3]. No new technique is an unalloyed blessing however, and in a few years, after waxing full bloom, this enthusiasm has now dimmed with the tarnished realisation that a quantitative interpretation of low-energy electron diffraction is at best a monumentally difficult task, or as in some recently expressed opinions, attainable only in concept if at all but in any case utterly beyond reasonable expectation.

The initial optimism was not entirely without foundation however. In principle at least, low-energy elecdon diffraction is simple. Elastic penetration of electrons in the energy range 10–1000 eV is limited to the outermost 2–4 atomic layers[4]. One then simply directs a monoenergetic beam of slow electrons onto a carefully prepared single crystal surface in high vacuum and observes the back-reflected diffraction pattern. Figure 3.1 is a schematic of one such device for accomplishing this purpose: the 3-grid, fluorescent-screen LEED display apparatus[1]. The crystal face under examination is placed at the centre of curvature of a series of spherically concentric grids. Electrons are emitted by the cathode at a negative variable voltage, $-V$, focused onto the crystal by the gun lenses, and are reflected back to the display screen. In the diffraction mode the second grid is biased at cathode potential to suppress the large inelastic component, and only the elastically scattered electrons which carry the diffraction information are displayed. Precise intensities can be measured by insertion of a Faraday collector.

With minor external modifications, changes in work functions and secondary-electron emission ratios can be measured, plasmon and other losses can be observed[5], and surface chemical compositions can be monitored

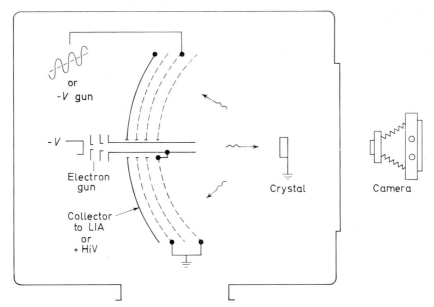

Figure 3.1 Schematic of the principal features of the 3-grid fluorescent-screen LEED apparatus
(From Ellis[28], by courtesy of Academic Press)

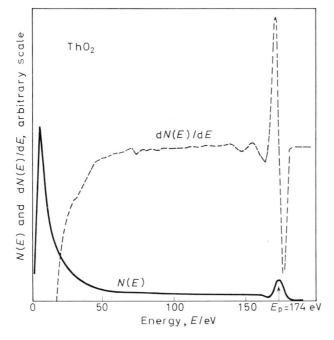

Figure 3.2 The energy distribution $N(E)$–v.–E and its derivative $dN(E)/dE$–v.–E of electrons scattered from the $ThO_2(111)$ face with a primary energy $E_p = 174$ eV

by the presence of Auger transitions in the secondary-electron spectrum[6]. Four or more grids can be used instead of three as shown in Figure 3.1, and the voltages may be arranged in various alternative manners depending upon the type of information desired. In the total inelastic mode, which includes both Auger and loss spectra, the primary beam energy is fixed at E_p, the second grid is ramped with a modulated sweep voltage, and the screen is used as a collector. The signal is processed with a lock-in amplifier (LIA) and displayed as an energy distribution $N(E)-v.-E$ or its derivative $dN(E)/dE-v.-E$ on an XY-recorder or oscilloscope[5,6]. A distinct advantage of the arrangement shown in Figure 3.1 is that both LEED and the inelastic spectral analysis can be performed on identically the same area of a possibly inhomogeneous crystal.

Figure 3.2 is an example of the type of data obtainable in a low-gain spectral analysis. Only 5–10% of the electrons are scattered elastically. Herein lies the major difficulty in LEED. It is ironic that the major obstacle in the interpretation of LEED should derive from the same principles that make it such a sensitive surface probe. Both the elastic and inelastic scattering cross-sections are so high that only a few atomic layers are probed and the intensities of the diffraction profile are not related in a simple manner to structure.

In the following two sections the question of LEED intensity analysis is explored first, followed by recent advances in the spectroscopy of inelastically scattered electrons. The concluding section treats briefly some attractive prospects which the related field of photoelectron spectroscopy offers to the study of surfaces.

3.2 ELASTICALLY SCATTERED LOW-ENERGY ELECTRONS

3.2.1 Difficulty of intensity analysis

If the dispersion curves of all elastic and inelastic scattering interactions were known completely for slow electrons propagating in a crystal lattice, then it would be a straightforward matter to compute surface structures from LEED intensities. These interactions are not known, but if they were such desirable parameters as atomic positions within the surface unit mesh could be obtained, as well as lattice expansions and contractions, configurations of fractional monolayers, and so on. The two-dimensional reciprocal net of course is presented directly in the diffraction array. But in spite of the large number of LEED observations – in his discussion in 1970, May[7] tabulated 414 references – according to Stern and Taub[8], as of July 1970 not a single periodic structure has been uniquely determined in a rigorous manner by low-energy electron diffraction techniques. For those accustomed to thinking of diffraction in terms of the reciprocal lattice the reasons for this deficiency can perhaps best be demonstrated by an illustration such as Figure 3.3.

3.2.2 Reciprocal lattice considerations

In Figure 3.3a the dots represent the reciprocal lattice points of an infinitely extended array of point-scattering centres. The incident propagation wave-vector is k_0, and the 60 degrees cone of LEED observation is also shown. In the kinematical formation which assumes only one scattering event per electron, photon or neutron, a diffraction beam emerges from the crystal in those directions where the Ewald sphere of radius $1/\lambda$ intersects a reciprocal lattice point[9]. As the crystal is rocked, or as the radius $1/\lambda$ is changed as in LEED where $1/\lambda = \sqrt{(V/150)}\,\text{Å}^{-1}$ and V is the variable crystal-to-cathode voltage difference, the intensity of a given Bragg beam would appear as a sharp spike, as indicated at the top. For a crystal two atomic layers thick the points elongate into rods as shown in figure 3.3b, and instead of sharp spikes the intensity profile becomes broadened. This effect is variously

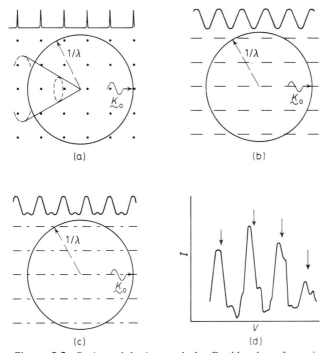

Figure 3.3 Reciprocal lattices and the Ewald sphere for point scattering centres and penetrating monochromatic radiation in (a) a perfect, infinite crystal, (b) a crystal two atomic layers thick, (c) a crystal three atomic layers thick and (d) typical LEED I–V plot

called the Fourier transform over crystal shape, the Laue interference function or crystal-form factor. For a crystal three atomic layers thick as shown in Figure 3.3c the beams are sharper than in Figure 3.3b but still broad, and between the principal Bragg reflections are secondary peaks of reduced intensity. Some aspect of this effect undoubtedly is present in LEED since penetration is of atomic dimensions. For comparison a typical experimental LEED intensity v. voltage curve observed for example with Ni or Ag is shown in Figure 3.3d. The calculated principal Bragg peaks shown by the arrows are displaced from the observed maxima by the so-called inner potential. Although Figure 3.3d bears some resemblance to the distribution at the top of Figure 3.3c in that both have broad principal maxima and weak secondary peaks, the correspondence is not at all acceptable for purposes of convolution even when corrected for non-point scattering. This modified kinematical formulation then is only qualitatively acceptable in low-energy electron diffraction, and is of no apparent value in the most pressing quantitative problem of structure analysis, that is, of assigning atomic positions within the unit cell. This limitation has been recognised, even if not fully appreciated, since the early days of LEED, but it has only been within recent years that theoretical progress of promising practical value has been made on dynamical formulation.

3.2.3 Dynamical approaches

The starting point of most dynamical treatments of diffraction is the Bethe theory[10] in which solutions are sought for the familiar Schroedinger equation in the crystal potential

$$\nabla^2 \psi((r)) + (2me/\hbar^2)[E + V(r)]\psi(r) = 0 \qquad (3.1)$$

The translational symmetry of the regular crystal lattice prescribes a periodic interaction potential, $V(r)$, which together with a given total energy E determine those modes that are excited in the bulk. In solid-state valence-band theory where E is below the vacuum level the solutions are in terms of the familiar Block waves. In low-energy diffraction, energies of the electrons injected into the surface are 10–1000 eV above the vacuum level and solutions for $\psi(r)$ are sought with particular emphasis on the forbidden regions in the band structure since electrons with energies that fall within a gap cannot propagate through the lattice and thus are reflected back into the Bragg beams[11]. In Equation 3.1, the form of the scattering potential $V(r)$ and the ratio $V(r)/E$ are critical parameters, and together with the manner in which $\psi(r)$ is solved as a sum over states are features which characterise the different dynamical treatments.

Boudreaux and Heine[11] have used the band-theory approach, whereas Stern and Taub[8] have emphasised the dispersion hypersurface, i.e. locus of allowed solutions in energy-momentum space, and inelastic Kikuchi effects. McRae[12] has utilised the Darwin theory in LEED and has predicted the scattering resonances observed in LEED. As recently as 1970, Kambe and Moliere[13] stated that theoretical treatment of LEED had to be developed further and was not included in their extensive article on the dynamical theory of electron diffraction. However, the multiple-scattering approach of Gafner[14], Duke and Tucker[15], more recently the quantum-field formulation of inelastic scattering by Laramore and Duke[16], and inelastic band treatment by Strozier and Jones[17] are in encouraging agreement with experimental diffraction intensity profiles and are of distinct value at this time in that they may help restore the diminished confidence in LEED as a useful quantitative technique.

It is outside the intent of this article to survey more fully the different formulations or to compare their relative usefulness. For a thorough presentation the reader may refer to the original articles. The above listing is by no means complete however, and in a number of laboratories workers are pursuing their own calculations quietly without preliminary publication. To a non-theoretician, familiar with but only partially exposed to, the mathematical difficulties, it would appear that to be useful in extracting structural information a formulation of necessity would include the experimental restraints that (a) only 5–10% of the electrons are scattered elastically, and (b) elastic penetration is so severely limited. Comprehensibility, ease of computation and applicability to a wide range of materials are added desirable qualities but they may be too much to expect since it is now abundantly clear that LEED presents one of the most difficult quantitative problems of contemporary mathematical physics.

In scattering theory, attenuation of amplitude is introduced by making $V(r)$ in equation (3.1) complex, thereby giving the so-called optical potential in analogy to adding an imaginary term to the dielectric constant in optics[13]. Unfortunately this violates the cherished principle in quantum mechanics that to have physical meaning, i.e. conservation of matter in this case, the Hamiltonian operator must be hermitian. This difficulty can be circumvented by including it as Yoshioka[18] did as described by Kambe and Moliere[13], ignoring it as in early LEED theories, or by not solving equation 3.1 for $\psi(r)$ in the first place. Computationally, the multiple-scattering approach in which the amplitude[9] $f(s) = \sum_i a_i^*(s, r)$ is summed over all probable paths including both single- and multiple-scattering events is appealing because it relates conceptually to the kinematical theory, damping and phase shifts are easily included as well as directional dependence of scattering amplitude, and computation is straightforward on large computers programmed to handle complex exponentials. It is limited in that $\psi(r)$ is not solved for explicitly; it thus has no first-principles theoretical foundation and dispersion effects are not calculated except as included in the set of empirically derived scattering parameters.

On the basis of the preceding discussions alone, one might conclude somewhat glumly that LEED is of no immediate practical value in surface studies, whereas indeed just the opposite is true in the author's opinion. It seems appropriate to conclude this section with a few examples in which LEED does yield apparently unambiguous results. Low-energy electron diffraction does show (i) whether or not a solid has an ordered single crystalline surface[1,2], (ii) if an adsorbed layer is ordered and bears a simple two-dimensional crystallographic relationship to the substrate and under what conditions ordering occurs[7,19], (iii) with low melting-point materials, that the surface does not premelt before the bulk[20], (iv) that the surface Debye–Waller effect is greater than in the bulk[21], (v) the faceting sequence of thermodynamically unstable planes[22], and (vi) the ordering of steps on high-index surfaces[23–25]. For these features a qualitative description may be adequate. Regrettably, other pertinent but quantitative questions such as atomic positions within the surface unit cell at fractional monolayer coverage of adsorbed gases, whether or not there is place exchange, the statistical structures of defect clusters in non-stoichiometric compounds, to name just a few, will receive no solid quantitative answers from the scattering profiles until good intensity data are analysed with an equally good theory.

3.2.4 Optical transforms

3.2.4.1 Assignment of the two-dimensional space group

An addition to the problem of intensity analysis is the frequent ambiguity in the very first step of structure determination, that is, in assigning a two-dimensional space group to the surface lattice. Complex structural features may be expected to occur on surfaces: random nucleation and growth of surface layers, ordering or disordering of steps and half-crystal sites on

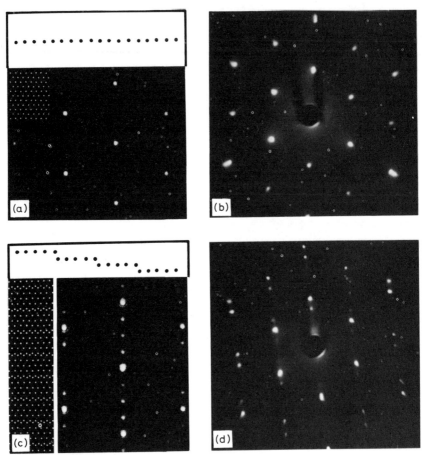

Figure 3.4 Simulated and observed LEED patterns; (a) optical transform of the hexagonal grating, *inset*, (b) observed LEED pattern from $UO_2(111)$ at $E_p = 128$ eV, (c) transform of a flat projection of the fcc(553) surface, *inset*, and (d) observed LEED pattern from $UO_2(553)$ at $E_p = 127$ eV.

high-index faces, ordered or statistically disordered arrays of antiphase domains, and adsorption which is statistical on the atomic scale. The principal diffraction effects of such features can be approximated quite readily by computing the Fourier integral, but a much simpler and faster way is with optical transforms[26, 27]. The method is particularly appealing first because there is no pretence of finality and second because the simulated Fraunhofer diffraction pattern from a two-dimensional grating corresponds to the kinematical pattern of the outermost layer of atoms, and is presented in the same format as with the fluorescent-screen LEED unit. It should be emphasised however that except for the azimuthal positions, beams in the analogue do not vary in intensity as in the LEED I–V curve. The application of optical transforms to LEED studies has been treated extensively elsewhere[28], but a few examples to illustrate the applicability are given below.

Figure 3.4a shows an optical simulation of the hexagonal array of which a magnified portion is shown in the upper left corner. As known and expected from the Fourier integral, the pattern is reciprocal to the lattice. For comparison Figure 3.4b is the experimental, LEED pattern from the f.c.c $UO_2(111)$ planar surface at $E_p = 128$ eV. Figure 3.4c simulates the pattern from an f.c.c.(553) surface with equally spaced steps of minimum height. The grating, of which a portion is seen to the left, is a flat projection of the sideview shown by the rows of dots at top. Figure 3.4d is an experimental LEED pattern from a carefully prepared $UO_2(553)$ face at $E_p = 127$ eV. Agreement of beam positions in Figure 3.4c and 3.4d is good which adds support to the previous contention[23] that Figure 3.4d represents a terraced surface with steps of minimum height. Figure 3.4d also indicates that elastic penetration of low-energy electrons is only a few atomic layers at most: for penetrating radiation such as x-rays, surface effects are wiped out and a pattern similar to Figure 3.4a would be seen. The transforms shown here are quite simple. Considerably more complicated arrays can be simulated with equal ease[28].

3.2.4.2 The low-energy electron microscope

Although the term 'recent advances' in the title of this article implies *fait accompli*, there may be some value in the inclusion of one experimental possibility that has not been perfected yet. The low-energy electron microscope has received only scant attention[29], but optical analogues have shown that a dark field microscope when perfected may possibly be a valuable extension to the study of the nucleation and growth of surface structures[28]. Certainly a bright field (BF) low-energy microscope presents a formidable problem of aberration in lens design, but in the dark field (DF) mode, which selects only one diffraction beam, that problem is minimal, i.e. $f/60$ instead of $f/1$. Furthermore the DF mode is more informative. Figure 3.5a is a photograph of an optical-transform grating which contains two $c(2 \times 2)$ nuclei on a square lattice. Figure 3.5b is the BF image, in Figure 3.5c only the (10) diffraction beam was imaged, and in Figure 3.5d all beams but the $(\frac{1}{2}\frac{1}{2})$ were excluded. In a practical microscope of limited resolution, the $(\frac{1}{2}\frac{1}{2})$ DF mode of Figure 3.5d is the most informative of those shown here and indicates the possibility that the initial nucleation stages of gas–solid reactions someday may be visualised. In addition, if ever such nuclei are imaged, by choice of DF modes it may be possible to distinguish between bonding positions on the surface[28], i.e. whether in centred-bond or bridge-bond positions, and thus be of considerable assistance in intensity analysis.

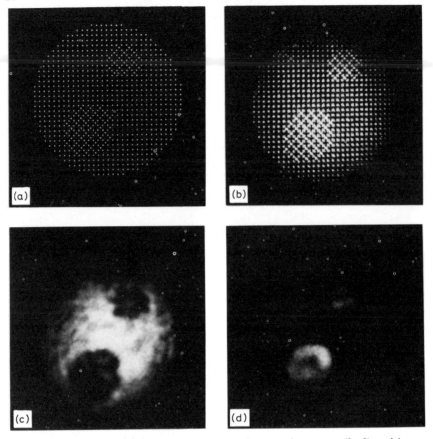

Figure 3.5 Optical simulation of the low-energy electron microscope, c(2 × 2) nuclei on a square lattice; (a) photograph of the grating, (b) simulated bright-field micrograph, (c) simulated (10) dark field micrograph, and (d) simulated ($\frac{1}{2}\frac{1}{2}$) dark field micrograph

The prospects are somewhat exciting, and if the analyses of even a few structures are aided in this fashion they will make the effort of construction worthwhile.

3.3 INELASTICALLY SCATTERED LOW-ENERGY ELECTRONS

3.3.1 Auger electron spectroscopy (AES)

In 1967 the volume, 'ESCA', by Siegbahn[30] and his co-workers was published, and Harris[31] wrote his famous GE report on 'Analysis of Materials by Electron-Excited Auger Electrons'. In the same year Weber and Peria[6] and Scheibner and Tharp[5] described a modulation technique by which the conventional retarding-grid LEED apparatus could be modified for energy analysis of secondary electrons, a technique which provided rapid and simple determinations of chemical constituents on surfaces. Since then the spectroscopy of electrons scattered inelastically from solid surfaces has become a booming field, and several instrumental advances, such as the cylindrical mirror analyser, have been made. The field of Auger electron spectroscopy (AES) has been the subject of several recent reviews[32-34], with listings of Auger energies, which do not need repetition here. Instead a general description with a few brief examples of more recent work will be discussed below including some new experimental approaches in the following sections.

Much of the LEED work carried out prior to the new spectroscopy has been repeated to evaluate the role that impurities play in surface structure. LEED of course requires single crystals: the reciprocal lattice rods of Figure 3.3 effectively smear the azimuthal intensities beyond recognition, and the familiar Debye–Scherrer rings of x-ray powder patterns are not seen at all

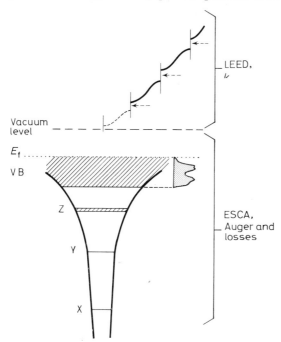

Figure 3.6 Schematic of energy states of interest in ESCA, Auger, and loss spectroscopy, and above in k-space the bands of interest in LEED

with microcrystalline solids. Electron spectroscopy in contrast is not a diffraction method, does not require single crystals or specimen temperatures below the Debye–Waller limitation, and thus has opened to investigation a wide range of polycrystalline materials. Figure 3.6 is a schematic diagram showing core levels X, Y and Z, the valence band VB, the Fermi level E_f, the vacuum level, and above in k-space the higher bands of interest in LEED. Incident exciting radiation promotes an electron from level X to a higher empty or partially filled level. The excited X^* state decays by an electron from a higher level falling into the empty state with the energy difference being carried away either by (i) x-rays which when analysed form the basis of the electron microprobe, or (ii) in a radiationless transition by ejection of an Auger electron from a higher, e.g. valence-band, level. The primary excitation can either be by x-rays as in photoelectron spectroscopy (PES), or by electrons. In this article, except for the concluding section on PES, only electron excitation is considered. Losses of discrete increments of energy by primary electrons are observed for plasmon excitation, intra-band transitions, and promotion of inner-shell electrons to the Fermi level. The valence band is drawn with broad features in Figure 3.6 to indicate that any process on the surface that alters the densities of states also alters the VB Auger spectrum.

That such an alteration can be pronounced is illustrated by one example in Figure 3.7 which shows the total inelastic spectrum from polycrystalline thorium metal with two different surface preparations. The lower dashed

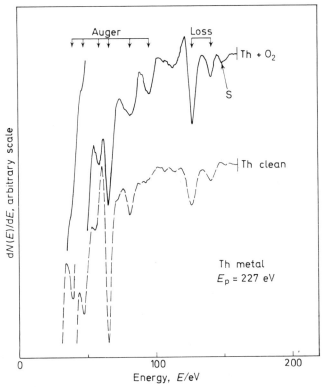

Figure 3.7 Auger and loss spectra observed for thorium metal under two different surface conditions

curve is the spectrum of a clean Th surface free of detectable impurities and the upper solid curve is from the same surface after exposure to oxygen. The curves have the same two loss peaks and the same six Auger transitions, but at significantly different intensities. The interpretation is that oxygen radically alters the densities of states in the Th valence band, but that the same states are present in both. Factors in a quantitative interpretation of the Auger intensities include dispersion effects on excitation and transition probabilities, densities of states and x-ray yields, as well as instrumental response and the presence of reactive trace impurities. These features as yet are largely unknown, but the possibilities have received some recent attention. For example, Amelio[35] derived a band structure of silicon from Auger intensity data in general agreement with the calculated densities of states.

By far the greatest application of Auger spectroscopy in surface studies has been to monitor surface cleanliness and to follow gas–solid reactions at

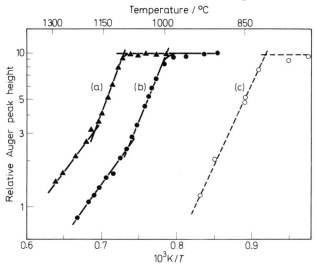

Figure 3.8 Bulk-to-surface equilibria of trace impurities on surfaces of polycrystalline thorium metal, logarithm of derivative Auger peak heights v. $1/T$; (a) sulphur, 15 p.p.m. in bulk by weight, (b) sulphur after removal of 34 saturated layers with Ar^+ bombardment, and (c) phosphorus (50 p.p.m.)

(From Ellis[40], by courtesy of the American Institute of Physics)

fractional monolayer coverages. Sickafus[36] for example correlated LEED patterns of Ni(110) with the presence of adsorbed C and S, Fiermans and Vennik[37] concluded that the electron-beam induced transition $V_2O_5 \rightarrow V_{12}O_{26}$ at $E_p = 1000$ eV probably was not the result of carbon contamination, and Dooley and Haas[38] observed C, S and O impurities on almost all refractory metals, to name just a few.

In their Auger study of fractured tungsten, Joshi and Stein[39] found that segregation of phosphorus in grain boundaries was the primary cause of embrittlement in powder specimens, and that C and O had no effect. They further found that the ductile–brittle transition temperature was directly related to the amount of phosphorus on the grain boundary, and that P segregates preferentially in the boundary rather than in the bulk. Such studies of internal surfaces is yet another powerful extension of the Auger tool.

Another recent example of the application of AES to surface studies is provided by the rapid bulk-to-surface equilibria of trace impurities in polycrystalline thorium[40], as shown in Figure 3.8. In Figure 3.8, the logarithms of the derivative Auger peak heights are plotted versus reciprocal absolute temperature. Curves (a) and (b) are for sulphur: curve (a) is from a fresh specimen with S present in the bulk at 15 ± 5 p.p.m. by weight, and curve (b) is from the same surface after 34 saturated layers of S had been removed by Ar^+ bombardments. Curve (c) shows a similar effect occurring with P(50 p.p.m.). The plots for S are extremely reproducible. They indicate that impurities return to the bulk in a thermodynamic fashion, that several equilibrium ranges exist for impurities on surfaces and furthermore that the thermodynamic enthalpies can be measured with reasonable accuracy.

Auger electron spectroscopy now appears to be firmly established as a surface analytical tool, It has drawbacks however. First, Auger emission is a complex, multi-electron process involving as many as three separate bound states, and the excited core level X* can decay in several alternative modes. Secondly, relative intensities are largely unknown. Thirdly, loss (or gain[41]) satellites have to be deconvoluted in a rigorous intensity analysis[42]. Fourthly, fine structure is masked by the inherently broad peaks. Fifthly, broad peaks at neighbouring energies tend to overlap, and with elements of high atomic number are hard to identify. Some of these difficulties are avoided in the methods given below.

3.3.2 X-ray appearance potential spectroscopy

In x-ray appearance potential spectroscopy[43] the secondary electrons are not analysed; they are dispensed with entirely. Rather, the excitation threshold for x-ray emission is measured. As indicated in Figure 3.9 the method is uncomplicated and thus has the appeal of being relatively inexpensive. Electrons from filament F excite x-rays in sample S which in

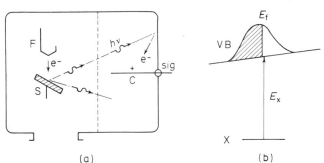

Figure 3.9 Schematic of the principal features of x-ray appearance potential spectroscopy; (a) detection system showing filament F, sample S, and the electronically isolated detection chamber C and (b) energy-level diagram showing core level X, valence band VB, fermi level E_f and excitation threshold energy E_X

turn are detected in the electronically isolated chamber C by the photoelectron current from the walls. The voltage difference between sample and filament can be oscillated, and as the energy sweeps back and forth across the threshold an a.c. signal is detected. As seen in Figure 3.9b, the threshold for excitation of level X is that energy E_x required to promote an electron to the empty states above the Fermi level, E_f. The signal at C then depends upon the distribution of unfilled states above E_f, excitation and transition probabilities for all processes, densities of all filled states below E_f from which electrons fall into X thereby creating a soft x-ray, dispersion and reproducibility of the photoelectric effect in the collector chamber, instrumental gain, and so on. Also the signal is weaker than in Auger spectroscopy but in principle, because the excitation process is simpler and the detected peaks sharper, the intensity information promises to be more valuable.

3.3.3 Ionisation spectroscopy

In ionisation spectroscopy, characteristic loss energies, E_X, corresponding to the excitation of core level X in Figure 3.9b are sought in the spectra of inelastically scattered electrons[44]. An energy analysis of scattered electrons is required of course, and the intensity of a given loss peak depends upon

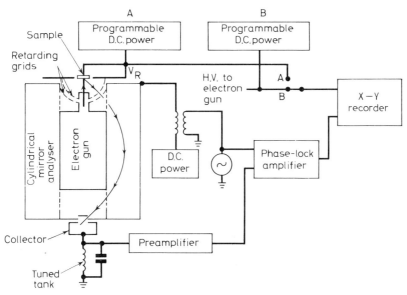

Figure 3.10 Simplified schematic of an ionisation and Auger spectrometer. Auger spectra are obtained with power supply A programmed, B fixed, switch on A. Loss spectra are obtained with power supply B programmed, A fixed, and switch on B
(From Gerlach[44], by courtesy of The American Institute of Physics)

excitation probabilities, densities of unfilled states above E_f, and the usual instrumental response characteristics in common with all measuring devices. In principle though, the intensity analysis would appear to be simpler than in either appearance potential or Auger spectroscopy because the complexities of x-ray and secondary-electron emission are avoided.

Experimentally the instrumentation is no more difficult than conventional total electron spectroscopy. Figure 3.10 is a schematic of Gerlach's latest arrangement[45a] in which sharper response is achieved by adding spherical retarding grids prior to a cylindrical mirror analyser. By means of a simple switch the analyser can function either in the total spectra mode, i.e. detect both Auger and loss peaks, or in the loss mode only. Figure 3.11 is an example of the type of data taken in second derivative, i.e. d^2N/dE^2–$v.$–E, loss mode for a carbon-contaminated palladium surface. Another advantage

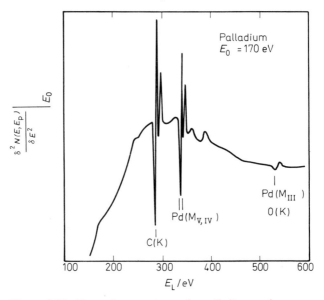

Figure 3.11 Energy-loss spectrum of a palladium surface contaminated with carbon
(From Gerlach[44], by courtesy of The American Institute of Physics)

which this spectroscopy has over Auger analysis is that the core levels are known better and the loss peaks in general are sharper.

In ESCA terminology (cf. next section), excitation to an empty level immediately above E_f is referred to as a 'shake-up' process because the electron is 'shaken-up' so to speak from a lower to a higher bound state. The 'shake-off' process corresponds to excitation to the vacuum continuum.

3.4 PHOTOELECTRON SPECTROSCOPY (PES)

Electron spectroscopy for chemical analysis, i.e. ESCA, has come to imply the high-resolution energy distribution analysis of photoelectrons, whereas in general the term also includes electron-excited emission as well. High-resolution photoelectron spectroscopy, i.e. PES, is included in this final section because it became apparent at the Asilomar Conference[45] that x-ray PES offers a prolific field in the study of solid surfaces. That soft x-ray photoelectrons with discrete energies are ejected from the surface region appears to be established[4, 45b]. Chemical shifts of a few eV in the inner-shell binding energies, corresponding to different chemical environments, are presently measured to within ±0.2 eV or better. If these shifts can be correlated with the thermodynamic ΔH values as proposed by Jolly[45c], then the high-temperature chemist investigating multicomponent solids will have a powerful tool to study phase stabilities and reaction energies in a manner reminiscent of the Engel–Brewer correlation[46]. It may become feasible to investigate binding states and energies of adsorbed species on surfaces for a direct comparison in the same ultra-high-vacuum system with LEED, FEM, and flash desorption. Finally, it may be feasible in some instances with elemental solids to obtain the occupied density-of-state profile of the valence band directly in the photoelectron spectrum, as shown in Figure 3.12. Figure 3.12 shows the adjusted VB photoelectron spectrum of

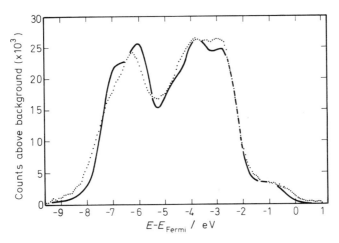

Figure 3.12 Gold valence bands, theory and experiment. The experimental spectrum, shown as points, was taken with monochromatised $Al(K_{\alpha 1,2})$ x-rays. A scattering correction was made; the average background was 2×10^4 counts. The theoretical band structure curve was obtained by broadening the relativistic band calculation of Connolly. The ordinate was adjusted by a scale factor; no adjustment of the abscissa was made
(Reproduced by permission of Professor D. A. Shirley).

gold taken by Shirley[45d] with $Al(K_\alpha)$ radiation. Agreement with the calculated bulk VB relativistic density-of-states curve is good enough to suggest that high-resolution x-ray PES may be a much more straightforward way of obtaining the occupied-band structure of pure elements than the convolution of Auger data for example. The agreement also suggests first that the band structure at least of Au at the surface is not far different from in the bulk and second that the cross-section for the x-ray photoejection of VB electrons is constant. Since the chemical and electronic properties of solid surfaces are determined by the allowed electron-energy states, the possibilities offered by soft x-ray PES certainly merit the thorough examination they undoubtedly will receive by surface chemists.

References

1. Germer, L. H. (1965). *Sci. Amer.*, **212**, 32
2. MacRae, A. U. (1963). *Science*, **139**, 379
3. Lander, J. J. (1970). *Progress in Solid State Chemistry*, **2**, 26 (New York: Pergamon)
4. Palmberg, P. W. and Rhodin, T. N. (1968). *J. Appl. Phys.*, **39**, 2425
5. Scheibner, E. J. and Tharp, L. N. (1967). *Surface Sci.*, **8**, 427
6. Weber, R. E. and Peria, W. T. (1967). *J. Appl. Phys.*, **38**, 4355
7. May, J. W. (1970). *Advan. Catal.*, **21**, 151
8. Stern, R. M. and Taub, H. (1970). *Crit. Rev. Solid State Sci.*, **1**, 221
9. Vainshtein, B. K. (1964). *Structure Analysis by Electron Diffraction*, 13 (New York: Macmillan)
10. Bethe, H. A. (1928). *Ann. J. Phys.*, **87**, 55
11. Boudreaux, D. S. and Heine, V. (1967). *Surface Sci.*, **8**, 426
12. McRae, E. G. (1971). *Surface Sci.*, **25**, 491
13. Kambe, K. and Moliere, K. (1970). *Advan. Struct. Res. Diffr. Meth.*, **3**, 53
14. Gafner, G. (1969). *4th International Materials Symposium: The Structure and Chemistry of Solid Surfaces*, 2 (Ed. by Somorjai, G.). (New York: Wiley)
15. Duke, C. B. and Tucker, C. W., Jr. (1969). *Phys. Rev. Lett.*, **23**, 1163
16. Laramore, G. E. and Duke, C. B. (1971). *Phys. Rev. B*, **3**, 3198
17. Strozier, J. A., Jr. and Jones, R. O. (1971). *Phys. Rev. B*, **3**, 3228
18. Yoshioka, H. (1957). *J. Phys. Soc. Japan*, **12**, 618
19. Estrup, P. J. and McRae, E. G. (1971). *Surface Sci.*, **25**, 1
20. Goodman, R. M. and Somorjai, G. A. (1970). *J. Chem. Phys.*, **52**, 6325
21. Woodruff, D. P. and Seah, M. P. (1969). *Phys. Lett.*, **30A**, 263
22. Ellis, W. P. (1968). *J. Chem. Phys.*, **48**, 5695
23. Ellis, W. P. and Schwoebel, R. L. (1968). *Surface Sci.*, **11**, 82
24. Henzler, M. (1970). *Surface Sci.*, **19**, 159
25. Rhead, G. E. and Perdereau, J. (1969). *Compt. Rend. Acad. Sci. C*, **269**, 1183
26. Taylor, C. A. and Lipson, H. (1964). *Optical Transforms* (London: Bell)
27. Campbell, B. D. and Ellis, W. P. (1968). *Surface Sci.*, **10**, 124
28. Ellis, W. P. (in the press). *Optical Transforms*, Ch. 7 on LEED (Ed. by Lipson, H.) (London: Academic)
29. Turner, G. and Bauer, E. (1966). *6th International Congress for Electron Microscopy. I*, 163 (Tokyo: Maruzen)
30. Siegbahn, K. (1967). *ESCA* (Uppsala: Almqvist & Wiksella)
31. Harris, L. A. (1968). *J. Appl. Phys.*, **39**, 1419. Also available as GE Report 67-C-201 (May, 1967)
32. Chang, C. C. (1971). *Surface Sci.*, **25**, 53
33. Sickafus, E. N. and Bonzel, H. P. (in press). *Progress in Surface and Membrane Science*, **4**, (Ed. by Danielli, J. F., Pankhurst, K. G. A. and Riddiford, A. C.). (New York: Academic)
34. Pentenero, A. (1971). *Catal. Rev.*, **5**, 199
35. Amelio, G. F. (1970). *Surface Sci.*, **22**, 301
36. Sickafus, E. N. (1970). *Surface Sci.*, **19**, 181
37. Fiermans, L. and Vennik, J. (1971). *Surface Sci.*, **24**, 541
38. Dooley, G. J. and Haas, T. W. (1970). *J. Vac. Sci. Technol.*, **7**, S90
39. Joshi, A. and Stein, D. F. (1970). *Met. Trans.*, **1**, 2543
40. Ellis, W. P. (in press). *J. Vac. Sci. Technol.*
41. Jenkins, L. H. and Chung, M. F. (1971). *Surface Sci.*, **26**, 151
42. Mularie, W. M. and Peria, W. T. (1971). *Surface Sci.*, **26**, 125
43. Park, R. L. and Houston, J. E. (1971). *Surface Sci.*, **26**, 664
44. Gerlach, R. L. (1971). *J. Vac. Sci. Technol.*, **8**, 599
45. Shirley, D. A., ed. (in press). *International Conference on Electron Spectroscopy, Asilomer, 7–10 Sept., 1971* (Amsterdam: North Holland). (a) Gerlach, R. L. (b) Steinhardt, R. G., Hudis, J. and Perlman, M. L. (c) Jolly, W. L. (d) Shirley, D. A.
46. Brewer, L. (1968). *Science*, **161**, 115

4
Aggregation in Surfactant Systems

PER EKWALL*
Stockholm, Sweden

INGVAR DANIELSSON** and **PER STENIUS****
Åbo Akademi, Finland

4.1	INTRODUCTION	98
	4.1.1 *Scope of the survey*	98
	4.1.2 *General properties of the aggregates*	99
	4.1.3 *The interfacial layers of polar groups*	100
4.2	SURFACTANTS IN AQUEOUS SOLUTION	101
	4.2.1 *The interactions between monomers and water*	101
	4.2.2 *The energetics of micelle formation*	102
	4.2.2.1 *Thermodynamic treatment of micelle formation*	102
	4.2.2.2 *The role of the hydrophobic factors in micelle formation*	103
	4.2.2.3 *The role of the hydrophilic groups in micelle formation*	104
	4.2.2.4 *The role of the counter-ions in micelle formation*	106
	4.2.2.5 *Pre-micellar association*	106
	4.2.3 *Changes in micelle structure above the c.m.c.*	106
	4.2.4 *Solubilisation*	107
4.3	LIQUID CRYSTALLINE PHASES	108
	4.3.1 *General remarks on the liquid crystalline phases*	108
	4.3.1.1 *Anhydrous liquid crystalline phases*	109
	4.3.1.2 *Lyotropic liquid crystalline phases*	109
	4.3.1.3 *The structure of the lyotropic mesophases*	112
	4.3.1.4 *The occurrence of the lyotropic mesophases*	112
	4.3.1.5 *The phase rule*	116

*Present address: Gråhundsvägen 134, 12362 Stockholm, Sweden.
**Present address: Department of Physical Chemistry, Åbo Akademi, Åbo (Turku), Finland.

	4.3.2	Liquid crystalline phases of type 1 and the transition from aqueous solution to mesophase		116
		4.3.2.1	The middle phase of type 1	117
		4.3.2.2	Mesophases in the region between the middle phase and the neat phase	119
		4.3.2.3	Transitions to mesophase from the micellar aqueous solution at the upper boundary of the solution region	120
		4.3.2.4	Transitions to mesophase from aqueous solutions of lower concentrations	121
	4.3.3	Lamellar liquid crystalline phases		122
		4.3.3.1	Mesophase B	122
		4.3.3.2	Neat phase	123
	4.3.4	Liquid crystalline phases of type 2 and the transition from organic solution to mesophase		125
		4.3.4.1	The middle phase of type 2	125
		4.3.4.2	Other mesophases of type 2	127
		4.3.4.3	Transitions to mesophase from micellar organic solutions	128
	4.3.5	Summary		128
4.4	SOLUTIONS OF SURFACTANTS IN ORGANIC SOLVENTS			129
	4.4.1	General aspects		129
	4.4.2	Surfactant solutions in non-polar organic solvents		130
		4.4.2.1	The dissolution process	130
		4.4.2.2	The aggregation	131
		4.4.2.3	A model substance	132
		4.4.2.4	Other surfactants	134
		4.4.2.5	Extreme solubilisation of water	134
	4.4.3	Surfactant solutions in polar organic solvents		135
		4.4.3.1	Carboxylic acid as solvent	135
		4.4.3.2	Alcohol as solvent	136
4.5	THE STAGEWISE AGGREGATION			138

4.1 INTRODUCTION

4.1.1 Scope of the survey

Aggregation in surfactant systems results in the formation in aqueous and non-aqueous solutions of micelles and of liquid crystalline phases, so-called mesophases. This aggregation is combined with the formation of internal boundaries within the homogeneous phases. These interfaces are formed by polar groups; in aqueous systems they separate the lipophilic regions from those with high water content. The spontaneous formation of these interfaces plays an important role in determining the properties of the surfactant systems.

In this survey attention will be drawn to the phase equilibria of surfactant systems, the phase structures and changes in them when the composition of the phases is altered, and to the factors responsible for these changes and the transition from one phase to another. The survey makes an attempt to describe the general picture of surfactant systems provided by the research of recent years.

4.1.2 General properties of the aggregates

The term 'micelle' in this survey is used only for aggregates of limited size in homogeneous solutions. The micelles may be defined as aggregates of amphiphile molecules whose lyophilic groups face outwards towards the surrounding solvent, while the lyophobic groups are oriented inwards and form the core of the aggregate. In aqueous solutions the core consists of the lipophilic parts of the molecules, while the hydrated polar groups lie in the surface of the micelle (micelles of the normal type, type 1). In organic solutions the polar groups form the core and the hydrocarbon groups of the molecules face outwards toward the solvent (micelles of the reversed or inverted type, type 2). This qualitative picture will distinguish the micelles from polymolecular chains, stoichiometrically defined aggregates, etc. where the cohesive forces are covalent or of a purely electrostatic nature.

The aggregation number of the micelles is often high. This leads to their formation being observed only above a defined concentration – the critical micelle concentration, c.m.c. and is particularly characteristic of aqueous solutions. At low surfactant concentrations above the c.m.c. the micelles appear to be spherical. When the concentration is increased, the properties of the solutions indicate a change in the micellar structure, probably primarily due to an interaction between the micelles, e.g. disturbances in the electrical double layers of the normal micelles of ionic surfactants. At still higher concentrations the crowding in the micellar solution may increase to a point where the micelles become cylindrical or lamellar and, ultimately, when a random orientation of the micelles becomes impossible, there is a transition from micellar solution to mesophase.

The aggregates in liquid crystalline phases, which often are of indefinite extent in one or two dimensions, will be referred to as mesophase aggregates. In them the surfactant molecules are arranged principally in the same manner as in the micelles.

The micelles are capable of incorporating – solubilising – compounds of the same nature as the interior of the micelles; lipophilic molecules are solubilised in micelles of aqueous solutions, while strongly polar molecules, especially those of water, are solubilised in the micelles in organic solutions. Molecules having an amphiphilic structure, such as long-chain alcohols and carboxylic acids, are built in between the molecules of the surfactants, with their lipophilic group in contact with the hydrocarbon part of the micelle and their hydrophilic group oriented towards the polar groups. The mesophase aggregates are capable of solubilising added compounds in a similar way. As the concentration of solubilisate in a micellar solution is increased, solubilisation in many systems is interrupted by the formation of a mesophase, but often continues within the aggregates of the mesophase and may proceed throughout the whole surfactant system.

Concerning the question of the balance that minimises the energy for a certain micellar size distribution two approaches, the phase separation and the mass action approaches, have been discussed[1-4].

The phase separation approach implies the precipitation of a colloidal 'pseudophase' in the solution at the c.m.c.[5-9]. It therefore implies that the activity of surfactant not bound to micelles is constant above the c.m.c. and that at the c.m.c. there are discontinuous changes in the properties of the solutions. Inconsistent with this approach is, however, the fact that, as careful measurements show, the activity of monomers decreases above the c.m.c.[10-14] and that around the c.m.c. the properties of the solutions change rapidly but continuously. The phase model has been criticised also on the grounds that it requires inclusion of the electric double layer so as not to imply, for ionic micelles in aqueous solution, the formation of a charged phase in the system. A realistic approach also requires account to be taken of the significant effects of the change in the total interfacial area of the micelles as the total concentration of surfactant increases[15].

According to the mass action approach the formation of micelles is considered to be an association to large aggregates in homogeneous solution[16-22, 44]; there is an equilibrium between these aggregates and monomers. This approach is consistent with the fact that the properties of the solutions change continuously at the c.m.c. which is a concentration range rather than a point; the interval is narrower the larger the micelles formed. This approach predicts correctly that the activity of the monomers changes above the c.m.c.[10-14], and enables bound counter-ions to be included in the equilibria in the case of ionic surfactants. The mass-action approach in its simplest form assumes that only two surfactant states occur in the solutions, namely monomers and micelles with a definite aggregation number. It is quite well established that the formation of micellar aggregates is preceded by dimerisation or formation of small polynuclear complexes and that the micelles are polydisperse. The formation of micelles thus involves a multiple equilibrium and the mathematical treatment can therefore be quite complex[21], especially in the case of ionic surfactants.

In the following, micelle formation in aqueous, as well as in organic solutions is discussed on the basis of the mass-action approach.

4.1.3 The interfacial layers of polar groups

It has become increasingly evident that water in the various phases of the surfactant systems does not function only as a dispersion medium between the amphiphile aggregates. Neither is it the influence of the surfactant molecules and aggregates on the structure of water, that is not directly bound to them, that is solely responsible for the effect of water on the aggregation. The water is also involved in interactions with the polar groups through ion–dipole or dipole–dipole attraction and through hydrogen bonds. As a result, some, often most, and sometimes all of it, is bound to the interfacial layers of polar groups and must be regarded as being an inherent part of the aggregates. These interactions are particularly strongly manifested in systems with a low water content.

Attention was drawn to the essential importance of this interaction of surfactants with water when it was found that the alkali soaps of the fatty acids dissolved in the liquid alkanols only in the presence of water, and that the minimum amount of water required was that needed for the hydration of alkali ions[23-30]. This was seen as conclusive evidence that for dissolution of these soaps the alkali ions must be hydrated. It was found to apply also to the formation of mesophases in the soap–alcohol systems[23-25, 27-30]. Further investigations indicated that the alkali ions are hydrated in all phases of these systems, and that they are linked to water molecules also when they are bound to the interfacial layers of polar groups of the micelles in the alcoholic and aqueous solutions as well as of the aggregates of the mesophases[24, 27, 31-35].

Subsequently it was shown that the micelles and the mesophase aggregates can bind considerably more water than is accounted for by hydration of the alkali ions. This was taken as evidence that the carboxylate groups bind water through ion–dipole interaction and/or hydrogen bonds; further support for this view was given by the fact that the region of existence of certain phases is limited to an upper water content corresponding approximately to the water-binding capacity of the amphiphiles; in other phases their properties are found to undergo well-defined changes when this water content is exceeded[24, 26-28, 32]. These conclusions have since been found to apply also to many other types of surfactants.

In many systems there is also an interaction between the polar groups of different amphiphile molecules: dipole–dipole attractions, direct hydrogen bonding between the groups, or hydrogen bonding of two polar groups to the same water molecule. In the case of ionic amphiphiles, of course, ion attraction and repulsion and, in some instances, coordination plays an important part. These different types of interactions in the interfacial layers of polar groups occur extensively in many amphiphile systems and may account for the formation of particular types of aggregates and phases.

This is the case for the acid soaps studied quite early on, where carboxylic and carboxylate groups are linked by hydrogen bonds which persist throughout the system in the different phases[26, 27, 32, 36-41]. A similar interaction between fatty acid soaps and alkanols has also long been known. More recently, these effects have been demonstrated in many other surfactant systems[27, 31, 42, 43].

These different interactions between the polar groups of a surfactant and an added amphiphilic compound, and between these groups and water, have of course an important bearing on the conditions in the interfacial layers of polar groups of the micelles and mesophase aggregates. These layers and consequently the interactions play an increasingly important role passing from aqueous solutions via mesophases to organic solutions. This survey will underline the fundamental significance of these phenomena, which have previously been given little attention.

4.2 SURFACTANTS IN AQUEOUS SOLUTIONS

4.2.1 The interactions between monomers and water

The statistical-mechanical theories developed by Aranow et al.[45, 46] on the one hand, and by Sheraga et al[47, 48] on the other, have assumed essential importance for the description of the mechanism of micelle formation and the factors governing it. Aranow assumes that the cause of aggregation lies in an increase in the internal mobility of the hydrocarbon chains, whereas Sheraga *et al.* ascribe principal importance to the sparing solubility of the hydrocarbon chains owing to hydrophobic hydration. Both models are able to describe experimental results, but are less successful in predicting them, especially as regards the influence of various end groups and solvent-modifying additives on the micellisation. In spite of this[49], the experimental results obtained in recent years, together with the ideas advanced in the theories mentioned, furnish a fairly convincing qualitative picture of micelle formation, which will be described below. In the light of the results of recent research[50] water would appear to be a structure held together by hydrogen bonds, with unbound water molecules in the 'cavities' of this structure. The water molecules reorient themselves in the structure in a time of $c.\ 10^{-11}$ s[51]. On addition of hydrocarbon there is 'increased structuring' of the water ('cage formation', 'iceberg formation', 'hydrophobic hydration'), which is best described as a marked increase in the re-orientation time[52]. This implies a reduction in the entropy of the system in relation to pure water + pure hydrocarbon. As a result, hydrocarbons are very sparingly soluble in water, in spite of the dissolution process being exothermic or weakly endothermic[53, 54]. Independent proof of the increased structuring has been provided by proton spin relaxation times and elastic neutron scattering[55, 56].

However, it is also possible to explain a large part of the change in entropy on dissolution of hydrocarbons in water by assuming that their internal torsional vibrations are inhibited[45]. Proof of this diminished mobility has been given by investigations of the hydrocarbon proton spin relaxation times[57].

It remains to differentiate quantitatively between these two effects. One has to assume that they cooperate, i.e. the hydrocarbon chain of an amphiphilic molecule reduces the mobility of the water and vice versa.

The interactions between the hydrophilic groups and water can be divided into at least the following groups: (a) Formation of solvation shells by ion–dipole attraction of water around the end groups, with simultaneous solvation of counter-ions. (b) Binding of water through electrostatic attraction of hydration counter-ions to the end groups (ion association)[58]. (c) Hydrogen bonding or dipole–dipole attraction of water to the end group[59]. (d) 'Increased structuring' of water by end groups containing alkyl groups[60–66].

In any systematic description of the properties of systems containing surfactants and water one must be aware of the relative importance of these different interactions. Few investigations so far have been carried out which aim at a quantitative differentiation between them.

4.2.2 The energetics of micelle formation

4.2.2.1 *Thermodynamic treatment of micelle formation*

The 'mass-action approach' seems to give a realistic description of micelle formation from both thermodynamic and kinetic aspects. In its simplest form the aggregation process is given by

$$pD^+ + nR^- = M^{(n-p)-} \quad (4.1)$$

where D^+ = counter-ion, R^- = surfactant monomer and M = micelle, for which the standard change in free energy per mole of monomer is given by

$$\Delta G_M^\circ = -\frac{RT}{n}\ln x_M - \frac{RT}{n}\ln x_D^p x_R^n - \frac{RT}{n}\ln \frac{f_M}{f_D^p f_R^n} \quad (4.2)$$

The reference states are chosen so that for each component the activity coefficient $f \to 1$ when $x \to 0$. This extrapolation cannot be done exactly for micelles, since they cannot be observed below the c.m.c. This difficulty, however, is generally of no importance. For concentrations not far above the c.m.c., normal values of p and n and activity coefficients as low as 0.6, the first and third term in equation (4.2) amount to only c. 1% of the second. To a close approximation, therefore, $\Delta G_M^\circ = -RT/n \ln x_R^n x_D^p$. Near the c.m.c. $x_R \approx x_D \approx$ c.m.c., so that

$$\Delta G_M^\circ = -[1+(p/n)] RT \ln (\text{c.m.c.}) \quad (4.3)$$

The approximations in equation (4.3) are of the same order of magnitude as normal errors in the determination of the c.m.c. With similar approximations, equation (4.3) gives the change in free energy for the process

$$R^- + (p/n) D^+ + M_n = M_{n+1} \quad (4.4)$$

where M_n is a micelle of the most probable size at the c.m.c.[22]. For a completely dissociated micelle, p/n in equation (4.3) is 0; this also applies to non-ionic micelles. In the case of complete ion binding, $p/n = 1$. Unfortunately, a large number of the energies of micelle formation given in the literature have been calculated with no regard of the fact that for ionic micelles p/n may have any value between the used extreme values 0 or 1. Moreover, micelles are polydisperse. This may be taken into account relatively easily for non-ionic micelles, for which the expression[20]

$$\Delta G_M^\circ = -RT(\ln x_2 - \frac{1}{\langle n \rangle} \ln \Sigma x_r) \quad (4.5)$$

is valid. Here, x_2 is the concentration of monomers and Σx_r the total concentration of micelles with the mean aggregation number $\langle n \rangle$.

Enthalpies of micelle formation have often been determined from the dependence of the c.m.c. on temperature, which is possible if the approximations in equations (4.3) or (4.5) hold, and the size of the micelles is independent of temperature. Conclusions drawn on the basis of thermodynamic properties therefore should be supplemented by as many independent experimental results as possible.

4.2.2.2 The role of the hydrophobic factors in micelle formation

The transition of a surfactant molecule from aqueous solution to a micelle seems to imply the following changes in the state of the hydrocarbon chain: (a) The 'increased structuring' of the water in the vicinity of hydrocarbon chains disappears—that is, the entropy of the water increases while some energy must be supplied to overcome the van der Waals–London attraction between hydrocarbon and water. (b) The internal freedom of the hydrocarbon chains is increased and energy is liberated through the attraction between adjacent chains.

The possibilities of distinguishing between these factors have recently been reviewed. On the basis of thermodynamic data, Mukerjee[67] concludes that the major part of the free energy change per —CH_2— group on transfer into a micelle, must derive from the interaction of the hydrocarbon chains between themselves. The effect of increasing the disordering of water, which certainly is real, is considerably smaller than this interaction. The main cause of micelle formation in any case is not, as it was long thought, that water 'expels. the hydrocarbon chains. Calorimetric determinations of the enthalpy of micelle formation[68] show that the reduction in ΔG_m° when the hydrocarbon chain length is increased is mainly an enthalpy effect.

This may seem to be in conflict with the view that the reduced water structuring is of great importance for micelle formation, which is based on the following arguments: (i) The reduction in free energy on micelle formation is mainly an entropy effect ($T\Delta S$ is 90–95% of ΔG). (ii) The enthalpy of micelle formation becomes more negative with a rise in temperature[69–77] which is ascribed to the reduction of the water structure with temperature. (iii) Below the c.m.c. the partial molal volume of the surfactant decreases up to the c.m.c.; micelle formation causes a positive volume change[104–108]. This can be interpreted as the result of hydrophobic hydration being diminished on micellisation. (iv) The 'structure-breaking' substance, urea, increases the c.m.c. when it is added to surfactant solutions, decreases the entropy of micelle formation and renders the enthalpy of micelle formation more negative[66, 78, 79]. (v) N.M.R. investigations of the water protons in surfactant solutions indicate that their mobility increases on micelle formation[55, 80].

It seems to be indisputable that the introduction of monomers from aqueous solution into micelles must imply a change in the water structure. Points (i)–(iii), however, may also be due to a change in the water structure around the polar end group. Conclusions from results of the type indicated in (iv) should be drawn with great care, since the c.m.c. is also greatly affected by substances that cannot be compared to urea as structure breakers[49, 77]. The spectroscopic results constitute direct proof of the increased mobility of water, but do not permit a quantitative estimation of the importance of this effect.

To summarise, although the increased ordering of the water around the hydrocarbon chains of molecule–disperse surfactants cannot be disregarded, the principal hydrophobic factor leading to micelle formation seems to be the increased mobility and mutual attraction of the hydrocarbon chains. There is patently a need for more experiments to distinguish quantitatively between these effects.

4.2.2.3 The role of the hydrophilic groups in micelle formation

The role of the hydrophilic groups in micelle formation is not only to keep the aggregates in aqueous solution, but also to restrict the size of the aggregates formed. This is clearly brought out in a comparison between the properties of ionic and non-ionic surfactants having the same hydrocarbon chain: the micellar weight of the ionic micelle is much lower than for the non-ionic micelle, the c.m.c. of the ionic surfactant is much higher than for the non-ionic surfactant, the micellar weight of ionic micelles rises with electrolyte concentration while the weight of non-ionic micelles are only slightly affected, and the micellar weight of non-ionic micelles increases rapidly with temperature while the effect is very slight or opposite on ionic micelles[81–88, 94–97].

The cause of these phenomena is obvious: because of the repulsion between the ionic end groups and their strong ion–dipole binding of water, a transfer of an ionic molecule to a micelle requires greater energy than a corresponding transfer of a non-ionic molecule containing the same hydrocarbon chain. The spherical shape of ionic micelles is determined by the repulsion between the charges and their radius by the hydrocarbon chain length and the bulkiness of the polar end group; their size is therefore not very dependent on temperature but highly dependent on electrolyte concentration. The repulsion between non-ionic end groups is governed mainly by the binding of water to them, which is highly temperature dependent. Hence the strong temperature and weak electrolyte dependence of non-ionic micellar weights and shapes[89–94, 98–100]. In line with this argument non-ionic micelles with small, strongly polar end groups have about the same properties as ionic micelles with the same hydrocarbon chain[85, 101–103].

A number of attempts to estimate the magnitude of the electrostatic effect have been made[1, 19, 22, 110, 111]. It is found that the electrostatic part of ΔG_m° is small and positive and that it is considerably decreased by the inclusion of counter-ions in the calculations. It has also been estimated from attempts to calculate the hydrocarbon part of ΔG_m° [22, 67, 112]. The agreement between electrostatic calculations and the latter method is fairly good and suggests that it really is the electrostatic effects, including the binding of counter-ions, that are mainly responsible for the contribution of the ionic end groups to the energy of micelle formation. Further evidence for this is provided by the fact that micelles with simple ionic end groups behave experimentally as colloidal particles surrounded by a Guoy–Chapman diffuse double layer with some counter-ions adsorbed in the Stern layer[70, 112–117].

Apart from the electrostatic effects, the question of the importance of the end group for micelle formation is complicated, since it will be dependent on all the factors mentioned in Section 4.2.2.2. It is often found that an increase in the hydrophilic nature of the end groups results in an increase in the c.m.c. and the size of the micelles[98, 118, 120]; an increase in the hydrophobic nature of the end group often lowers the c.m.c. and reduces the micelle size[119]. Such results can, however, usually be obtained only when, for example, an end group may grow without undergoing any change in its nature (e.g. by increasing the number of oxyethylene groups in a series of polyoxyethylene glycol monoalkyl ethers). Although there is some correlation between increasing hydrophobicity of the end group, increasing micelle size and decreasing c.m.c., an important part is often played by specific possibilities for individual end groups to form hydrogen bonds and by the bulkiness of the hydrocarbon part of the end group[121–122].

4.2.2.4 *The role of the counter-ions in micelle formation*

A wide range of methods have been applied to study the effect of counter-ions on micelle formation[111, 123, 124]. They give, on the whole, mutually consistent results. The nature of the binding is not quite clear; it seems, however, that it is possible to define the bound counter-ions as those that belong to the kinetic micellar unit.

A number of studies[41, 110, 123] in recent years have shown that nearly all counter-ions bound to the micelles must be hydrated; however in the case of alkylated ammonium halides n.m.r. line-width studies[41] suggest that these may bind counter-ions without any water between the counter-ion and polar end group. In view of the binding in hydrated form it is evident why there is an increase in the micelle size for a particular cation in the series $Cl^- < Br^- < I^-$ and for a particular anion in the series $Na^+ < K^+ < Cs^+$; it would seem to be generally the case that the more weakly hydrated a counter-ion, the larger the micelles formed by the amphiphilic ions. The more weakly hydrated ions can be absorbed more readily in the micelle surface, so that the repulsion between the end groups is diminished[125].

4.2.2.5 *Pre-micellar association*

Investigation of pre-micellar association offers possibilities to separate the primary effects of the interaction between monomeric surfactants and water to some extent from the effect of the formation of micelles proper at the c.m.c. of ordinary association colloids. A good review of the work in this field up to 1967 has been presented by Mukerjee[67].

While there is hardly a dispute any longer that micelle-forming surfactants also build aggregates that are much smaller than micelles, their actual size, structure and stability are still a matter for discussion[104, 108, 126-128]. A systematic investigation of short-chain carboxylates[129, 130] has shown that association occurs with salts with more than four carbon atoms in the chain. When the concentration is increased, the changes in volume, the enthalpy effects and the activities display close similarities with those shown by alcohols having the same number of carbon atoms in the chain; when association occurs, deviations become apparent. It should therefore be possible to study interactions of monomeric surfactants with water through analogy with compounds having an amphiphilic structure but where the difference between hydrophilic and hydrophobic groups is not distinct enough for micelles to be formed.

Most studies suggest that the pre-micellar aggregates, too, are held together mainly by forces of attraction between hydrocarbon chains in conjunction with a reduction in the water structure on association (hydrophobic bonding). The possibility that binding of counter-ions plays a role cannot, however, be ruled out.

4.2.3 Changes in micelle structure above the c.m.c.

Above the c.m.c. both ionic and non-ionic micellar solutions display properties that, in a limited concentration range, change in a way which indicates that the size, spherical shape, water binding, counter-ion binding and other properties of the micelles remain constant.

At higher concentrations, however, there are changes in the concentration dependence of the properties. For a number of ionic association colloids these changes have been found to appear within a quite well defined concentration range, the 'second critical concentration'[23, 24, 26, 27, 30, 33, 131–137, 220]. These changes are probably due to interactions between the electrical double layers of the micelles. The counter-ion and water binding of the micelles increases at this concentration limit, but the spherical shape is retained.

When the concentration is raised further, there are, over a relatively wide but well defined concentration range ('the third critical concentration'), further marked changes in the dependence of the properties on concentration. This interval would seem to be reached when the volume fraction of hydrated micellar substance approaches the value at which spheres in simple cubic packing make contact with each other, i.e. the volume fraction 0.52. This crowding in the solution would seem in some cases to result in rapid deformation of the micelles, which tend to assume a cylindrical shape, while the deformation of micelles of other surfactants proceeds slowly or is non-existent despite the reduction in mobility[23, 24, 26, 27, 30, 33, 131–137, 220].

The solvation water incorporated in the micelles must play an important part in producing the crowding that is assumed to be the prime cause of the structural changes.

In solutions of non-ionic micelles there is no electrostatic repulsion between the micelles. This means that general methods for studying macromolecules (sedimentation, diffusion, viscosity, Rayleigh–Gans light scattering, etc.) may be applied with some confidence[13, 102, 103, 139]. Octyl- and dodecyl-amine oxides form spherical micelles up to extremely high concentrations[139]. The more complex and bulkier polar group in alkyl polyoxyethylene glycol ethers causes formation of cylindrical micelles at concentrations not much above the c.m.c.[81, 82, 90, 94, 99, 119]. It is not clear, however, whether this is due to elongation of spherical micelles containing large amounts of bound water or to an increased interaction between smaller cylindrical or spherical micelles.

At the upper limit of the solution region most surfactants are in equilibrium with a mesophase.

4.2.4 Solubilisation[15, 140]

The maximum amount of solubilised substance per mole of surfactant varies with the concentration of the solution parallel with the changes in the micellar structure. This is particularly so in the cases where the whole solubilising capacity of the micelles is used and the solubilisation is not interrupted by the formation of a mesophase; the ability of many association colloids to solubilise hydrocarbons of low molecular weight (mol/mol) increases at both the second and the third critical concentration[26–30].

Recent n.m.r. studies have confirmed that completely non-polar compounds are incorporated in the hydrocarbon core of the micelles[142–144]. Other studies have, however, been interpreted as showing that the first molecules of aromatic hydrocarbons to solubilise are located fairly close to the surface of the micelle while as the concentration of solubilisate is raised there is an increased penetration into the micelle[145, 146].

The distance that amphiphilic solubilisates are located from the interfacial layer of polar groups of the micelles will be determined by the specific possibilities for different solubilisates to interact with different polar groups[141, 147–152]. Solubilisates with end groups of higher polarity[13] are solubilised in such a way that the hydrophilic group is located between the polar groups of the surfactant. This may result in a reduction in the water binding of the micelles, as shown by, for example, vapour pressure measurements[224].

If the polar layer consists of voluminous polyoxyethylene glycol groups, it is also possible for the solubilised substances to be confined within this layer[153].

In some cases 'solubilisation' can occur below the c.m.c. too. For instance, pre-micellar aggregates of short-chain carboxylates are able to solubilise small amounts of fatty acids and alkanols. This solubilisation is often observed above a well-defined concentration, designated 'the limiting association concentration', l.a.c.[30]. When saturated with solubilisate, these solutions often are in equilibrium with different mesophases (see Section 4.3.2.4).

Non-polar solubilisates induce an increase in size of the micelles, which however retain their spherical form. If large amounts of solubilisate are incorporated, interactions between the micelles can occur and promote the transition to mesophases. Solubilisates with polar groups affect the shape of the micelles in a much more complex way. The interaction with the micelle end groups and the reduction of the charge density in the ionic micelles may liberate a part of the bound counter-ions and at high solubilisate/surfactant molar ratios lead to a deformation of the micelles. Often this causes a transition to a mesophase even at low molar ratios (see Section 4.3.2.4).

4.3 LIQUID CRYSTALLINE PHASES

4.3.1 General remarks on the liquid crystalline phases

The research on the liquid crystalline phases of the surfactants has so far been concerned mainly with demonstrating various phases and the extent of their region of existence, analysing their structure and examining the conditions for their formation and transition to other phases.

As a result of the fundamental work by Luzzati and his co-workers since the end of the 1950s the x-ray diffraction technique has been the principal one for determining the structure of lyotropic liquid crystalline phases[154–155]. Beside the long familiar 'soap boiler's neat soap' and 'middle soap', the occurrence at relatively low water contents of several other mesophases was demonstrated in binary aqueous surfactant systems, and their structure was examined, among them certain types of optically isotropic liquid crystalline phases[156–159]. The significance of the temperature and the water content for the appearance of lyotropic mesophases in the binary systems has been examined, and the influence of molecular structure on the structure of the phases has been elucidated. The high-temperature phases of the anhydrous soaps have been studied[160–166].

Systematic investigations by Ekwall and co-workers at about the same period were concerned with the phase equilibria at constant temperature in binary and ternary systems of surfactant, water, and an organic solvent. By separation in the pure form their nature as independent phases was proved; their composition, properties and structure were also examined[23, 25, 28, 167-170]. The effect of the molecular structure of the components on the phase equilibria was studied. Particular attention was devoted to the binding of water to the polar groups and the interaction between the polar groups of the various components. The existence of new mesophase structures and of mesophases with normal and reversed structure in the same system was demonstrated[23-25, 171-174].

During the same period many other investigators have contributed to our knowledge of the lyotropic mesophases, usually in binary systems[175-193]. Winsor has developed general aspects of the transition between the various phases[194]. Among the new investigational techniques brought into use during this period molecular spectroscopy has proved especially promising[35, 195-201]. Electron microscopy also deserves mention, even though the pre-treatment of the material for the studies has in some cases cast doubt as to whether the demonstrated mesophase structure is actually the same as that found in the untreated system[202-207].

The research in this field has previously been summarised in a number of surveys by the Luzzati group[208-210, 227], Winsor[194, 211] and Ekwall[30, 212].

In this review attention in the first place will be paid to the structure of the mesophases and the changes in it with composition, and to the factors governing the formation of a mesophase and its transition into other mesophases. In particular, the present knowledge of the conditions in interfacial layers of polar groups and the binding of water in them will be considered.

4.3.1.1 Anhydrous liquid crystalline phases

A number of crystalline surfactants show thermotropic mesomorphism[160-166]. The alkali soaps above c. 130 °C pass through a series of transformations to mesophases, which are lamellar in so far as the polar groups of the soap molecules are disposed in double layers. The lateral extent of these layers varies; they may be sheets of indefinite extent, ribbon- or lath-shaped with definite width but indefinite length, or discs of definite lateral extent in two dimensions. The space between the sheets, ribbons and discs is occupied by disordered paraffin chains. The soaps with bivalent cations transform to mesophases in which the polar groups are clustered in rod-like regions, either indefinitely long or of definite length, and linked to form two- and three-dimensional networks.

Some anhydrous surfactants occur in the liquid crystalline state even at room temperature. The lack of solid crystallinity is in most cases obviously associated with chain branching and a bulky hydrocarbon part of the molecule.

4.3.1.2 Lyotropic liquid crystalline phases

Added water, and in some cases also organic solvents, transforms most crystalline surfactants into the liquid crystalline state. It is these lyotropic

Table 4.1 Common lyotropic liquid crystalline phases

Designation	Type	Basic structure	Description of the supposed structure	Optical properties	Alphabetical notation Luzzati	Winsor	Ekwall
Layer structures							
Structural arrangement displaying Bragg spacing ratio $1:\frac{1}{2}:\frac{1}{3}$							
1. Neat phase 'soap boiler's neat soap'		Lamellar, double layers	Coherent double layers of amphiphile molecules with the polar groups in the interfaces with the intervening layers of water molecules	Anisotropic	L LL L_α	G	D
2. Single layered neat phase		Lamellar, single layers	Coherent single layers of amphiphile molecules oriented with the polar groups towards opposite interfaces with the intervening layers of water molecules	Anisotropic			D_s
3. Mucous woven phase		Lamellar, double layers	Coherent double layers of amphiphile molecules with the polar groups in the interfaces with the intervening layers of water molecules	Slightly anisotropic			B
Particle structures							
Structural arrangement displaying Bragg spacing ratio $1:1/\sqrt{3}:1/\sqrt{4}$							
4. Middle phase, normal	1	Two-dimensional hexagonal	Indefinitely long, mutually parallel rods in hexagonal array, with a hexagonal or circular cross-section; the rods consist of more or less radially arranged amphiphile molecules	Anisotropic	H_1	M_1	E
5. Middle phase, reversed	2			Anisotropic	H_2	M_1	F

6. Hexagonal complex phase, normal	1	Two-dimensional hexagonal	Indefinitely long, mutually parallel rods in hexagonal array, possibly consisting of a double layer of amphiphile molecules surrounding a core of hydrated polar groups and water and located in a water continuum	Anisotropic	H_c	H_c

Particle structures
Structural arrangement displaying Bragg spacing ratio $1 : \frac{1}{2} : \frac{1}{3}$

7. Rectangular phase, normal	1	Two-dimensional orthorhombic	Indefinitely long, mutually parallel rods in orthorhombic array, and with a rectangular cross-section	Anisotropic	R	R
8. Square phase, normal 'white phase'	1	Two-dimensional tetragonal	Indefinitely long, mutually parallel rods in tetragonal array and with a square cross-section	Anisotropic		C
9. Square phase, reversed	2			Anisotropic		K

Structural arrangement displaying cubic symmetry

10. Optical isotropic mesophase, normal 'Viscous isotropic phase', normal	1	Cubic, body centred Space group $Ia3d$	Short rod-like elements (axial ratio near 1) joined in threes at each end to a three-dimensional network; two networks interwoven but unconnected	Isotropic	Q V_1	I'_1
11. Optical isotropic mesophase, reversed 'Viscous isotropic phase', reversed	2			Isotropic	Q V_2	I'_2
12. Optical isotropic mesophase, normal	1	Cubic, body centred Possibly space group $Pm3n$	Possibly, spherical aggregates caged within a network of short rod-like elements joined in threes at one and in fours at the other end	Isotropic	F_{1h}	I''_1
13. Optical isotropic mesophase, reversed	2	Cubic, space group unknown	Structure unknown	Isotropic		I'''_2

Optically isotropic liquid phases are designated S (by Winsor), and L (by Ekwall); S_1 and L_1 denote aqueous solutions, S_2 and L_2 solutions in organic solvents.

liquid crystalline phases, and especially the aqueous ones, that will be dealt with below.

Just as in the case in the micelles, the surfactant molecules with their polar groups are more or less firmly anchored in the interface of the mesophase aggregates, while the hydrocarbon parts of the molecules possess so great a freedom that they can be considered to be in a semi-liquid state. In some systems at low water-contents and low temperatures, a 'gel phase' may appear with a somewhat more ordered state in the hydrocarbon parts of the aggregates, where the paraffin chains are considered to be stiff and parallel[186, 214, 215]. This may also be effected by the incorporation of a compound with a rigid hydrocarbon skeleton[216]. In contrast to the micelles in solution the mesophase aggregates are arranged in a structure of a higher order.

4.3.1.3 The structure of the lyotropic mesophases

The various mesophases differ from each other by virtue of their consistency, gross appearance and microscopic appearance between crossed polaroid plates. The inner structure of the phases, i.e. the shape and mutual arrangement of the mesophase aggregates varies. There are structures with lamellar, network, rod-shaped and possibly also spherical aggregates.

Most of the lyotropic liquid crystalline structures with particle-shaped or network-shaped aggregates have been found to occur as two complementary types.

Type 1 (the normal type), mesophase aggregates in a water continuum; they have a hydrocarbon core surrounded by an interfacial layer of hydrated polar groups.

Type 2 (the reversed or inverted type), mesophase aggregates in a hydrocarbon continuum; they have a core of unhydrated (in anhydrous systems) or of hydrated polar groups and water (in aqueous systems) which is surrounded by a layer of hydrocarbon chains.

Intermediate between these two types are the mesophases with a lamellar structure, composed of coherent layers of amphiphile molecules, with the hydrocarbon parts innermost and the polar groups in the interface with intervening layers of water molecules. Various mesophases of this type seem to differ from each other by virtue either of the arrangement of the molecules as such in the amphiphile layers, of the hydrocarbon chains in the innermost of these layers, or of a different orientation of the polar groups in the interfacial layers.

A survey of various mesophases and of their structure is given in Table 4.1. The various lyotropic mesophases have been given alphabetical notation (Ekwall's notation will be used where appropriate in this survey).

4.3.1.4 The occurrence of the lyotropic mesophases

In binary systems (with organic solvents or with water) the lyotropic mesophases occur chiefly at relatively high surfactant contents; exceptions are known. Mesophases with anhydrous organic solvents, studied up to the present, are of type 2. With water some surfactants form type 1 and others type 2 mesophases. The former group includes surfactants that in aqueous solution occur as typical association colloids, that is, those whose solubility in water is so high that the c.m.c. is exceeded and the various types of micelles are found. To the other group belong the amphiphiles that have a lower solubility in water, but that are able to take up water while swelling, and surfactants with a bulky hydrocarbon part, which obviously for this reason neither crystallise nor form normal micelles and mesophase aggregates[30, 212].

In ternary systems consisting of surfactant, water and an organic solvent, both types of lyotropic mesophases can occur. In both micellar aqueous and

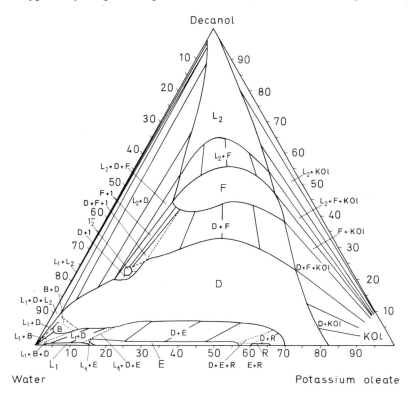

Figure 4.1 Phase equilibrium diagram at 20 °C of the ternary system potassium oleate–n-decanol–water. Alphabetical notations: see Table 4.1 (From Elkwall et al.[219], by courtesy of Academic Press, Inc.)

organic solutions these are usually in equilibrium with each other via a three-phase zone; its position differs widely, according to the system, but the mesophases invariably appear in the direction towards higher surfactant contents, reckoned from this zone (see Figures 4.1 and 4.2). The type 1 mesophases occur mainly in the parts of the system with higher water and surfactant contents, and are often in equilibrium with the aqueous solution, while the type 2 mesophases are encountered mainly in parts of the system with higher contents of organic solvent and surfactant, and often in direct equilibrium with the organic solution. Lamellar mesophases occur chiefly in the concentration ranges between the regions of existence of the above-men-

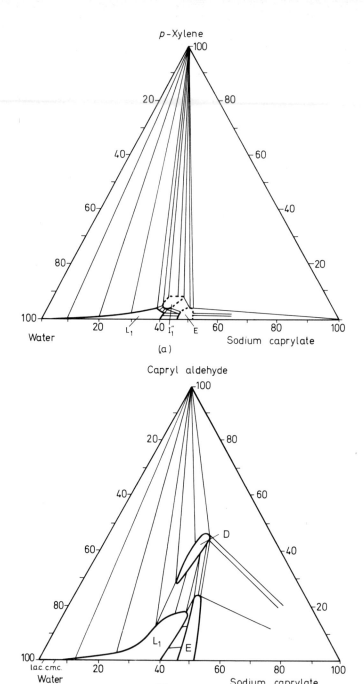

Figure 4.2 (a) Phase equilibria diagram at 20°C of the ternary system sodium caprylate–*p*-xylene–water.
(b) Phase equilibria diagram at 20°C of the ternary system sodium caprylate–capryl aldehyde–water.

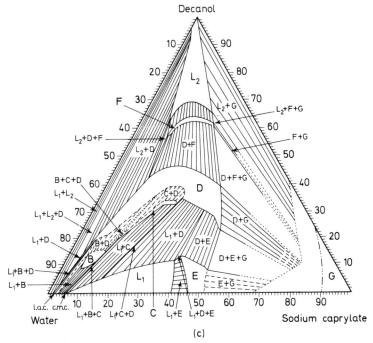

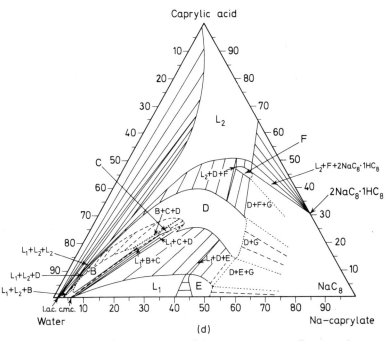

(c) Phase equilibria diagram at 20 °C of the ternary system sodium caprylate–n-decanol–water.
(d) Phase equilibria diagram at 20 °C of the ternary system sodium caprylate–caprylic acid–water. Alphabetical notations: see Table 4.1
(From Ekwall, P., Mandell, L. and Fontell, K. (1969). *Mol. Cryst.*, **8,** 157, by courtesy of Gordon and Breach, Inc.)

tioned types. The solutions in equilibrium with lyotropic mesophases have generally proved to contain micelles or other aggregates[23, 30, 212, 222].

4.3.1.5 The phase rule

Investigations of the phase equilibria have confirmed the validity of the phase rule in systems with lyotropic liquid crystalline phases. Some typical phase diagrams for three-component systems are given in Figures 4.1–4.4. The extent of the two-phase zones that separate the regions of homogeneous

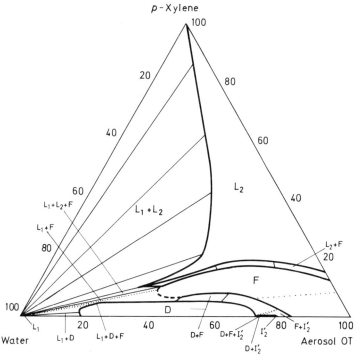

Figure 4.3 Phase equilibria diagram at 20 °C of the ternary system di-2-ethylhexyl sodium sulphosuccinate (Aerosol OT)–p-xylene–water. Alphabetical notations: see Table 4.1
(From Ekwall, P., Mandell, L. and Fontell, K. (1970). *J. Colloid Interface Sci.*, **33**, 215, by courtesy of Academic Press, Inc.)

phases has been shown to vary widely. The direction of the tie lines in these zones (see Figures 4.1 and 4.2, a–d) show at which compositions two phases are in equilibrium; the location of the phase boundaries gives information about some of the factors that govern the existence of the phases.

4.3.2 Liquid crystalline phases of type 1 and the transition from aqueous solution to mesophase

If the water content of the micellar aqueous solutions of the association colloids is lowered, there is sooner or later a transition from homogeneous solution – via a two-phase zone – to homogeneous liquid crystalline phase; in rare cases an optical isotropic mesophase I_1'' is formed, but usually it is the two-dimensional hexagonal mesophase of type 1, the normal middle phase, that occurs. On further reduction of the water content this phase changes, via a new two-phase zone, to another mesophase – often the lamellar

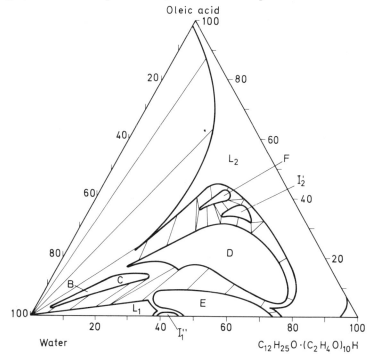

Figure 4.4 Phase equilibria diagram at 20 °C of the ternary system decaoxyethyleneglycol monolauryl ether–oleic acid–water. Alphabetical notations: see Table 4.1
(From Ekwall, P., Mandell, L. and Fontell, K. (1969). *Mol. Cryst.*, **8**, 157, by courtesy of Gordon & Breach, Inc.)

neat phase. Between them, other type 1 phases often occur – a complex hexagonal (H_c), a rectangular (R) and/or an optical isotropic mesophase (I_1') (see Table 4.1 and Figures 4.1, 4.4 and 4.5).

4.3.2.1 The middle phase of type 1

The relationship between the lattice parameter and the volume fraction of amphiphilic substance would seem to indicate that the cross-section of the rod-shaped aggregates is circular at high water contents, but probably deformed towards a hexagonal shape when the volume fraction of hydrated amphiphile exceeds 0.91 [212]. As the water content of the phase is lowered, there is a slow decrease in the interfacial area per polar group; the rate is independent of chain length, but varies with the type of polar group and the temperature[210, 217]. Even at the lowest water contents the calculated values of the interfacial area per polar group of the alkali soaps is so high (39–40 $Å^2$) that the alkali ion and the water of the phase may well be located between the carboxylate groups in the interface of the rod-shaped aggregates. For a more or less cylindrical aggregate there is a minimum value for the interfacial area per polar group and this increases with the chain length of the soap. The minimum content of water of the middle soap is probably determined by the fact that the space between the end groups in the interface must remain completely occupied by hydrated counter-ions and water molecules if the structure in question is to remain stable. Accordingly, the region of existence of the middle phase is displaced towards higher water contents, with increasing chain length of the surfactant, when its polar end group is kept constant. On the other hand, for surfactants of the same type and same chain length the region is displaced towards lower water contents when the size of the end group is increased.

It seems probable that the increase of the areas, when the middle phase takes up more water, is due to the fact that a part of this water, too, is located between the carboxylate groups in the interface. What factors restrict this uptake of water, and thus the extent of the middle phase region towards high water contents, is still unknown. The interfacial area per polar group here rises to c. 60 $Å^2$. The charge density in the interface has thus fallen to a level where it is no longer certain that all the counter-ions remain bound to the surface. The possibility that a Donnan distribution of liberated ions here contribute to the uptake of water — as is the case for certain other phases — cannot be ruled out[212].

The middle phase is in most systems able to incorporate various additional lipophilic and amphiphilic compounds without prejudice to the basic structure of the phase. The capacity for incorporation differs with their character (Figures 4.2, a–d).

In n-alkanols sparingly soluble in water the maximum amount that the middle phase of a given alkali soap is able to solubilise is independent of the chain length of the alcohol; for sodium caprylate at 20 °C it is 0.33 moles of alcohol (n-decanol – n-pentanol) per mole of soap[212, 218]. This has been regarded as proof that it is the interaction between the carboxylate group of the soap and the hydroxyl group of the alcohol that determines the number of alcohol molecules that can be inserted between the soap molecules of the rod-shaped aggregates. For other (less well examined) surfactants similar values have been recorded (0.27–0.44 moles per mole of surfactant)[212]. The calculated mean interfacial area per polar group decreases rapidly, but proportionally to the increase in the molar ratio of alcohol to soap. This decrease is due not only to the smaller space needed for the alcohol group and the interaction between the polar groups, but obviously in the first place to expulsion of the hydrated alkali ions that had been interposed between the carboxylate groups[212]. The vapour pressure of the phase then rises, as is the case when alcohol is solubilised in micelles of the aqueous solution[224].

The solubilisation of fatty acids follows the same mechanism; maximally a molar ratio of 0.26 has been noted. Additives with the less strongly polar aldehyde, nitrile and methyl ester groups can be solubilised in considerably greater quantities (maximum molar ratios of 0.75, 0.94 and 1.43, respectively, have been recorded); the mean interfacial area per polar group decreases to a lesser degree. These solubilisates, too, appear to be inserted between the molecules of the surfactant — at least so long as the amount solubilised is not too high — but the interaction between the polar groups would seem to be weaker[212].

For non-polar lipophilic compounds (hydrocarbons, chlorohydrocarbons) which are incorporated as a whole in the hydrocarbon environment of the core, the extent of the solubilisation seems to be determined by the size of the hydrocarbon core of the rod-aggregates (0.25 and 1.24 moles of *p*-xylene per mole of sodium caprylate and potassium oleate, respectively). This solubilisation increases the diameter of the aggregates, whereas the interfacial area per carboxylate group remains unaltered or undergoes a small increase[212].

4.3.2.2 Mesophases in the region between the middle phase and the neat phase

Between the middle phase and neat phase a two-dimensional hexagonal mesophase, H_c (see Table 4.1) has been demonstrated in a number of systems.

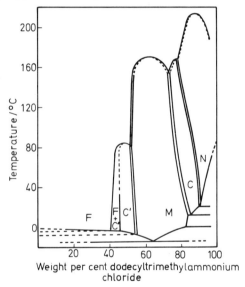

Figure 4.5 Temperature *v.* concentration diagram for the binary system dodecyltrimethylammonium chloride–water. $F = L_1$, aqueous solution; $C' = I_1''$, optical isotropic mesophase of type 1; $M = E$, middle phase of type 1; $C = I_1'$, optical isotropic mesophase of type 1; $N = D$, neat phase. (From Balmbra *et al.*[181], by courtesy of Macmillan Co.)

The phase has hitherto not been isolated from the neighbouring phases in the pure state, and its existence has therefore been questioned[154, 155, 158, 206–208, 215, 219].

In the same region another mesophase, R, composed of rod-shaped particles of rectangular cross-section has been observed in a number of systems (see Table 4.7). Judging from the observations so far the solubilising capacity of this phase seems to be very low[154, 155, 208, 215, 219].

In this region, in the immediate vicinity of the neat phase in many systems, there also occurs the optically isotropic mesophase I'_1, considered to consist of two interwoven networks of short rod-like elements[157, 158, 178, 219] (see Table 4.1 and Figure 4.5) and capable of solubilising organic additives[218, 219].

4.3.2.3 Transitions to mesophase from the micellar aqueous solution at the upper boundary of the solution region

From investigations of the phase equilibria, together with the study of the changes in micellar structure in concentrated aqueous solutions, it has been possible to get an idea of the factors which govern the transition from micellar solution to mesophase. In Section 4.2.3 it was intimated that one reason for this transition is to be sought in the increasing crowding in the micellar solution with concentration, which promotes a deformation of the spherical micelles. On the other hand, it has been shown that the randomly oriented, more or less cylindrical, micelles of the concentrated solution in equilibrium with the middle phase have practically the same composition as the long rod-shaped mesophase aggregates in hexagonal array[23, 169]. At the phase transition there thus seems to have been only an increase in length of the micelles. This phase transition has been related to the fact that the volume fraction of hydrated micellar substance at the upper boundary of the solution region attains a critical value above which a random orientation of the micelles becomes impossible[26, 27, 131, 220]. This critical volume fraction should – in accordance with a theory advanced by Flory for the transition from a solution of rod-like, stiff randomly oriented molecules or particles to a tactoid, anisotropic phase with largely parallel rods[221] – be a function of the axial ratio of the micelles. Qualitatively this has been proved to be the case. The concentration at which the phase transition occurs as a rule is lower the greater the chain length of the surfactant and the more anisometric the micelles[212]. For lower alkali soaps (C_7–C_{12}) the upper limit of the solution region falls from 2.8 to 1.7 m at 20 °C, at the same time as the critical volume fraction of the hydrated micellar substance remains relatively unchanged (0.85–0.70) and the axial ratio remains low (7–9). For the higher soaps (C_{12}–C_{18}) the volume fraction of the hydrated micellar substance, and hence the limit of the solution region, are attained earlier (1.4–0.7 m at 100 °C), obviously because the anisometry of the micelles progresses at ever lower concentrations. In octyl-, octylmonomethyl- and octyldimethyl-ammonium chloride solutions, where the hydration is low and the micelles appear to be little deformed, the limit of the solution region lies at quite high concentrations (52, 53, 58 % respectively[212]), while in the cetyltrimethyl-ammonium bromide solution, with its special, extremely marked, hydration and rapidly progressing anisometry of the micelles, it is located at extremely low concentration (26 %)[137].

In the light of this theory many displacements of the boundary of the solution region resulting from solubilisation can be explained. A solubilisate that promotes anisometry of the micelles would presumably displace the boundary downwards, while a solubilisate that inhibits or decreases the deformation would displace the boundary upwards. Examples of effects in both direction are known[212].

Systems are, however, known where the middle phase can exist only at relatively high surfactant concentrations, but the volume fraction of the hydrated micellar substance even at a much lower concentration reaches a value above which the solution structure cannot persist, although the micelles still appear to be more or less spherical. In these cases there may be a transition to an optical isotropic mesophase of type I''_1 with spherical aggregates in a clathrate cage (or possibly consisting of spherical aggregates alone) (decaoxyethyleneglycolmonolauryl ether[30, 212], dodecyl trimethylammonium chloride[181] (Figures 4.4 and 4.5)). Conditions for a transition to this mesophase can, however, also be created by solubilisation of a non-polar organic compound in an aqueous solution of an ionic association colloid which normally produces middle phase (sodium caprylate and water + a hydrocarbon or a chlorinated hydrocarbon (Figure 4.2a))[30, 174].

4.3.2.4 Transitions to mesophase from aqueous solutions of lower concentrations

A transition from the micellar solution to a liquid crystalline phase can be produced by solubilisation also at concentrations much lower than that at which the volume fraction of hydrated micellar substance attains the critical value. This occurs when the solubilisation leads to major changes in the interfacial layer of the micelles. The paraffin chain alcohols and fatty acids produce such changes by the incorporation of their molecules in the micelle with the polar groups located in the interface. The primary effects are that the density of charged groups in the interface, the electrokinetic potential and the water binding of the micelle decrease and that the interaction between the polar groups leads to a reduction in the mean interfacial area per group, which can result in a flattening of rod-shaped and spherical micelles. To these effects under certain circumstances should be added an increase of the hydrocarbon content of the micellar substance. When these changes increase above a certain limit the solution structure cannot persist and a transition to a new phase takes place[23, 24, 29, 169, 212, 222].

For most association colloids these changes result in a transition to lamellar neat phase at concentrations lower than those at which the middle phase forms. For fatty acid soaps they also lead to the formation of the two-dimensional tetragonal mesophase of type 1, C (see Table 4.1 and Figure 4.2c) even between the c.m.c. and the 2nd c.c.; this liquid crystalline phase has hitherto been observed only in ternary systems including an alkanol or a fatty acid. The phase is characterised by the fact that the molar ratio of solubilisate to soap is nearly constant throughout its region of existence. This has been regarded as evidence that the polar groups of the two organic components are bound to each other (probably via a water molecule) forming small molecule groups (consisting in the alcoholic systems of 3 moles of alcohol and 2 moles of soap); in the rod-shaped aggregates these groups are presumably arranged in a structure with the polar groups distributed over the surface, the rods having a more or less square cross-section. The mean interfacial area per polar group is constant throughout the phase (38 Å^2 in the sodium caprylate–decan-1-ol–water system); the charge density is low (one charge to c. 76–83 Å^2), the counter-ions thus being loosely bound. The water content of the mesophase is high (40–65 %), only part of which can be bound to the polar groups. This has been ascribed to liberation of a part of the counter-ions of the soap and their diffusion into the 'inter-micellar' solution between the rods; this would result in a Donnan distribution and hence provide the conditions for the retention of the water in the phase. When the water content falls below a certain value, and at the latest when all the unbound water has disappeared, there is a transition from this mesophase to the lamellar neat phase. It is easy to separate this phase in pure form from the aqueous solution with which it is in equilibrium, but because of its high water content it is extremely difficult to separate it in the unaltered form from the neat phase by high-speed centrifugation[23, 24, 28, 29, 169, 212, 222].

Although below the c.m.c. there are no micelles in the solutions of association colloids, paraffin-chain alcohols and fatty acids dissolve in them in small quantities with the formation of small aggregates composed of surfactant and solubilisate. On addition of more solubilisate the aqueous solution becomes saturated with these aggregates and a liquid crystalline product is formed. This leads to a decrease in the concentration of association colloid in the aqueous solution. This decrease may proceed until a definite concentration, which is found to coincide with the l.a.c., is reached (see page 108). Below this concentration solubilisation has not been observed and here the aqueous solution is in equilibrium with a solution where alkanol is the solvent (Figure 4.2 c and d). The process leading to formation of mesophases between the l.a.c. and the c.m.c. thus differs from that resulting in the formation of mesophase C between the c.m.c. and the 2nd c.c. and of neat phase above the 2nd c.c., and also from that resulting in the formation of middle phase at the upper boundary of the solution region. At the l.a.c. the lamellar neat phase always forms, and in many systems the formation of this mesophase continues at higher concentrations. In a number of systems, however, another liquid crystalline lamellar phase, denoted B, forms between the l.a.c. and the c.m.c.[23, 24, 28, 29, 212, 219, 222].

4.3.3 Lamellar liquid crystalline phases

4.3.3.1 Mesophase B

The lamellar mesophase B, which is formed on adding alcohol or fatty acid to dilute surfactant solutions between the l.a.c. and the c.m.c., has not been observed in binary surfactant–water systems (Figure 4.2 c and d). The amount of solubilisate in the phase varies over quite a wide range – c. 1.2–2.7 moles per mole of surfactant[169, 219]. The phase is composed of water layers alternating with double amphiphile layers; the arrangement of the polar groups in the interfaces is thought to be other than in the neat phase. The calculated mean interfacial area per polar group is constant ($c.$ 25–27 Å^2) throughout the phase region; the charge density is low — one charge to $c.$ 68–83 Å^2 [28, 212, 219]. The water content of the phase is extremely high, $c.$ 70 to over 90%. Only a small part of this water can be bound to the polar groups; the reason that so much water can be retained in the phase is, according to Ekwall, that a Donnan distribution is produced by counter-ions liberated from the amphiphile layers entering the water layers [24, 28, 222].

The water content and the density of phase B differ very little from those of the phases with which it is in equilibrium; this makes it almost impossible to separate it from them.

4.3.3.2 Neat phase

The neat phase is the most common of the lamellar mesophases. It occurs in many binary surfactant–water systems at quite high surfactant concentrations, 90–70%. It is able to incorporate various lipophilic and amphiphilic compounds, and its region of existence then extends towards the centre of the phase diagram for the ternary system. The capacity of the phase for taking up water is often only slightly altered by solubilisation, but if the added compound is a paraffin chain alcohol or a fatty acid it may be increased markedly. The phase region may then extend over a wide area throughout the ternary system, down towards the water corner, that is, to 80–90% of water (Figure 4.1). The neat phase can be in equilibrium — via two- and three-phase zones — with almost all other phases occurring in aqueous amphiphile systems.

Most, but not all, surfactants need a certain minimum amount of water to produce neat phase. For the alkali metal soaps this has proved to be the amount needed for more or less complete hydration of the alkali ions, namely 2–3 moles for potassium, 5–6 moles for sodium and 7–8 moles for lithium soaps [212, 223]. Throughout the neat-phase region the alkali ions remain hydrated — that is to say, they are bound to water molecules both when they are interposed in the interface of the amphiphile layers and when they are dispersed in intercalated water layers. The alkylammonium halides also require a minimum amount of water (0.5–2 moles per mole of surfactant) to form neat phase. At low water contents in these systems the counter-ions tend to be bound directly to the positive ammonium group — apparently without intermediate water molecules; as the water content is raised, however, hydration of the anions seems to increase.

In different surfactant systems the mechanism for the uptake of water is different. In many systems (monoglycerides, diethyleneglycol monolaurate, hexanolamino-oleate, the dialkylsulphosuccinates AOT and MA) it is accompanied by one-dimensional swelling; the slope of the curve of log d $v.$ log $(1/\phi_a)$ is unity (d = long spacing, ϕ_a = volume fraction of the amphiphile). This means that all the additional water is intercalated between the double layers; the calculated interfacial area per polar group is almost independent of the water content. This situation is not altered by solubilising of non-ionic lipophilic or amphiphilic compounds in the phase [212, 213, 223].

In contrast, the uptake of water produces almost no swelling of the neat phase of most typical ionic association colloids. The addition of water results in some reduction in the thickness of the double layers and a marked rise in the calculated interfacial area per polar group. The water molecules penetrate the layers between the amphiphile molecules, forcing them apart and reducing the thickness of the layer so that the phase as a whole becomes non-expanding[212, 223]. The increase in the interfacial area per polar group is independent of the chain length of the surfactant and dependent only on the type of polar group and temperature[210, 217]. Even at the lowest water contents this area is so great (c. 26 Å^2) that the hydrated counter-ion can be situated between the ionised end groups of the surfactant molecules. The large increase in interfacial area accompanying the continued uptake affords a proof that most of this water, too, is interposed between the groups. This seems to be connected with the high water-binding capacity of the ions and ionised groups. The maximum amount of water that the neat phases of the alkali soaps are able to take up would seem not to exceed that which can be bound to the alkali ions by ion–dipole attraction, and to the carboxylate groups by hydrogen bonds. The maximum values of the interfacial area per polar group are of the order of 40–46 Å^2 the charge density thus remaining rather high.

When a long-chain alkanol is solubilised in the neat phase of these ionic surfactants, the uptake of water by the phase changes; as the alkanol content is increased, the water uptake gradually approaches that of the neat phase of the first-mentioned surfactants – that is to say, the water finally will be taken up with one-dimensional swelling. As a result of the insertion of the alcohol molecules between surfactant molecules in the double layers, hydrated counter-ions are displaced and the mean interfacial area per polar group is reduced. As the amount of alcohol incorporated is increased, the slope of the curve of log d v. log $(1/\phi_a)$ increases from zero for the pure surfactant to close on unity. This means that the alcohol causes an increasingly large fraction of the additional water molecules to remain outside the amphiphile layers, within the intercalated water layers. A limit is ultimately reached above which further water cannot be incorporated between the polar groups of the amphiphile layers, but is intercalated between these layers. The mean interfacial area per polar group has now fallen to 24–25 Å^2. This value is unaffected by a change in water content or any further increase in the alcohol to soap ratio[28, 168, 212, 219, 223].

At this limit the properties of the neat phase are modified in many respects. The rheological behaviour of the phase changes, the vapour pressure that has risen from a low value to one close to that for pure water now remains relatively constant, and the capacity of the phase to incorporate water increases suddenly, with the result that the neat-phase region in the triangular diagram projects in a long salient down towards the water apex (Figure 4.2c). The reason for these changes has been sought in the fact that up to the limit in question all the water of the phase has been bound to the polar groups of the amphiphiles, whereas, above the limit the additional water, which is intercalated between the double layers with one-dimensional swelling, seems to behave as free bulk water. The sudden increase in the ability to incorporate water has been ascribed to the liberation of counter-ions of the soap as a result of the decrease in the charge density in the interface of the double layers to about one charge to 72–75 Å^2. This liberation creates the conditions for a Donnan distribution of counter-ions, which in turn enables increasing amounts of water to be taken up until the vapour pressure reaches the value characteristic of the more water-rich phase in equilibrium with the neat phase[23, 28, 168, 212, 219, 224, 225]. The same factors that have been considered to govern the uptake of water in phases B and C may thus account for the increased water uptake in the alcohol-rich part of the neat phase of the soap–alcohol system.

While the extension of the neat-phase region towards higher water contents in the alcohol- and water-rich part of the region thus seems to be governed by the Donnan distribution of liberated counter-ions, in other parts it seems to be limited by the maximal water-binding capacity of the polar groups. Towards lower water contents the extension of the region may be limited partly by the hydration need of the alkali ion of the soap, partly by the fact that the interfacial area per polar group cannot decrease below a value of 23–24 $Å^2$. Towards higher alcohol contents the region of existence seems to be limited by the fact that the amphiphile layers are not able to solubilise more than c. 4 moles of alcohol per mole of soap[28, 30, 212, 218].

In the alkali-soap systems the above type of neat phase, with its large capacity for taking up water, is formed by solubilising paraffin chain alcohols from n-decanol to n-pentanol[218]. The alcoholic neat phase of alkyl sulphates and sulphonates is of the same type. This is, with some modifications, also the case for the alcoholic neat phases of the alkylammonium halides. In all these cases the interaction between the polar group of surfactant and alcohol seems to be the critical factor for the formation of the actual type of neat phase and especially for its stabilisation at high water contents. The importance of this interaction is confirmed by the fact that this type of neat phase also appears in systems of soap, fatty acid and water where there is a corresponding interaction between the carboxylic group of the fatty acid and the carboxylate group of the soap. On the other hand, this type has not been encountered in ternary systems where there is no such interaction between the solubilisate and the ionic association colloid; when the solubilisate in such systems is a hydrocarbon the region of the neat phase would not seem to extend to higher water contents than can be bound by the polar groups of the colloid. The same applies when the solubilisate is a compound with a weak polar group, such as a methyl ester, nitrile or aldehyde[30, 212].

A counterpart to the extreme uptake of water accompanying solubilising of an alcohol or a fatty acid in the neat phase of ionic surfactants is obtained, as has quite recently been shown, when small amounts of an ionic association colloid are dissolved in the neat phase of lecithin[226].

4.3.4 Liquid crystalline phases of type 2 and the transition from organic solution to mesophase

4.3.4.1 The middle phase of type 2

Among the few surfactants in the pure form that are known only in the liquid crystalline state, di-2-ethylhexyl sodium sulphosuccinate (Aerosol OT, AOT) is the one that has been most thoroughly examined. It has a reversed two-dimensional hexagonal structure, with the long parallel rods of surfactant molecules in hexagonal array. In the rods the AOT molecules are oriented with their hydrophilic sodium sulphonate groups facing inwards and the bulky branched hydrocarbon chains outwards. x-Ray diffraction studies show that the molecules are tightly packed and that the ion pairs in the core of the rods must lie very closely together, held by dipole–dipole forces obviously in such a way that the sodium ions coordinate with oxygen atoms from different sulphonate groups. In the outer parts of the rods, too, the packing seems to be so tight that the cross-section of the rods is more or less hexagonal. This liquid crystalline phase can incorporate both anhydrous organic solvent and water, and hence pass continuously into the lyotropic liquid crystalline state. It can dissolve considerable amounts of, for instance, anhydrous hydrocarbon, without any essential modification of its two-dimensional hexagonal structure. The hydrocarbon molecules penetrate mainly between the AOT rods and perhaps also into the outer parts of their hydrocarbon shell; the main change seems to be that the rods assume a more circular circumference. On the other hand, pure AOT is capable of dissolving up to 16% of water without the phase undergoing any modifications of its basic structure; than the sodium sulphonate groups in the rod cores are hydrated and the core undergoes swelling. The same occurs if the mesophase contains hydrocarbon, but then a considerably greater amount of water can be solubilised (up to more than 50% in the presence of c. 8–10% of p-xylene). AOT thus affords an example of non-aqueous and aqueous phases with type 2 two-dimensional hexagonal structure and also illustrates the formation of this type of mesophase in one-, two- and three-component systems (Figure 4.3)[197, 206, 213].

Most other surfactants that behave as swelling amphiphiles give rise to the mesophase in question only in the presence of a minimum amount of water; this is the case for, for instance, monoglycerides[183, 186] and dipalmitoyl lecithin[157]. It also applies to the phases of this type that occur in many ternary systems of typical association colloids (alkali soap–alkanol–water systems, etc.[29, 212]). It is also the case in some chemically less well defined systems with three or more components (some bio-lipids and water, and some commercial products with hydrocarbon and water)[227–229].

It is obvious that amphiphile molecules with a laterally bulky hydrocarbon part and a relatively small polar group are particularly liable to yield this reversed phase (Figure 4.3)[213]. In some systems an increase of the hydrocarbon volume, or a decrease of the mean interfacial area per polar group by the solubilising of a hydrocarbon, a long-chain alkanol or a fatty acid is sufficient to give this mesophase (Figure 4.1, 4.2(c), 4.2(d) and 4.4).

In most systems the interfacial area per polar group increases more rapidly for the initial than for the subsequent amounts of water solubilised; sometimes the increase ceases completely after solubilisation of a certain amount of water. The upper limit of the water solubilisation seems in some systems to be determined by the maximum water-binding capacity of the polar groups, while in others this can be exceeded[212]; and example of this is provided by the reversed middle phase of the AOT systems containing hydrocarbon, which shows a long salient towards higher water contents. In this the interfacial area per polar group is independent of the water content, and the charge density in the interface approaches values (one charge to c. 64 Å^2)

that should allow the liberation of counter-ions and an uptake of water regulated by the Donnan distribution (Figure 4.3)[213].

In spite of the swelling of the core through the solubilisation of water, the thickness of the hydrocarbon layer of the rod-shaped aggregates remains fairly constant throughout the region of existence of the phase; it is slightly less than the length of the surfactant's hydrocarbon chain. The molecules of an added hydrocarbon are chiefly located between the rods; in the majority of systems containing water the phase can no longer exist when the ratio between hydrocarbon and surfactant falls below a certain value. On the other hand, if the added component is an n-alkanol or a fatty acid, only a portion of it is located between the rods, the remainder being solubilised in them with the polar groups directed towards the core. The latter amount seems to be regulated by the interaction between the polar groups of the organic components, and therefore varies from one system to another; in addition it often varies with the water content of the phase (e.g., in the reversed middle phase of the sodium caprylate–decanol–water system (Figure 4.2(c)). When the amount of the added substance in the phase is decreased, the molar ratio of the aggregates remains unchanged, while the amount outside them diminishes; the phase can no longer exist when this latter amount has fallen to zero[28, 212].

4.3.4.2 Other mesophases of type 2

The liquid crystalline phase ascribed to the reversed two-dimensional tetragonal structure, K, (see Table 4.1) has so far been observed only in one system (namely, potassium caprate–n-octanol–water). It appears in the same concentration region between the organic solution and the lamellar neat phase as that where the reversed two-dimensional hexagonal mesophases usually occur in ternary systems, but at slightly higher surfactant concentration. In this phase there would seem to be no alcohol between the rod-shaped aggregates; the region of existence of this phase is extremely small[219].

Optically isotropic mesophases of the reversed type are quite common and usually appear in the concentration range between the reversed middle phase and the neat phase. Their detailed structure varies; those considered to have a structure of short rods joined in threes at each end to form two three-dimensional networks (designated I'_2) have been observed at fairly high surfactant contents (binary aqueous systems of bio-lipids[157, 226], binary and ternary systems of Aerosol OT[213] and of decaoxyethylene glycol monolauryl ether with oleic acid[29, 174] (Figures 4.3 and 4.4)). This structure may be regarded as an intermediate stage between a structure of long rod-shaped aggregates of type 2 and a lamellar structure, which occurs when the water content is increased. The incorporation of water in the rod core leads to its swelling and to a looser packing in the hydrocarbon layer surrounding the core; the latter trend is counteracted if the hydrated cores merge to some extent, as occurs when the rods are joined in threes. This structure however, seems to be possible only at particular molar and volume proportions of water to amphiphile.

In contrast, the water content of the reversed optical isotropic mesophase found in a number of monoglyceride–water systems varies within quite wide limits; its structure has not been definitely established (possibly I'_2)[183, 186].

In a number of ternary systems optically isotropic mesophases of the reversed type occur (for example, I'''_2 in Figure 4.1), which probably have a different structure from that ascribed to phase I'_2 [219].

4.3.4.3 Transitions to mesophase from micellar organic solutions

In many systems the reversed mesophases are in equilibrium, via two and three-phase zones with solution of the surfactants in organic solvents. At the transition from solution to mesophase the randomly-oriented reversed micelles seem, without any change in composition, to pass over to the mesophases F and K with their rod-shaped aggregates of reversed type arranged in an hexagonal or tetragonal array; only the length of the aggregates appears to increase[29, 169, 212, 213, 219]. As in aqueous solutions, this transition probably takes place as soon as the volume fraction of the micellar substance exceeds a critical value, above which a solution with randomly-oriented micelles can no longer exist. Here, too, the critical value is probably a function of the anisometry of the micelles; since the change of the micelles towards anisometry appears to take place more slowly in the organic solutions with their uncharged micelles than in aqueous solutions of ionic surfactants; the critical volume is often not reached until quite high concentrations.

4.3.5 Summary

In the transition from one liquid crystalline structure to another there is a change in the ratio of the total interfacial area available for polar groups, to the total volume occupied by the hydrocarbon parts of the organic components; within each phase this ratio can vary only within certain limits.

The interfacial area per polar group is, in many phases or parts of a phase, independent of the water content; this is the case when the water is intercalated between the aggregates of the mesophase. In other phases or parts of a phase, on the other hand, this area varies with the water content, because the water is built in between the polar groups. Also the interaction between polar groups of different organic components can result in a change in the mean interfacial area per polar group.

The ratio of the total interfacial area to the total volume occupied by the hydrocarbon can thus be influenced both by the water content and by the proportions of the organic components. These are two of the reasons why a change in the composition of a system leads to a series of liquid crystalline phases which replace each other in a definite sequence. Another reason is that these phases cannot exist with more than a certain maximum amount of water even when the area per polar group is unaffected by the water content. In many systems of ionic surfactants, at least, this water content is governed by a Donnan distribution of liberated counter-ions; other factors are probably also involved.

The experience to date suggests that in the case of binary systems of surfactant and water a change solely in the water content cannot produce a transition from a type 1 to a type 2 mesophase or vice versa. So far as the commonest mesophases are concerned, we have for typical association colloids the following sequence of phases as the *water content is diminished*:

$$L_1 \rightleftharpoons E \rightleftharpoons H_c \rightleftharpoons R \rightleftharpoons I'_1 \rightleftharpoons D \rightleftharpoons \text{crystalline surfactant}$$

For many typical swelling amphiphiles and surfactants with similar properties we have the following sequence of phases as the *water content is increased*:

$$\text{crystalline amphiphile} \rightleftharpoons F \rightleftharpoons I'_2 \rightleftharpoons D \rightleftharpoons L_1$$

In each case the *transition from one mesophase to another* ceases at the lamellar neat phase.

In ternary systems on the other hand, a transition between type 1 and type 2 mesophases is quite common. It occurs especially in systems where the third component is a paraffin chain alcohol or a fatty acid, and usually proceeds via the lamellar neat phase. When ratio of water to surfactant is kept constant but *the ratio of the third component to surfactant is increased*, among others, the following sequences have been observed:

$$L_1 \rightleftharpoons B \rightleftharpoons D \rightleftharpoons F \rightleftharpoons L_2 \,; L_1 \rightleftharpoons C \rightleftharpoons D \rightleftharpoons F \rightleftharpoons L_2 \,;$$
$$E \rightleftharpoons D \rightleftharpoons F \rightleftharpoons L_2 \,; R \rightleftharpoons D \rightleftharpoons F \rightleftharpoons L_2$$

The transition from one type to the other is made possible by the changes in the interfacial layers of polar groups resulting from the changes in the proportion of the two organic components; the interaction between their polar groups contributes to the formation of special mesophase structures. According to the above results the area per polar group cannot solely determine the phase transitions, as Winsor pointed out[194]. Obviously, other factors are also involved.

Continued systematic investigations on the dependence of the phase equilibria and phase structure on the molecular structure of the organic components may elucidate the conditions of the polar groups and the binding of water in lyotropic mesophases in general.

4.4 SOLUTIONS OF SURFACTANTS IN ORGANIC SOLVENTS

4.4.1 General aspects

Many surfactants are oil soluble; they dissolve in lipophilic solvents such as hydrocarbons and chlorinated hydrocarbons, or in amphiphilic liquids such as fatty acids, alcohols etc. In some cases they dissolve in the anhydrous solvent, while in others a solution forms only in the presence of water. In both cases the solutions so formed are capable of taking up more water without losing their homogeneous nature. Solutions of these types will be dealt with in this section. The research field before 1965 has been summarised in a number of surveys[230–233].

Most surfactants occur in an aggregated form in these solutions. When micelles are formed, they are of the 'reversed' or 'inverted' type with a core consisting of non-hydrated or hydrated polar groups surrounded by a layer formed of hydrocarbon groups. The cohesive forces in these micelles are due to the interaction – direct, or indirect, via water molecules – between the polar groups of the surfactant. In many systems there is, as the concentration is increased, a transition from these reversed micellar solutions to mesophases of the reversed type.

Questions of particular relevance are: (a) whether the micelles are composed solely of surfactant molecules, possibly with loosely attached, more or less mechanically retained, molecules of solvent, or whether the solvent actively participates in the association process; (b) the effect of water on the formation of micelles, where in the micelles the water is located, and which factors govern the water solubilisation.

Different types of mechanism of micelle formation are known[32, 213]: (a) The solvent takes no active part in the association, it acts solely as a dispersion medium between the micelles. In anhydrous solution some surfactants occur only in molecular form, but micelles are formed when water is added; others give rise to micelles even in the anhydrous solution and these persist when water is added. (b) The solvent participates actively in the micelle formation. Some surfactants dissolve in the anhydrous solvent to form small molecularly dispersed aggregates with a definite stoichiometric ratio of the two components. When water is added, these primary aggregates are hydrated and associate to micelles. Other surfactants require a minimum amount of water to go into solution; hydrated aggregates of surfactant and solvent molecules in varying proportions then appear and these form micelles.

Intermediate mechanisms appear to exist, and new variations will no doubt be discovered. Thus, the nature of the association process varies largely, not only depending on variations in the chemical structure of the surfactants and the interaction between the polar groups of the surfactant and water, but also depending on the non-polar or polar nature of the organic solvent.

4.4.2 Surfactant solutions in non-polar organic solvents

A list of micelle-forming surfactants in non-polar solvents was published in 1955 by Singleterry[230]. Among the groups of chemically pure surfactants that have been examined in this respect during the last years are fatty acid salts of bi- and tri-valent metals, aryl–alkyl carboxylates, dialkyl–aryl sulphonates, dialkyl sulphosuccinates, alkylated amine and ammonium salts, alkyl and aryl derivatives of polyoxyethylene glycols, monoglycerides and lecithin. Among these, the aluminium salts of the fatty acids and the phenylstearates of alkali and alkaline-earth metals give, at low concentrations in a water-free environment, long chain-like aggregates, held together by intermolecular bonds involving free hydroxyl groups or coordinate linkages between metal ions and carboxylate oxygen. Some aggregates are spontaneously re-formed if broken by stirring; their length appear to vary randomly. They are to some extent more reminiscent of high-polymer fibrous aggregates than of surfactant micelles, and it is therefore doubtful whether their solutions should be regarded as micellar systems. Unlike these substances, other surfactants associate under suitable conditions in non-polar solvents, to form micelles of limited size. It is with these that this survey is particularly concerned.

4.4.2.1 The dissolution process

Because of the low dielectric constants of non-polar solvents ionic, surfactants are undissociated in them, as has been confirmed by conductivity and n.m.r.

measurements. The aggregation can be ascribed to dipole–dipole attraction between the ion pairs or other polar groups, hydrogen bonds between them, and, in some cases, also to coordination of end groups around a central ion.

The solvent capacity of the anhydrous non-polar solvents may be attributed to dispersion forces between their molecules and the hydrocarbon parts of the surfactant molecules, evidence for which is found in vapour-pressure measurements, and from them derived values of the heat of solution[234]. A semiquantitative theory for the interaction between surfactant and the non-polar solvents supposes an equilibrium between surfactant micelles of definite size and a solvent swollen surfactant phase[235]. The solubility phenomena observed, therefore, are those characteristic of binary liquid systems and the solubility parameters rather than the dielectric constants or the dipole moments, appear to be the quantities that gives a correct measure of the solvent power of the solvents. The dissolution seems to proceed in two stages as the system is warmed: swelling of the surfactant lattice by solvent penetration and then, when, at a critical solution temperature the forces holding the surfactant lattice together are overcome, a swelling without limit or a disaggregation[236]. The importance of the solubility parameter for the swelling, dissolution and size of the formed aggregates has been demonstrated for many surfactants[232, 234, 236–238]

4.4.2.2 The aggregation

In many systems it has been possible to demonstrate a minimum concentration below which no association occurs. This is often extremely low; it is usually regarded as the c.m.c. of the solution, even though the aggregates formed are very small.

Results obtained with methods entailing the addition of a polar substance should be regarded with caution, especially if the added compound is water, since this can produce aggregation even where it does not occur in the water-free environment[240]. Care is also indicated in drawing conclusions regarding aggregation from the critical solution temperature. There are cases where the solution continues to be molecularly dispersed even at concentrations reached above this temperature[241, 242]. The increase in solubility may instead have its explanation in changes in the solid phase of the type indicated above[235] or in a transition from solid crystalline phase to mesophase[243].

The fluorescence depolarisation technique gives a valuable supplement to the previously known methods for determining the size of the micelles; it gives their volumes[244–246].

The aggregates formed at first in the non-polar solvents would seem in general to be relatively small, the aggregation number being of the order of 4–30. When the aggregates are nearly spherical, as they seem to be at low concentrations, the low aggregation numbers are due to steric factors. The space in a spherical micelle core permits only a limited number of polar groups; this also applies to the number of bulky hydrocarbon groups in the outer parts of the micelle.

Monodispersity seems to be usual in the solutions[238, 245]. The relationship between the aggregation number and the molecular structure of surfactants has recently been thoroughly examined by Kitahara[238]. The aggregation number in benzene solutions of straight-chain dialkyl sodium sulphosuccinates (from dibutyl to didodecyl compounds) decreases rapidly (from 15 to 7) with increasing total number of carbon atoms of the alkyl chains until this exceeds 16, but thereafter slowly. A homologue with straight chains has a higher aggregation number than one with branched chains. Similar results were obtained with quaternary dialkyldimethylammonium halides in various non-polar solvents.

As it is known from earlier work that the aggregation number is low for quaternary alkylammonium salts having only short alkyl groups, compounds with a moderately large number of carbon atoms in one or two alkyl chains seem to have the highest aggregation number. For a surfactant to form stable micelles in a non-polar solvent its hydrocarbon part apparently must be so large that the polar core is adequately screened from the solvent. After the optimal size has been reached any further increase in chain length and especially in bulkiness of the hydrocarbon parts will result in a reduction in aggregation number. In the case of ionic surfactants this often decreases also when the counter-ion radius is increased.

With reference to the fact that many surfactants are salts, Fowkes states that the polar interactions in the core of the micelle should be seen against a background of Lewis acid–base theory; this serves as a useful aid in estimating the magnitude of polar–polar and polar–non-polar interactions. He gives examples where a variation in micelle size can be accounted for on this basis[232].

The above results apply in anhydrous solutions or in the presence of extremely small quantities of water. The micellar solutions can, as a rule, solubilise considerable amounts of water; this is bound to the polar groups of the surfactant molecules by ion–dipole or dipole–dipole attraction; the water can in addition, because of its tendency to form hydrogen bonds and its bifunctional nature, at the same time bind several surfactant molecules simultaneously and hence contribute to the cohesion in the micelles.

4.4.2.3 A model substance

Solutions of di-2-ethylhexyl sodium sulphosuccinate (AOT) in non-polar solvents are among the most thoroughly examined. Because of its liquid crystalline structure in pure form the solubility in hydrocarbons is high (85% in p-xylene). The possibility of relating the properties of the AOT solutions to the mesophase structure of the pure substance has provided a better insight into the conditions in the solutions.

In Section 4.3.4.1 the structure of the pure surfactant with its parallel rods of AOT molecules in hexagonal array was described. It was pointed out that when this mesophase dissolves a hydrocarbon the hydrocarbon molecules are mainly intercalated between the rods and only to a minor part penetrate the outer parts of the hydrocarbon shell. Only when the hydrocarbon content exceeds a critical value is the structure destroyed, the mesophase going into solution, where the AOT rods are divided into shorter micelles dispersed in the hydrocarbon, apparently without any change in the mutual arrangement of the polar groups in the core and the cohesive forces between them (Figure 4.3)[213].

Aggregation has been shown to begin at relatively low concentrations[237, 247]. The solutions are monodisperse from c. 5% to at least 20% with an aggregation number that varies slightly with the type of solvent (20–29 in hydrocarbons, 17 and 14 respectively in tetrachloromethane and ethylene bromide[239]). Viscosity measurements suggest a weak anisometry, a slight solvation, or both.[213, 239]. While Peri proposes a lamellar structure for these micelles[239], Ekwall et al. — referring to the rod-like aggregates of the pure substance — ascribed to them a cylindrical structure[213]. In p-xylene solutions where the aggregation number is 23, the micelles would have the form of short cylinders with a diameter of 22–24 Å and an axial ratio of 1.4–1.8. Any solvation of the micelles, as well as their shape and size, remain practically constant up to a concentration of c. 73% of AOT. A further increase in concentration results in crowding which appears to deform the micelles into true cylinders the length of which seems to increase rapidly. At a content of 85% AOT there is then a transition to the mesophase with hexagonally arranged cylindrical aggregates with the same diameter and structure as the micelles. This interpretation of the experimental findings implies that from the pure substance down to the c.m.c. we have the same aggregating forces and the same arrangement of surfactant molecules within the aggregates; only their length and mobility vary[213].

Pure AOT dissolves up to 5–6 moles of water per mole of surfactant without modification of the mesophase structure, the sodium sulphonate group being hydrated and the core swelling. When the mesophase contains a hydrocarbon, it is able to solubilise up to 36 moles of water per mole of AOT. The micellar hydrocarbon solutions, too, can solubilise considerable quantities of water, probably by the same mechanism as that obtaining in the mesophase (Figure 4.3)[213].

From viscosity measurements it is evident that anisometry of the micelles of the last mentioned solutions does not appear until the volume fraction of the micellar substance exceeds the critical value of 0.74. That here, too, the deformation resulting from the crowding results in a cylindrical shape of the micelles is seen from the fact that these water-containing solutions are in equilibrium with the two-dimensional hexagonal mesophase with its cylindrical aggregates along the entire boundary of the solution region towards higher AOT concentrations (Figure 4.3)[213].

It is well established that the first water solubilised is bound particularly firmly through hydration of the sodium ions by 5–6 moles of water. This is shown by measurements of the partial molar volume, the vapour pressure, the enthalpy of solubilisation ΔH_s and n.m.r.[213, 248, 249, 270]. Kitahara has adduced direct proof that this water is bound to the cation[270].

The hydration of the cations leads to some separation of cations and anions and this process must affect the derived value of ΔH_s; these values may also be affected by the water binding to the sulphonate groups.

The special properties of solubilised water are also evident from the fact that the first solubilised 1.5–1.6 moles of water per mole of AOT in tetrachloroethylene are not available for dissolving sodium chloride[250] or a water-soluble food dye[247]. The dissolving capacity of the rest of the solubilised water is also reduced; for the dye it is only 5–10% of the solubility in free bulk water. These results probably apply to the water-containing surfactant systems in general; the water bound to the polar groups of the mesophase aggregates and the micelles in these systems has a reduced solvent capacity – a point that is usually overlooked.

4.4.2.4 Other surfactants

The above picture of the AOT solutions would seem to apply also to the solutions of many other surfactants in non-polar solvents, e.g. to other dialkylsulphosuccinates, dialkyl–aryl sulphonates, fatty acid metal soaps, alkylated amine and ammonium salts, and monoglycerides. In the aggregation in the monoglyceride solution, hydrogen bonding between the polar groups plays an important part[251]. Most of these surfactants have hitherto usually been examined only in a restricted concentration range above the c.m.c.

Lecithin solutions differ in certain respects from the general picture. In these the micelles would appear to have a lamellar structure at higher concentrations consisting of bimolecular leaflets with the polar groups facing each other and with an aggregation number of 70–80 [252].

The non-ionic polyoxyethylene glycol derivatives also display deviant properties; they do not seem to form micelles in anhydrous solvents, but only on addition of water. In the latter solutions the polar groups of separate surfactant molecules are bound together by the water, evidently by hydrogen bonds between the water molecules and the ether oxygen[240-242, 260]. In the water-containing solutions, the conditions are in many respects reminiscent of those in the solutions of other surfactants.

4.4.2.5 Extreme solubilisation of water

It is indisputable that the binding of the water to the polar groups governs the solubilisation in hydrocarbon solutions of surfactants, but it is clear that other factors must also be involved. This is evident from, for instance, the extent of the solution region in the AOT–p-xylene–water system (region L_2 in Figure 4.3). At high concentrations this region extends a salient towards higher water contents, and here the water content of the micelles can rise to 40 moles mol^{-1} of AOT, which is much more than the polar groups can bind directly. Here the micelles may be regarded as small drops of water surrounded by a monomolecular layer of hydrated AOT molecules[213].

Extremely water-rich micellar solutions of this reversed type occur in many other surfactant systems, too[27, 32, 38]. As was shown in Section 4.3, there also occur liquid crystalline phases with high water content, for instance the middle phase of type 2 in the AOT–p-xylene–water system (mesophase F in Figure 4.3[213]) and lamellar mesophases in many systems[23, 24, 27, 28, 212, 253] (for instance the D, C and B phases in Figure 4.2 c and d). They seem to occur when the charge density in the interfacial layer of polar groups falls so low that a part of the counter-ions can be liberated and give rise to a Donnan distribution. Just as in the water-rich part of phase F of the AOT system, the charge density of the micelles in the salient of the xylene solutions is probably so low that conditions may exist for a Donnan distribution of liberated counter-ions and a consequent increase in the water content of the micelles.

As has recently been shown, an increase in temperature seems to produce exceptionally high water contents in the hydrocarbon solutions of several surfactants (amongst others AOT) at considerably lower surfactant concentrations than those in Figure 4.3. When the temperature is raised above a certain value a marked increase in the water-solubilising capacity of the micelles is reported. The temperature stability of the solutions so formed seems as a rule, to be limited, and when the temperature is raised further there is a separation into two phases[254-260].

From experimental data it seems as if the increase in water solubilisation would have its principal cause in the effect of the temperature on the interaction between micelles and solvent; the temperature at which the increase in solubilisation of water begins differs from one solvent to another. The penetration of solvent molecules into the hydrocarbon shell of the micelle is presumably promoted by the rise in temperature, and this will consequently lead to a loosening of the micelle structure and lowering of the packing density in the interfacial layer of polar groups. One must ask if this may lead to a decrease in the charge density and a liberation of counter-ions in micelles of ionic surfactants?

Kitahara has shown that the temperature at which increased water solubilisation is elicited is affected by the presence of electrolytes, and that it is the cations that are active in the case of an anionic surfactant[259, 260], and the anions in the case of a cationic surfactant[257]. He is inclined to ascribe this to the different adsorptivity of the ions at the interfacial layer of polar groups.

4.4.3 Surfactant solutions in polar organic solutions

Organic solvents with one or more polar groups tend to interact with the polar groups of the surfactant. Fowkes has referred to Lewis acid–base theory for an explanation of these interactions and has given several examples of the use of this theory for a qualitative understanding of the effect of polar solvents[232].

In the light of today's experimental experience it would, however, seem more appropriate in classification to take note of the fact that there is often permanent aggregation between a polar solvent and a surfactant, which results in the formation of mixed micelles. In some cases this interaction takes place via molecular compounds with stoichiometric composition, while in others the composition of the aggregates varies. In the case of interaction between surfactants and polar solvents and especially in micelle formation, water can also play a critical part.

4.4.3.1 Carboxylic acid as solvent

The fatty acid sodium soaps dissolve in anhydrous fatty acid forming a molecular compound with a soap:fatty acid molar ratio of 1:2, the components being held together by strong hydrogen bonds between the acid and the ionised carboxylate group[27, 32, 38, 39-41]. The bonds remain even when a non-polar solvent is added, and in this solution the composition of the aggregates is 2 sodium soap:4 fatty acid[261]. The sodium ions are firmly bound to the carboxylate groups[41]; apparently they coordinate oxygen atoms from several carboxyl groups.

Added water is solubilised in these fatty acid solutions. At small excesses of fatty acid the molar ratio of the hydrated acid soap is 1 soap:2 fatty acid:x water and this compound has a large water-solubilising capacity. The water is bound particularly firmly up to 5–6 moles per mole of soap above which level there is a loosening of the bonds between the sodium ions and the carboxylate–carboxylic groups and instead there is apparently a binding between hydrated cations and the increasingly hydrated soap–fatty acid complexes. Light-scattering measurements show that when the water content exceeds 6–8 moles per mole of soap the primary aggregates of the acid soap combine to form micelles of the reversed type whose core is apparently held together by water molecules which are bound to the sodium ions by ion–dipole attraction and to the carboxylate and carboxylic groups by hydrogen bonds. This micellar structure of the solution seems to persist until the acid soap cannot bind more water and 'free bulk water' appears in the solution (at 40–50% (w/w) of water in the sodium caprylate–caprylic acid–water system). Under suitable conditions the system remains a homogeneous solution up to still higher water contents (Figure 4.2(d)); after a transition region the hydrated acid soap forms micelles of normal type with a hydrocarbon core in aqueous intermicellar solution. In this system we thus have a continuous transformation in homogeneous solution from a micellar structure of type 2 to one of type 1[29, 32, 39, 40, 262].

Potassium and lithium soaps appear to behave in essentially the same way[263]. AOT also dissolves in anhydrous fatty acid with the formation of aggregates of 2 AOT:2 fatty acid[213]; these aggregates too, are able to solubilise large quantities of water, with the formation of micellar solutions.

Non-ionic surfactants, as polyoxyethylene glycol ether derivatives and monoglycerides react with anhydrous fatty acid in a similar way forming aggregates in the molar ratio 1:1, which are held together by hydrogen bonds[263].

Many surfactants react with a carboxylic acid even when the surfactant is dissolved in anhydrous non-polar solvent. It was shown long ago by Honig and Singleterry that small additions of phenylstearic acid break the strong bonds in the chain aggregates of alkali and alkali-earth metal phenylstearates in hydrocarbon solutions, forming practically molecular disperse solutions; on addition of water a new aggregation occurs[230, 264]. Kaufman has shown that acetic acid is solubilised in the micellar hydrocarbon solutions of alkali, alkaline-earth and zinc dinonylnaphthalene sulphonates[265, 266]. As the acid carboxylate soaps have proved to be stable in a hydrocarbon environment, it seems conceivable that the dinonylnaphthalenesulphonate, too, primarily, forms a molecular compound, which then – for example, on addition of water – gives rise to micelles.

4.4.3.2 Alcohol as solvent

Alkali metal soaps are sparingly soluble in anhydrous alkanols and the solubility is not appreciably dependent on temperature up to *c.* 50 °C. However the solubility is considerably increased on addition of some water (Figures 4.1 and 4.2(c))[23, 24, 29, 32, 42, 169, 218, 219]; irrespective of the chain length of the alkanol in the range C_1–C_{10}, the minimum amount of water required to produce a solution of the soap in the alcohol is roughly that for hydration of the alkali ion, namely 2–3 moles per mole of soap for potassium, 5–6 moles for sodium, and 7–8 moles for lithium soaps. This suggests that an alkali metal soap dissolves only when its alkali ion is completely hydrated. Above a certain soap concentration the hydrated soap associates with the alcohol to form micelles containing soap, alcohol and water. Their alcohol content rises with the molar of water to soap (the limits lie between 1 and 4 moles of alcohol per mole of soap and seem to vary slightly with the nature of the alkali metal ion). In decanolic solutions of sodium caprylate the aggregation number of the micelles was found to be 30–40 soap molecules. In these reversed micelles the alkali ions are firmly bound, especially so long as the water to soap ratio is low. The water is firmly bound at low water contents; however, when the boundary towards higher water contents is reached the vapour pressure gradually rises to a value near that of pure water. The amount of water in the micelles at this boundary seems to be restricted to about the maximum that can be bound to the cations by ion–dipole attraction and to the carboxylate and alcohol groups by hydrogen bonds.

Sodium alkyl sulphates and sulphonates also dissolve in alkanols only in the presence of at least the amount of water required for hydration of the alkali ion. Here, too, the solution process is associated with an interaction with alcohol, and formation of mixed micelles of the above type[30].

The bile-acid salts produce in decanol, when water is present, similar mixed micelles containing 3–4 moles of alcohol per mole of the salt. In these systems there is a continuous transformation in homogeneous solution from an alcoholic solution with mixed micelles of the reversed type to an aqueous solution with mixed micelles of the normal type[267].

Alkylated ammonium halides are, as a rule, more soluble in anhydrous alkanols than the above[30, 138]. To judge from the CTAB–hexanol–water systems the surfactant dissolves in small amounts as ion pairs in the anhydrous alkanol, and only in the presence of the minimum amount of 4–5 moles of water per mole of CTAB is there an interaction with the alcohol and an aggregation that leads to the formation of reversed mixed micelles[138]. In the micelles with fairly low water content the halogen ions are extremely firmly bound, but even in those with a high water content the bonds are comparatively strong. The alcohol content of the micelles is close to the molar ratio 1:1, while the water content varies widely (from 4 to 40 moles per mole of CTAB). The water content thus reaches much higher amounts than can be bound to the polar groups; this has been considered to be a consequence of the structuring effect of quaternary ammonium ions on water[30, 268].

AOT dissolves in anhydrous decanol forming rather small aggregates; this points to an interaction with the alcohol[213]. Peri has shown that when alcohol is added to the micellar solution of AOT in nonane, the micelles are broken down to smaller aggregates[239] and Kitahara has found that micelles did not occur in anhydrous methanol[237].

Singleterry et al. reported some time ago that the originally extremely large aggregates in water-free hydrocarbon solutions of alkali and alkaline earth arylstearates are broken down on addition of small amounts of alcohol[230, 264]. Kaufman *et al.* have recently found that the aggregation number of the micelles in benzene solutions of barium dinonylnapthalene sulphonate is reduced on addition of methanol and that in pure nonanol no micelle formation occurs[269].

These last examples point to an extremely powerful reaction between alcohol and some surfactants without the participation of water. It is not known whether mixed aggregates form here nor whether these have a stoichiometric or varying composition. When water is added, however, mixed micelles of the reversed type, with a core of hydrated polar groups, are probably also formed in most of these systems.

4.5 THE STAGEWISE AGGREGATION

When the composition of the surfactant systems is gradually varied from dilute aqueous solutions to pure surfactant or, in ternary systems, to dilute organic solutions, a number of well defined equilibrium states are passed through that may be regarded as stages of a continuous process of aggregation. These stages are defined by variations in attractive forces, extents of aggregation, aggregate sizes, the packing densities in the aggregates and the crowding of the aggregates in the system.

The aggregation can begin in dilute aqueous solutions with an association to pre-micelles, continue with the formation of spherical micelles with constant properties in the region above c.m.c., and then, when the water content is lowered, continue via a region where the properties of the spherical micelles have undergone a change, to another region where the micelles are modified toward anisometry. There may then follow a transition to a liquid crystalline phase — for example, type 1 middle phase (see Table 4.1) and further, perhaps one or several other type 1 mesophases with rod-shaped aggregates, and the lamellar neat phase. When the water content is diminished further, solid crystalline surfactant is formed and this may be regarded as the terminal stage of the aggregation process.

It is also possible to start with a dilute surfactant solution in an organic solvent, and then, as the concentration is increased, a largely corresponding series of micellar and liquid crystalline aggregation stages are passed through, but now of the reversed type; in these systems, however, the number of links is usually fewer.

Finally, in ternary systems of surfactant, water and an organic solvent, there may be a transition from normal (type 1) micellar and liquid crystalline structures, via lamellar to reversed (type 2) liquid crystalline and micellar structures. This implies first a stepwise increase in aggregation, which reaches its highest degree in the lamellar mesophase, and then a stepwise reduction.

The first stages of aggregation — formation both of the pre-micelles and micelles — may be regarded as ordinary association processes in solution. At higher concentrations, space factors are involved and the resulting interactions of various kinds produce further changes in the structure of the aggregates.

As a consequence of the increasing crowding in the micellar solution, the micelles are altered and finally a transition to a liquid crystalline phase is produced. At still higher concentrations (lower water content) the decrease in the interfacial area per polar group in these mesophase aggregates may cause a transition from one mesophase to another.

Three main classes of factors play a critical role in the aggregation: (a) the interactions of the lipophilic parts of the aggregate-forming molecules, (b) the interaction between the polar groups and (c) environmental factors.

Under the first two headings are also included the interactions between surfactant and solvent which lead to the incorporation of the latter with the aggregates. The last group of factors (c) includes not only the part played by the solvent insofar as it functions as an inert dispersion medium, but also one whose structure is disturbed or strengthened by the dissolved surfactant. These purely environmental factors are manifested most strongly in the solutions; their importance diminishes towards the mesophases and may be disregarded in some of them. On the other hand, that part of the solvent, and especially of the water, that is bound to the surfactant aggregates plays a role throughout the system.

In aqueous solutions the interaction between the lipophilic parts of the molecules is, together with purely environmental factors, regarded as being the most important cause of aggregation, whereas in organic solutions it is the interaction between the polar groups that is, besides the environmental factors, the prime factor. In the liquid crystalline phases the interaction between the lipophilic parts and that between the polar groups in the molecules are both of importance; as the water content of these phases is reduced and as a transition from normal to reversed structure occurs, the conditions in the interfacial layers of polar groups assume increasing importance.

The aggregation in stages, and the transition from normal to reversed structures, or vice versa, are among the most characteristic features of the surfactant systems.

The formation of interfacial layers of polar groups in homogeneous phases as a result of this aggregation may play an important part in biological systems. This may also be the case with the ability of the amphiphile systems to incorporate considerable amounts of compounds which are more or less sparingly soluble in water or in non-polar organic solvents. These phenomena, in combination with the fact that in many cases extremely small amounts of additives may change the mesophase and solution structures considerably, seem to make desirable an increased attention to the physiological risks for our external and internal environment implied in the extensive use of various manufactured chemicals.

References

1. Anacker, E. W. (1970). *Cationic Surfactants,* 203. (New York: Marcel Dekker)
2. Hall, D. G. and Pethica, B. A. (1967). *Nonionic Surfactants,* 516. (New York: Marcel Dekker)
3. Pethica, B. A. (1960). *Vortraege Originalfassung Intern. Kongr. Grenzflaechenahtive Stoffe, 3 Cologne,* **1,** 212
4. Shinoda, K., Nakagawa, T., Tamamushi, B. and Isemura, T. (1963). *Colloidal Surfactants* 1. (New York: Academic Press)
5. Stainsby, G. and Alexander, A. E. (1950). *Trans. Faraday Soc.,* **54,** 577

6. Matijević, E. and Pethica, B. A. (1957). *Trans. Faraday Soc.*, **54**, 587
7. Shinoda, K. and Hutchinson, E. (1962). *J. Phys. Chem.*, **66**, 577
8. Shinoda, K. (1967). *Chem. Phys. Appl. Surface Active Subst., Proc. Int. Congr., 4th 1964*, **2**, 527. (New York: Gordon and Breach)
9. Nakayama, H., Shinoda, K. and Hutchinson, E. (1966). *J. Phys. Chem.*, **70**, 3502
10. Elworthy, P. H. and Mysels, K. J. (1966). *J. Colloid. Interface Sci.*, **21**, 331
11. Elworthy, P. H., Gyane, D. O. and Macfarlane, C. B. (1969). *J. Pharm. Pharmacol.*, **21**, 6A
12. Ekwall, P., Eikrem, H. and Stenius, P. (1967). *Acta Chem. Scand.*, **21**, 1639
13. Corkill, J. M. and Goodman, J. F. (1969). *Advan. Colloid Interface Sci.*, **2**, 297
14. Abu-Hamdiyyah, M. and Mysels, K. J. (1967). *J. Phys. Chem.*, **71**, 48
15. Shinoda, K. (ed.) (1968). *Solvent Properties of Surfactant Solutions*. (New York: Marcel Dekker)
16. Grindley, J. and Bury, C. R. (1929). *J. Chem. Soc.*, **131**, 679
17. Debye, P. (1949). *Ann. N.Y. Acad. Sci.*, **51**, 575
18. Phillips, J. N. (1955), *Trans. Faraday Soc.*, **51**, 561
19. Overbeek, J. T. G. and Stigter, D. (1956). *Rec. Trav. Chim.*, **75**, 1263
20. Corkill, J. M., Goodman, J. F., Walker, T. and Wyer, J. (1969). *Proc. Roy. Soc. (London), A*, **312**, 243
21. Vold, M. J. (1950). *J. Colloid Sci.*, **5**, 506
22. Emerson, M. F. and Holtzer, A. (1965). *J. Phys. Chem.*, **69**, 3718
23. Ekwall, P., Danielsson, I. and Mandell, L. (1960). *Kolloid-Z.*, **169**, 113
24. Ekwall, P. (1963). *Finska Kemistsamfundets Medd.*, **72**, 59
25. Mandell, L. and Ekwall, P. (1967). *Chem. Phys. Appl. Surface Active Subst., Proc. Int. Congr., 4th 1964*, **2**, 659
26. Ekwall, P. (1965). *Wiss. Z. Friedrich-Schiller-Univ. Jena, Math.-Naturwiss. Reihe*, **14**, 184
27. Ekwall, P. (1967). *Svensk Kemisk Tidskr.*, **79**, 605
28. Fontell, K., Mandell, L., Lehtinen, H. and Ekwall, P. (1968). *Acta Polytech. Scand. Chem. Met. Ser.*, **74:3**, 1
29. Ekwall, P. and Mandell, L. (1968). *Acta. Chem. Scand.*, **22**, 699
30. Ekwall, P., Mandell, L. and Fontell, K. (1969). *Molec. Crystals*, **8**, 157
31. Ekwall, P. and Solyom, P. (1967). *Acta. Chem. Scand.*, **29**, 1619
32. Ekwall, P. (1969). *J. Colloid Interface Sci.*, **29**, 16
33. Ekwall, P. and Holmberg, P. (1965). *Acta Chem. Scand.*, **19**, 455, 573
34. Solyom, P. and Ekwall, P. (1968). *Chim. Phys. Appl. Prat. Aq. Surface, C. R. Congr. Int. Deterg., 5th*, **2**, 1041
35. Lindman, B. and Ekwall, P. (1968). *Molec. Crystals*, **5**, 79
36. Ekwall, P. (1937). *Kolloid-Z.*, **80**, 77
37. Friberg, S., Mandell, L. and Ekwall, P. (1966). *Acta. Chem. Scand.*, **20**, 1632
38. Ekwall, P. and Mandell, L. (1969). *Kolloid-Z.*, **233**, 938
39. Ekwall, P. and Solyom, P. (1969). *Kolloid-Z.*, **233**, 945
40. Friberg, S., Mandell, L. and Ekwall, P. (1969). *Kolloid-Z.*, **233**, 955
41. Lindman, B. and Ekwall, P. (1969). *Kolloid-Z.*, **234**, 1115
42. Ekwall, P. and Mandell, L. (1967). *Acta Chem. Scand.*,**21**, 1612
43. Gillberg, G. and Ekwall, P. (1967). *Acta Chem. Scand.*, **21**, 1630
44. Mijnlieff, P. F. (1970). *J. Colloid Interface Sci.*, **33**, 255
45. Aranow, R. H. and Witten, L. (1960). *J. Phys. Chem.*, **64**, 1643
46. Aranow, R. H. (1963). *J. Phys. Chem.*, **67**, 556
47. Poland, D. C. and Sheraga, H. A. (1965). *J. Phys. Chem.*, **69**, 2431
48. Poland, D. C. and Sheraga, H. A. (1966). *J. Colloid Interface Sci.*, **21**, 273
49. Holtzer, A. and Emerson, M. F. (1969). *J. Phys. Chem.*, **73**, 26
50. Frank, H. S. (1970). *Science*, **169**, 635
51. Walrafen, G. E. (1968). *Hydrogen Bonded Solvent Systems*, 9 (London: Taylor and Francis)
52. Franks, F. (1968). *Hydrogen Bonded Solvent Systems*, 31. (London: Taylor and Francis)
53. Glew, D. (1962). *J. Phys. Chem.*, **66**, 605
54. Reid, D. S., Quickenden, M. A. J. and Franks, F. (1969). *Nature (London)*, **224**, 1294
55. Clifford, J. and Pethica, B. A. (1965). *Trans. Faraday Soc.*, **61**, 182
56. Franks, F., Ravenhill, J., Egelstaff, P. A. and Page, D. I. (1970). *Proc. Roy. Soc. (London) A.*, **319**, 189
57. Danielsson, I., Lindman, B. and Ödberg, L. (1969). *Suomen Kemistilehti*, **42**, 209

58. Conway, B. E. (1970). *Physical Chemistry*, **9A**, 92. (New York: Academic Press)
59. Elworthy, P. H. and Florence, A. T. (1966). *Kolloid-Z.*, **208**, 157
60. Lindman, B., Wennerström, H. and Forsén, S. (1970). *J. Phys. Chem.*, **74**, 754
61. Lindman, B., Forsén, S. and Forslind, E. (1968). *J. Phys. Chem.*, **72**, 8205
62. Lindenbaum, S., Leifer, L., Boyd, G. E. and Chase, J. W. (1970). *J. Phys. Chem.*, **74**, 761
63. Hertz, H. G., Lindman, B. and Siepek, V. (1969). *Ber. Bunsenges. Physik. Chem.*, **73**, 542
64. Wen. W-Y. and Hung, J. H. (1970). *J. Phys. Chem.*, **74**, 170
65. Arnett, E. M., Ho, M. and Schaleger, I. (1970). *J. Amer. Chem. Soc.*, **92**, 7039
66. Corkill, J. M., Goodman, J. F., Harrold, S. P. and Tate, J. R. (1967). *Trans. Faraday Soc.*, **63**, 240
67. Mukerjee, P. (1967). *Advan. Colloid Interface Sci.*, **1**, 241
    967). *Advan. Colloid Interface Sci.*, **1**, 241
68. Corkill, J. M., Goodman, J. F., Harrold, S. P. and Tate, J. R. (1966). *Trans. Faraday Soc.*, **62**, 994; (1967). **63**, 773
69. Pilcher, G., Jones, M. N., Espada, L. and Skinner, H. A. (1969). *J. Chem. Thermodyn.*, **1**, 381; (1970). **2**, 1, 333
70. Corkill, J. M., Goodman, J. F. and Tate, J. R. (1966). *Trans. Faraday Soc.*, **60**, 966
71. Adderson, J. E. and Taylor, H. (1967). *Chem. Phys. Appl. Surface Active Subst.*, *Proc. Int. Congr., 4th 1964*, **4**, 613. (New York: Gordon and Breach)
72. Adderson, J. E. and Taylor, H. (1970). *J. Pharm. Pharmacol.*, **22**, 523
73. Robins, D. C. and Thomas, I. L. (1968). *J. Colloid Interface Sci.*, **26**, 407
74. Elworthy, P. H. and Florence, A. T. (1964). *Kolloid-Z.*, **195**, 23
75. Hudson, R. A. and Pethica, B. A. (1967). *Chem. Phys. Appl. Surface Active Subst.*, *Proc. Int. Congr. 4th 1964*, **2**, 631. (New York: Gordon and Breach)
76. Markina, Z. N., Tsikurina, N. N., Kostova, N. Z. and Rehbinder, P. A. (1964). *Kolloidn. Zh.*, **26**, 76
77. Emerson, M. F. and Holtzer, A. (1967). *J. Phys. Chem.*, **71**, 3320
78. Schick, M. J. (1964). *J. Phys. Chem.*, **68**, 3585
79. Benjamin, L. (1966). *J. Colloid Interface Sci.*, **22**, 386
80. Pethica, B. A. and Clifford, J. (1964). *Trans. Faraday Soc.*, **60**, 1483
81. Attwood, D. (1969). *Kolloid-Z.*, **232**, 785
82. Attwood, D. (1968). *J. Phys. Chem.*, **72**, 339
83. Benjamin, L. (1964). *J. Phys. Chem.*, **68**, 3375
84. Herrmann, K. W. (1964). *J. Phys. Chem.*, **68**, 1540
85. Herrmann, K. W. (1962). *J. Phys. Chem.*, **66**, 295
86. Herrmann, K. W. (1966). *J. Colloid Interface Sci.*, **22**, 352
87. Swarbrick, J. and Daruwala, J. (1969). *J. Phys. Chem.*, **73**, 2627
88. Tokiwa, F. and Ohki, K. (1967). *J. Phys. Chem.*, **71**, 1343
89. Schick, M. (1962). *J. Colloid Sci.*, **17**, 801
90. Schick, M., Atlas, F. M. and Eirich, S. R. (1962). *J. Phys. Chem.*, **66**, 1326
91. Doren, A. and Goldfarb, J. (1970). *J. Colloid Interface Sci.*, **32**, 67
92. Kuriyama, K. (1962). *Kolloid-Z.*, **181**, 144
93. Becher, P. (1962). *J. Colloid Sci.*, **17**, 186
94. Balmbra, R. S., Clunie, J. S., Corkill, J. M. and Goodman, J. F. (1962). *Trans. Faraday Soc.*, **58**, 1661
95. Elworthy, P. H. and McDonald, C. (1964). *Kolloid-Z.*, **195**, 23
96. Chinnikova, A. K., Markina, Z. N. and Rehbinder, P. A. (1970). *Kolloid. Zh.*, **30**, 592
97. Heilweil, I. J. (1964). *J. Colloid Sci.*, **19**, 105
98. Elworthy, P. H. and Macfarlane, C. B. (1964). *J. Chem. Soc.*, 907
99. Elworthy, P. H. and MacDonald, C. H. (1964). *Kolloid-Z.*, **195**, 16
100. Goldfarb, J. and Sepulveda, L. (1969). *J. Colloid Interface Sci.*, **31**, 454
101. Mukerjee, P. and Mysels, K. (1971). *Critical Micelle Concentrations of Aqueous Surfactant System*, Nat. Stand. Ref. Data Ser., Nat. Bur Stand., **36**
102. Courchene, W. L. (1964). *J. Phys. Chem.*, **68**, 1870
103. Herrmann, K. W., Brushmiller, J. G. and Courchene, W. L. (1966). *J. Phys. Chem.*, **70**, 2909
104. Franks, F. and Smith, H. T. (1962). *J. Phys. Chem.*, **68**, 3581
105. Hamann, S. D. (1962). *J. Phys. Chem.*, **66**, 1359
106. Tuddeham, R. F. and Alexander, A. J. (1962). *J. Phys. Chem.*, **66**, 1839

107. Shinoda, K. and Soda, T. (1963). *J. Phys. Chem.*, **79**, 2072
108. Franks, F., Quickenden, M. J., Ravenhill, J. R. and Smith, H. T. (1968). *J. Phys. Chem.*, **72**, 2668
109. Mukerjee, P. (1962). *J. Phys. Chem.*, **73**, 2054
110. Stigter, D. (1964). *J. Phys. Chem.*, **68**, 3603
111. Stigter, D. (1967). *J. Colloid Interface Sci.*, **23**, 379
112. Mukerjee, P. (1969). *J. Phys. Chem.*, **73**, 2054
113. Tartar, H. V. (1962). *J. Colloid Sci.*, **17**, 243
114. Tokiwa, F. and Ohhi, K. (1966). *J. Phys. Chem.*, **70**, 3437
115. Tokiwa, F. and Ohhi, K. (1967). *J. Colloid Interface Sci.*, **23**, 456
116. Tokiwa, F. and Ohhi, K. (1968). *Bull. Chem. Soc. Japan*, **41**, 2828
117. Attwood, D. (1969). *Kolloid-Z.*, **235**, 1193
118. Elworthy, P. H. and Macfarlane, C. B. (1962). *J. Chem. Soc.*, **52**, 537
119. Corkill, J. M., Goodman, J. F. and Walker, T. (1967). *Trans. Faraday Soc.*, **63**, 759
120. Carless, J. E., Challis, R. A. and Mulley, B. A. (1964). *J. Colloid Sci.*, **19**, 201
121. Geer, R. D., Eylar, E. W. and Anacker, E. W. (1971). *J. Phys. Chem.*, **75**, 369
122. Anacker, E. W. and Geer, R. D. (1971). *J. Colloid Interface Sci.*, **35**, 441
123. Mukerjee, P., Mysels, K. J. and Kapauan, P. (1967). *J. Phys. Chem.*, **71**, 4166
124. Anacker, E. W. and Westwell, A. E. (1964). *J. Phys. Chem.*, **68**, 3490
125. Anacker, E. W. and Ghose, H. M. (1968). *J. Amer. Chem. Soc.*, **90**, 3161
126. Clunie, J. S., Goodman, J. F. and Symons, P. G. (1967). *Trans. Faraday Soc.*, **63**, 754
127. Stead, J. A. and Taylor, H. (1970). *Aust. J. Pharm.* **51**, 51
128. Proust, J. and Ter-Miniassan-Saraga, L. (1970). *Compt. Rend.*, **270**, 1354
129. Stenius, P. and Zilliacus, C. H. (1971). *Acta Chem. Scand.*, **25**, 2232
130. Danielsson, I. and Stenius, P. (1971). *J. Colloid Interface Sci.*, **37**, 264
131. Ekwall, P. (1967). *Chem. Phys. Appl. Surface Active Subst.*, *Proc. Int. Congr.*, 4th 1964, **1**, 656 (New York: Gordon and Breach)
132. Ekwall, P., Eikrem, H. and Mandell, L. (1963). *Acta Chem. Scand.*, **17**, 111
133. Ekwall, P., Lemström, K-E., Eikrem, H. and Holmberg, P. (1967). *Acta Chem. Scand.*, **21**, 1401
134. Ekwall, P., Eikrem, H. and Stenius, P. (1967). *Acta Chem. Scand.*, **21**, 1639
135. Stenius, P. and Ekwall, P. (1967). *Acta Chem. Scand.*, **21**, 1643
136. Ekwall, P. and Stenius, P. (1967). *Acta Chem. Scand.*, **21**, 1767
137. Ekwall, P., Mandell, L. and Solyom, P. (1971). *J. Colloid Interface Sci.*, **35**, 519
138. Ekwall, P., Mandell, L. and Solyom, P. (1971). *J. Colloid Interface Sci.*, **35**, 266
139. Herrmann, K. W. (1964). *J. Phys. Chem.*, **68**, 1540
140. Elworthy, P. H., Florence, A. T. and Macfarlane, C. B. (1968). *Solubilization by Surface-Active Agents and its Application in Chemistry and the Biological Sciences.* (London: Chapman and Hall)
141. Riegelman, S., Allawala, N. A., Hronoff, M. K. and Strait, L. A. (1958). *J. Colloid Sci.*, **13**, 208
142. Nakagawa, T. and Tori, K. (1964). *Kolloid-Z.*, **194**, 143
143. Nakagawa, T. and Inoue, H. (1967). *Chem. Phys. Appl. Surface Active Subst.*, *Proc. Int. Congr.*, 4th 1964, **2**, 569. (New York: Gordon and Breach)
144. Inoue, H. and Nakagawa, T. (1966). *J. Phys. Chem.*, **70**, 1108
145. Eriksson, J. C. and Gillberg, G. (1966). *Acta Chem. Scand.*, **20**, 2019
146. Eriksson, J. C. (1963). *Acta Chem. Scand.*, **17**, 1478
147. Mulley, B. A. and Metcalf, A. D. (1962). *J. Colloid Sci.*, **17**, 523
148. Donbrow, M. and Rhudes, C. T. (1964). *J. Chem. Soc. Suppl.*, 6166
149. Lange, H. and van Raay, H. (1963). *Kolloid-Z.*, **189**, 55
150. Donbrow, M., Malyneux, P. and Rhodes, C. T. (1967). *J. Chem. Soc. A.*, 561
151. Chinnikova, A. V., Markina, Z. N. and Rehbinder, P. A. (1968). *Kolloid.*, **30**, 782
152. Markina, Z. N., Rybakova, E. V., Chinnikova, A. V. and Rehbinder, P. A. *Akad. Nauk. SSSR*, **179**, 75
153. Shinoda, K., Nakagawa, T., Tamamushi, B.-I. and Isemura, T. (1963). *Colloid*, 141. (New York: Academic Press)
154. Luzzati, V., Mustachi, M., Skoulios, A. and Husson, F. (1960). *Acta Crystallogr.*, **13**, 660
155. Husson, F., Mustachi, H. and Luzzati, V. (1960). *Acta Crystallogr.*, **13**, 668
156. Luzzati, V. and Reiss-Husson, F. (1966). *Nature (London)*, **210**, 1351
157. Luzzati, V., Gulick-Krzywicki, G. and Tardieu, A. (1968). *Nature (London)*, **218**, 1021

158. Luzzati, V., Tardieu, A., Gulick-Krzywicki, G., Riva, E. and Reiss-Husson, F. (1968). *Nature (London)*, **220**, 485
159. Tardieu, A. and Luzzati, V. (1970). *Biochim. Biophys. Acta*, **11**, 219
160. Skoulios, A. E. and Luzzati, V. (1961). *Acta Crystallogr.*, **14**, 278
161. Skoulios, A. E. (1961). *Acta Crystallogr.*, **14**, 419
162. Gallot, B. and Skoulios, A. E. (1962). *Acta Crystallogr.*, **15**, 826
163. Spegt, P. A. and Skoulios, A. E. (1963). *Acta Crystallogr.*, **16**, No. 13, suppl. A 88
164. Spegt, P. A. and Skoulios, A. E. (1963). *Acta Crystallogr.*, **16**, 301
165. Spegt, P. A. and Skoulios, A. E. (1964). *Acta Crystallogr.*, **17**, 198
166. Spegt, P. A. and Skoulios, A. E. (1962). *Acta Crystallogr.*, **21**, 892
167. Ekwall, P., Danielsson, I. and Mandell, L. (1960). *Vortraege Originalfassung Intern. Kongr. Grenzflaechenaktive Stoffe, 3. Cologne*, **1A**, 193
168. Mandell, L., Fontell, K. and Ekwall, P. (1967). *Advan. Chem. Ser.*, **63**, 89
169. Mandell, L. and Ekwall, P. (1968). *Acta Polytech. Scand. Chem. Met. Ser.*, **74 I**, 1
170. Mandell, L., Fontell, K., Lehtien, H. and Ekwall, P. (1968). *Acta Polytech. Scand. Chem. Met. Ser.*, **74 II**, 1
171. Fontell, K., Ekwall, P., Mandell, L. and Danielsson, I. (1962). *Acta Chem. Scand.*, **16**, 2294
172. Ekwall, P., Mandell, L. and Fontell, K. (1968). *Acta Chem. Scand.*, **22**, 373
173. Ekwall, P., Mandell, L. and Fontell, K. (1968). *Acta Chem. Scand.*, **22**, 697
174. Fontell, K., Mandell, L. and Ekwall, P. (1968). *Acta Chem. Scand.*, **22**, 3209
175. Mulley, B. A. (1961). *J. Pharm. Pharmacol.*, **13**, 205 T
176. Mulley, B. A. and Metcalf, A. D. (1962). *J. Colloid Sci.*, **17**, 523
177. Mulley, B. A. and Metcalf, A. D. (1964). *J. Colloid Sci.*, **19**, 501
178. Clunie, J. S., Corkill, J. M. and Goodman, J. F. (1965). *Proc. Roy. Soc. (London)*, **A285**, 520
179. Clunie, J. S., Corkill, J. M., Goodman, J. F., Symons, P. C. and Tate, J. R. (1967). *Trans. Faraday Soc.*, **63**, 2839
180. Corkill, J. S. and Goodman, J. F. (1969). *Advan. Colloid Interface Sci.*, **2**, 297
181. Balmbra, R. R., Clunie, J. S. and Goodman, J. F. (1969). *Nature (London)*, **222**, 1159
182. Clunie, J. S., Goodman, J. F. and Symons, P. C. (1969). *Trans. Faraday Soc.*, **65**, 287
183. Lutton, E. S. (1965). *J. Amer. Oil Chemists Soc.*, **42**, 1068
184. Lutton, E. S. (1966). *J. Amer. Oil Chemists Soc.*, **43**, 28
185. Herrmann, K. W., Brushmiller, J. G. and Courchene, W. L. (1966). *J. Phys. Chem.*, **70**, 2969
186. Larsson, K. (1967). *Z. Physik. Chem. (Frankfurt)*, **56**, 173
187. Small, D. M., Bourgés, M. and Dervichian, D. G. (1966). *Nature (London)*, **211**, 5051
188. Small, D. M., Bourgés, M. and Dervichian, D. G. (1966). *Biochim. Biophys. Acta*, **125**, 563
189. Small, D. M. and Bourgés, M. (1966). *Molec. Crystals*, **1**, 541
190. Bourgés, M., Small, D. M. and Dervichian, D. G. (1967). *Biochim. Biophys. Acta*, **137**, 157
191. Bourgés, M, Small, D. M. and Dervichian, D. G. (1967). *Biochim. Biophys. Acta*, **144**, 189
192. Lawson, K. D., Mabis, A. J. and Flautt, T. J. (1968). *J. Phys. Chem.*, **72**, 2058
193. Park, D., Rogers, J., Toft, R. W. and Winsor, P. A. (1970). *J. Colloid Interface Sci.*, **32**, 84
194. Winsor, P. A. (1968). *Chem. Rev.*, **68**, 2
195. Lawson, K. D. and Flautt, T. J. (1965). *J. Phys. Chem.*, **69**, 4256
196. Lawson, K. D. and Flautt, T. J. (1968). *J. Phys. Chem.*, **72**, 2066
197. Gilchrist, C. A., Rogers, J., Steel, G., Vool, E. G. and Winsor, P. A. (1968). *J. Colloid Interface Sci.*, **25**, 409
198. Drakenberg, T., Johansson, Å. and Forsén, S. (1970). *J. Phys. Chem.*, **74**, 4528
199. Johansson, Å. and Lindman, B. (1972). *Liquid Crystalline Systems*, in the press. (London: Van Nostrand)
200. Johansson, Å. and Drakenberg, T. (1971). *Thesis, Univ. of Lund*, to be published
201. Persson, N. G. and Johansson, Å. (1971). *Thesis, Univ. of Lund*, to be published
202. Stockenius (1962). *J. Cell Biol.*, **12**, 221
203. Balmbra, R. R., Clunie, J. S. and Goodman, J. F. (1965). *Proc. Roy. Soc. (London)*, **A285**, 534
204. Balmbra, R. R., Clunie, J. S. and Goodman, J. F. (1967). *Molec Crystals*, **3**, 281
205. Bucknall, D. A. B., Clunie, J. S. and Goodman, J. F. (1969). *Molec Crystals*, **7**, 215
206. Balmbra, R. R., Bucknall, D. A. B. and Clunie, J. S. (1970). *Molec Crystals*, **11**, 173
207. Eins, S. (1970). *Molec Crystals*, **11**, 173

208. Luzzati, V. and Husson, F. (1962). *J. Cell Biol.*, **12**, 207
209. Reiss-Husson, F. and Luzzati, V. (1967). *Advan. Biol. Med. Phys.*, **11**, 87
210. Skoulios, A. (1967). *Advan. Colloid. Interface Sci.* **1**, 79
211. Winsor, P. A. (1972). *Liquid Crystalline Systems,* in the press. (London: Van Nostrand)
212. Ekwall, P. (1972). *Liquid Crystalline Systems,* in the press. (London: Van Nostrand)
213. Ekwall, P., Mandell, L. and Fontell, K. (1970). *J. Colloid Interface Sci.*, **33**, 215
214. Vincent, J. M. and Skoulios, A. (1966). *Acta Crystallogr.*, **20**, 431
215. Vincent, J. M. and Skoulios, A. (1966). *Acta Crystallogr.*, **20**, 444
216. Ekwall, P. and Mandell, L. (1961). *Acta Chem. Scand.*, **15**, 1407
217. Gallot, B. and Skoulios, A. (1966). *Kolloid-Z.*, **208**, 37
218. Mandell, L., to be published
219. Ekwall, P., Mandell, L. and Fontell, K. (1969). *J. Colloid Interface Sci.*, **31**, 508, 550
220. Solyom, P. and Ekwall, P. (1968). *Chim. Phys. Appl. Prat. Aq. Surface, C. R. Congr. Int. Deterg. 5th*, **2**, 1681
221. Flory, P. J. (1961). *J. Polymer Sci.*, **49**, 205
222. Ekwall, P., Danielsson, I. and Mandell, L. (1960). *Vortraege Originalfassung Intern. Kongr. Grenzflaechenaktive Stoffe, 3. Cologne*, **1A**, 189
223. Ekwall, P., Mandell, L. and Fontell, K. (1968). *Acta Chem. Scand.*, **22**, 1543
224. Ekwall, P., Vapour pressure measurements, to be published
225. Solyom, P. and Ekwall, P. (1969). *Rheologica Acta*, **8**, 316
226. Gulick-Krzywicki, T., Tardieu, A. and Luzzati, V. (1969). *Molec. Crystals*, **8**, 285
227. Luzzati, V. (1968). *Biological Membranes*, 71
228. Rivas, E. and Luzzati, V. (1969). *J. Molec Biol.* **41**, 261
229. Friberg, S., Mandell, L. and Fontell, K. (1969). *Acta Chem. Scand.*, **23**, 1085
230. Singleterry, C. R. (1955). *J. Amer. Chem. Soc.*, **32**, 446
231. Pilpel, N. (1963). *Chem. Rev.*, **63**, 221
232. Fowkes, F. M. (1967). *Solvent Properties of Surfactant Solutions,* 65. (New York: Marcel Dekker)
233. Kitahara, A. (1970). *Cationic Surfactants,* 289. (New York: Marcel Dekker)
234. Kitahara, A. and Ishikawa, T. (1967). *J. Colloid Interface Sci.*, **24**, 189
235. Little, R. C. and Singleterry, C. R. (1964). *J. Phys. Chem.*, **68**, 3453
236. Little, R. C. (1966). *J. Colloid Interface Sci.*, **24**, 189
237. Kitahara, A., Kobayashi, T. and Tachibana, T. (1962). *J. Phys. Chem.*, **66**, 363
238. Kon-No, K. and Kitahara, A. (1971). *J. Colloid. Interface Sci.*, **35**, 636
239. Peri, J. B. (1969). *J. Colloid Interface Sci.*, **29**, 6
240. Gonick, E. (1946). *J. Colloid. Sci.*, **1**, 393
241. Shinoda, K. and Arai, H. (1965). *J. Colloid. Interface Sci.*, **20**, 93
242. Kitahara, A. (1965). *J. Phys. Chem.*, **69**, 2788
243. Martin, E. P. and Pink, R. C. (1948). *J. Chem. Soc.*, 1750
244. Singleterry, C. R. and Weinberger, L. A. (1951). *J. Amer. Chem. Soc.*, **73**, 4574
245. Kaufman, S. and Singleterry, C. R. (1957). *J. Colloid Sci.*, **12**, 465
246. Ford, T. F., Kaufman, S. and Nichols, O-D. (1966). *J. Phys. Chem.*, **70**, 3726
247. Wentz, M., Smith, W. H. and Martin, A. R. (1969). *J. Colloid. Interface Sci.*, **29**, 36
248. Mathews, M. B. and Hirschbom, E. (1953). *J. Colloid. Sci.*, **8**, 86
249. Kitahara, A., Watanabe, K. Kon-No, K. and Ishikawa, T. (1969). *J. Colloid Interface Sci.*, **29**, 4
250. Aebi, C. M. and Wiebush, J. R. (1959). *J. Colloid. Sci.*, **14**, 161
251. Debye, P. and Coll, H. (1962). *J. Colloid. Sci.*, **17**, 220
252. Elworthy, P. H. and Mcintosh, D. S. (1964). *J. Phys. Chem.*, **68**, 3448
253. Ekwall, P., Mandell, L. and Fontell, K. (1969). *J. Colloid. Interface Sci.*, **29**, 639
254. Shinoda, K. and Ogava, T. (1967). *J. Colloid. Interface Sci.*, **24**, 56
255. Shinoda, K. and Saito, H. (1968). *J. Colloid. Interface Sci.*, **26**, 70
256. Frank, S. V. and Zografi, F. (1969). *J. Colloid. Interface Sci.*, **29**, 27
257. Kon-No, K. and Kitahara, A. (1970). *J. Colloid. Interface Sci.*, **33**, 124
258. Kon-No, K. and Kitahara, A. (1970). *J. Colloid. Interface Sci.*, **34**, 221
259. Kon-No, K. and Kitahara, A. (1971). *J. Colloid. Interface Sci.*, **37**, 469
260. Kon-No, K. and Kitahara, A. (1971). *J. Colloid. Interface Sci.*, in the press
261. Gillberg, G., Lehtinen, H. and Friberg, S. (1970). *J. Colloid. Interface Sci.*, **33**, 40
262. Ekwall, P. and co-workers, to be published
263. Friberg, S., Mandell, L. and Ekwall, P. (1968). *Chim. Phys. Appl. Prat. Aq. Surface, C. R. Congr. Int. Deterg., 5th*, **2**, 1121

264. Honig, J. G. and Singleterry, C. R. (1956). *J. Phys. Chem.*, **60,** 1108, 1114
265. Kaufman, S. (1964). *J. Phys. Chem.*, **68,** 2814
266. Kaufman, S. (1967). *J. Colloid Interface Sci.*, **25,** 401
267. Fontell, K. (1968). *Chim. Phys. Appl. Prat. Aq. Surface, C. R. Congr. Int. Deterg., 5th,* **2,** 1033
268. Lindblom, G., Lindman, B. and Mandell, L. (1970). *J. Colloid Interface Sci.*, **34,** 262
269. Fryar, A. J. and Kaufman, J. (1969). *J. Colloid. Interface Sci.*, **29,** 444
270. Kon-No, K. and Kitahara, A. (1971). *J. Colloid. Interface Sci.*, **35,** 409

5
Theory of Homogeneous Nucleation from the Vapour

G. M. POUND
Stanford University, California

KAZUMI NISHIOKA and JENS LOTHE
Blindern University, Oslo

5.1	INTRODUCTION	148
	5.1.1 *General*	148
	5.1.2 *'Classical' nucleation theory*	149
	5.1.3 *The capillarity approximation and statistical mechanical considerations*	151
	5.1.3.1 *Early work*	151
	5.1.3.2 *Recent developments in statistical mechanical considerations of the capillarity approximation*	154
	5.1.4 *Recent experimental results on homogeneous nucleation from the vapour*	155
	5.1.5 *Some important areas in homogeneous nucleation from the vapour that are omitted in this review due to limitations of space*	155
	5.1.5.1 *Theoretical work on homogeneous nucleation from the vapour near the critical point*	155
5.2	GENERAL CLASSICAL PHASE-INTEGRAL APPROACH TO STATISTICAL MECHANICAL CONSIDERATIONS AND THE CAPILLARITY APPROXIMATION IN HOMOGENEOUS NUCLEATION FROM THE VAPOUR	156
	5.2.1 *General*	156
	5.2.2 *The treatment by Reiss and Katz for the equilibrium concentration of nuclei*	156
	5.2.3 *The treatment by Reiss, Katz and Cohen for the equilibrium concentration of nuclei*	159
	5.2.3.1 *Outline of the derivation*	159
	5.2.3.2 *Outline of an alternative derivation*	161
	5.2.4 *The Reiss–Kikuchi treatment for the equilibrium concentration of nuclei*	164

		5.2.4.1	The Reiss treatment	164
		5.2.4.2	Kikuchi's extension of the Reiss treatment	167
		5.2.4.3	Summary	167
	5.2.5		The Kikuchi treatment for the equilibrium concentration of nuclei	168
	5.2.6		Summary	171
5.3	CLASSICAL PHASE-INTEGRAL APPROACHES USING CRYSTAL MODELS AND NORMAL-MODE COMPUTER CALCULATIONS			171
	5.3.1		General	171
	5.3.2		The replacement partition function in a linear chain	172
	5.3.3		The replacement partition function in crystals	174
	5.3.4		Normal-mode computer calculations	179
		5.3.4.1	General	179
		5.3.4.2	The work of Burton	179
		5.3.4.3	The work of Nishioka et al.	180
		5.3.4.4	The work of Abraham and Dave	182
		5.3.4.5	The work of Daee et al. and Kassner et al.	185
	5.3.5		A few conclusions	185
5.4	SOME GENERAL CONCLUSIONS			186
ACKNOWLEDGEMENT				187

5.1 INTRODUCTION

5.1.1 General

The main purpose of this chapter is to appraise the major developments of the past 5 years in the theory of homogeneous* nucleation from the vapour. Inevitably the choice of topics will be affected by the interests of the present authors. Also limitations of space will preclude discussion of many important contributions. Furthermore, in the interests of conciseness, complete derivations will not be given in connection with the lengthier theoretical developments. Rather, an attempt will be made to describe the physics of the situation, the approach and the salient results. Mathematical details will be given only when they are necessary for a critical examination of the validity of the results. Also, an attempt will be made to retain the nomenclature of the original papers rather than to standardise the notation. In the past years a number of reviews[1-4] of nucleation have been published and hence no historical account will be given in the present article. On the other hand, a brief exposition of the 'classical' theory for homogeneous nucleation of liquid from vapour must now be presented in the interests of pedagogy. One notes that this traditional approach is still the one that is most used in all branches of nucleation, with the possible exception of heterogeneous nucleation from the vapour on to substrates.

---

*Uncatalysed by such things as surfaces, ions or impurities.

5.1.2 'Classical' nucleation theory

It was of course early recognised that a correct approach to nucleation theory might start from evaluation of the free energy of formation of the nuclei of critical size. Furthermore, it was clear that this free energy of formation might best be expressed in terms of the binding potential energy and partition functions of the molecules of the critical cluster. However, such calculations are difficult and are only now becoming a possibility due to the advent of high-speed computers, as will be discussed in following sections. Accordingly a simplification, variously called the droplet model or the capillarity approximation, was introduced to accomplish the calculation. In this capillarity approximation, the free energy of formation of the sub-critical embryos and nuclei of the stable phase is given in terms of its macroscopic thermodynamic properties. Thus the Helmholtz free energy of formation of a spherical droplet of radius r from a supersaturated vapour is written[1-4] in terms of the bulk free energy difference ΔF_v and surface tension σ as

$$\Delta F^0 = \tfrac{4}{3}\pi r^3 \Delta F_v + 4\pi r^2 \sigma \tag{5.1}$$

in which

$$\Delta F_v = -\frac{kT}{v_1} \ln(p/p_0) \tag{5.2}$$

where p is the partial pressure in the supersaturated vapour and p_0 is vapour pressure and v_1 is molecular volume of the bulk liquid. A plot of equation (5.1) is shown in Figure 5.1, where a maximum is observed to occur at r^*.

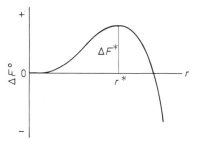

Figure 5.1 Plot of equation (5.1)

By differentiation of equation (5.1) one obtains the radius of the nucleus of critical size

$$r^* = -\frac{2\sigma}{\Delta F_v} \tag{5.3}$$

and its free energy

$$\Delta F^* = 16\pi\sigma^3/3\Delta F_v^2 \tag{5.4}$$

which is the free-energy barrier for formation of the critical nucleus, a result first obtained by Gibbs[5]. Of course, when one substitutes experimental values in equation (5.3) he finds that r^* is of the order of ·10 Å and that the critical nucleus contains only c. 100 molecules, rendering application of the droplet model to nucleation tenuous indeed. Much has been written but little is known of the corrections to the macroscopic value for σ which are necessitated by the high curvature of the small clusters.

In any case, the equilibrium concentration of critical nuclei containing n^* molecules was given incorrectly in the classical treatment by

$$c(n^*) = c(1)\exp(-\Delta F^*/kT) \quad (5.5)$$

or more generally for clusters of size class n

$$c(n) = c(1)\exp(-\Delta F^0/kT)$$

in which $c(1)$ is the concentration of monomer in the supersaturated vapour. The correct form of the equilibrium relationship is, as usual, obtained by minimisation of the Helmholtz free energy of the system at constant temperature and volume[6]; this result will be presented later in the discussion.

Farkas[7] and Volmer and Weber[8] postulated that the stationary nucleation rate would be given by the product of the equilibrium concentration of critical nuclei and the impingement frequency $\beta S(n^*)$ of vapour monomer on the critical nucleus

$$J = \beta S(n^*)c(n^*) \quad (5.6)$$

where β is the impingement flux of monomer, $p/\sqrt{2\pi mkT}$ in the present example, and $S(n^*) = 4\pi r^{*2}$ is the surface area of the critical nucleus.

Of course, this treatment neglects depletion of critical nuclei by growth and the effect of the reverse process of re-evaporation. Proper evaluation of these factors may proceed through treatment of the problem as generalised diffusion in size space n and this was accomplished by Becker and Doering[9] and, more succinctly, by Zeldovitch[10,11] and Frenkel[12]. Proceeding under the reasonable assumption of one-at-a-time condensation or evaporation of molecules, in which case the mean square displacement is unity, the result for the net current of molecules between size classes in the transient case is

$$J(n) = -\beta S(n)c(n)\frac{\partial [f(n,t)/c(n)]}{\partial n}$$

$$= -\beta S(n)\frac{\partial f(n,t)}{\partial n} - \frac{\beta S(n)}{kT}f(n,t)\frac{\partial \Delta F^0}{\partial n} \quad (5.7)$$

in which $f(n,t)$ is the actual concentration of clusters containing n molecules at time t and $\beta S(n)$ is the generalised diffusion coefficient. Integration of the time-independent form of (5.7) yields for the stationary nucleation rate

$$J = Z\beta S(n^*)c(n^*) \quad (5.8)$$

where the Zeldovitch factor

$$Z = (\Delta F^*/3\pi kTn^{*2})^{\frac{1}{2}} \quad (5.9)$$

Z, which expresses the effect of the generalised diffusional nature of the nucleaation process, lies between 10^{-1} and 10^{-2} for the usual experimental cases and hence, due to the inaccuracies inherent in both nucleation theory and experiment, is generally not of much practical importance.

The divergence of (5.7) for critical nuclei ($n = n^*$) may be approximately integrated to yield expressions for the induction period required for establishment of stationary conditions*. This induction period is of the order of microseconds in fluid phases, and hence the stationary nucleation rate is usually of greatest practical interest. Also, the generalised diffusion analysis has been extended to a two-dimensional space in size and energy[6], and it is found that dissipation of the heat of condensation is rapid and hence that the assumption of isothermal conditions is valid. Accordingly the difficulties with equations (5.6) or (5.8) for the stationary state appear to be thermodynamic rather than kinetic. Specifically, the trouble with the rate equations would appear to lie in the use of equations (5.4) and (5.5) for $c(n^*)$.

Equation (5.8) is the 'classical' nucleation rate equation, and it was found to describe the data for homogeneous nucleation of droplets from vapours quite well. In the first place it predicts a sharp increase, from essentially zero to a high value, of nucleation rate in a small increment of increase of imposed supersaturation. This phenomenon is actually observed in cloud chamber and other kinds of experiments, and the supersaturation, p/p_0, at which the sudden increase in rate of nucleation occurs, is called the critical supersaturation. The quantitative agreement of equation (5.8) with the critical supersaturations observed in the cloud-chamber measurements of Volmer and Flood[13, 14] is shown in Table 5.1. For many years this agreement stood as the basis of our understanding of nucleation.

Table 5.1 Critical supersaturations calculated from the 'classical' theory compared with the observations of Volmer and Flood

Vapour	Number of molecules in stable nucleus	Nucleus radius $r/Å$	$(p/p_0)_{cr}$ calculated	measured
Water, 275.2 K	80	8.9	4.2	4.2 ± 0.1
Water, 261.0 K	72	8.0	5.0	5.0
Methanol, 270.0 K	32	7.9	1.8	3.0
Ethanol, 273.0 K	128	14.2	2.3	2.3
n-Propanol, 270.0 K	115	15.0	3.2	3.0
Isopropyl alcohol, 265.0 K.	119	15.2	2.9	2.8
n-Butyl alcohol, 270.0 K	72	13.6	4.5	4.6
Nitromethane, 252.0 K	66	11.0	6.2	6.0
Ethyl acetate, 242.0 K	40	11.4	10.4	8.6 to 12.3

5.1.3 The capillarity approximation and statistical mechanical considerations

5.1.3.1 Early work

The necessity for consideration of the contributions from free translation and rotation to the free energy of formation of liquid nuclei from the vapour was recognised by Frenkel[12], Rodebush[15, 16], Kuhrt[17], Lothe and Pound[18], Hirth[19] and Dunning[20, 21].

---

*For example, see Ref. 6.

Frenkel[12] presented an essentially correct derivation of the free energy of formation of embryos and nuclei in 1946. However, he evidently did not evaluate the numerical magnitude of the external partition functions, which yield a contribution of $c. -45kT$ to the free energy of formation in typical cases, corresponding to an increase in equilibrium concentration of nuclei by a factor of roughly 10^{20}.

Rodebush[15,16] estimated the size of the contribution from free translation and rotation correctly, but did not attempt a rigorous derivation. He concluded that these factors must somehow be included in the 'classical' capillarity approximation.

Kuhrt[17] entered the problem from a discussion of the Mayer cluster theory of real gases. He gave an essentially correct derivation of the expression for the equilibrium concentration of liquid embryos and nuclei. However, he erroneously concluded that the translational and rotational contributions are fortuitously offset by removal of the entire free energy, including the binding potential energy, of two molecules. The reason for removal of the free energy of two molecules has to do with conservation of degrees of freedom and will be discussed in the following in connection with the replacement partition function. Suffice it to say here that the principal error in Kuhrt's work lies in inclusion of the binding potential energy in the required correction term, which is now known as the replacement free energy. This has been discussed in detail by Feder et al.[6].

Lothe and Pound[18] and Hirth[19] discussed the translational and rotational contributions to the free energy of formation of clusters in the vapour. They pointed out that these amount to $c. -45kT$ in typical cases. Further, they also recognised that all degrees of freedom are of course already included in the 'classical' capillarity approximation. Accordingly, if the external partition functions of translation and rotation are to be considered, six degrees of freedom that the molecules have as a part of the bulk phase must be deactivated in the isolated cluster or nucleus. Lothe and Pound[18] approximated this positive contribution, now called the replacement free energy, to the free energy of formation of a cluster by the molecular entropy of the bulk liquid, which is $c. 5 kT$ for many liquids. Therefore the net statistical mechanical contribution to the free energy of formation of typical nuclei would be of the order of $-40 kT$, corresponding to a factor of increase in equilibrium concentration of critical nuclei or stationary nucleation rate of $c. 10^{17}$. This amounts to a 35% decrease in critical supersaturation and hence would destroy the good agreement between theory and experiment noted in Table 5.1.

Dunning[20,21] took a similar approach, except that he considered that the entropy of the six degrees of freedom deactivated in the bulk *might* correspond to almost free translation and rotation of a 'cluster' as part of the bulk phase. In this case, the free translation and rotation of the isolated cluster in the vapour would be almost exactly offset by the replacement free energy, and one would return to the classical result. He thus introduced the important concept of a mathematical cluster in the bulk phase, an idea that was to be used extensively by subsequent investigators.

In 1966 Lothe and Pound[22] defined the replacement partition function as describing the modes of motion an isolated cluster does not have because it is no longer a part of the bulk phase. Further, they reasoned that all internal configurations accessible to a mathematical cluster of n molecules in the bulk phase are also accessible to the isolated cluster. Thus, the only motions not accessible to the isolated cluster are *the six translational and rotational motions of the cluster in bulk for which the relative positions of the molecules remain fixed.* These latter rigid-body motions then give rise to the replacement partition function. Lothe and Pound's crude estimates of this positive contribution to the free energy of formation of an isolated cluster, which were made by treating the imbedded cluster as an Einstein oscillator, also gave values of $c.\ 5\,kT$, again corresponding to a replacement partition function of $c.\ 10^3$.

Thus in 1966 the emerging picture of nucleation theory in terms of the equilibrium concentration of critical nuclei, as represented by equation (5.8), was concisely summarised by the following derivation[6] for the equilibrium concentration $c(n)$ of clusters containing n molecules. The Helmholtz free energy of a cluster containing n molecules in a system containing $c(n)$ clusters per unit volume may be written as

$$F = n\mu_1 + \sigma S(n) - \frac{kT}{c(n)} \ln\left[\frac{\lambda^{c(n)}}{c(n)!}\right] - kT \ln q_R + kT \ln q_{rep} \quad (5.10)$$

where μ_1 is the chemical potential of the bulk liquid, $S(n)$ the embryo surface area,

$$\lambda = (2\pi n m k T)^{\frac{3}{2}}/h^3 \quad (5.11)$$

the translational partition function, and

$$q_R = (2kT)^{\frac{3}{2}}(\pi I^3)^{\frac{1}{2}}/\hbar^3 \quad (5.12)$$

the rotational partition function in which the moment of inertia is approximated by that of a sphere of uniform density

$$I = \tfrac{2}{5}nmr^2 \quad (5.13)$$

q_{rep} is the replacement partition function described in the above discussion. Equation (5.10) may be rewritten using Stirling's formula as

$$F = F^0 - kT \ln[\lambda/c(n)] - kT - kT \ln q_R + kT \ln q_{rep} \quad (5.14)$$

in which

$$F^0 = n\mu_1 + \sigma S(n) \quad (5.15)$$

Considering all size classes, the total Helmholtz free energy of the unit volume is

$$\sum_{n=1}^{\hat{n}} c(n)F = \sum_{n=1}^{\hat{n}} \{c(n)F^0 - c(n)kT \ln[\lambda/c(n)] - c(n)kT$$
$$- c(n)kT \ln q_R + c(n)kT \ln q_{rep}\} \quad (5.16)$$

where $\hat{n}$ is taken as only somewhat larger than the critical molecular number[6] n^*. This total free energy is minimised at constant temperature and volume using the maximum term approximation subject to the constraint of fixed total number of molecules

$$N = \sum_{n=1}^{\hat{n}} nc(n) \tag{5.17}$$

Thus

$$\delta\left[\sum_{n=1}^{\hat{n}} c(n)F - \mu N\right] = 0 \tag{5.18}$$

in which the Lagrangian multiplier μ is the equilibrium chemical potential of the supersaturated vapour. It then follows, since

$$\frac{\partial N}{\partial c(n)} = n \tag{5.19}$$

that

$$\left\{F^0 - kT\ln\left[\frac{\lambda}{c(n)}\right] - kT\ln q_R + kT\ln q_{rep} - n\mu\right\}\delta c(n) = 0 \tag{5.20}$$

and the equilibrium distribution becomes

$$c(n) = c(1)\exp\left\{-\left[\Delta F^0 - kT\ln\left(\frac{\lambda q_R}{c(1)q_{rep}}\right)\right]/kT\right\} \tag{5.21}$$

For the critical nuclei,

$$c(n^*) = c(1)\exp\left\{-\left[\Delta F^* - kT\ln\left(\frac{\lambda^* q_R^*}{c(1)q_{rep}^*}\right)\right]/kT\right\} \tag{5.22}$$

in sharp contrast to the 'classical' expression of equation (5.5). $\lambda^*/c(1)$ is c. 10^{11} and q_R^* c. 10^9 in typical cases, and q_{rep}^* was thought to be 10^3, as described in the preceding discussion.

5.1.3.2 Recent developments in statistical mechanical considerations of the capillarity approximation

In the past 5 years the principal efforts in homogeneous nucleation theory have centred about:

(a) classical phase-integral approaches applied to both liquid and crystalline clusters and

(b) normal-mode computer calculations of the 'exact' free energy of isolated crystalline clusters and comparison of these values with those obtained from the capillarity approximation.

As one might expect, in view of the greater precision of the model for the crystalline state, the results for microcrystallites are fairly well accepted. As of this writing they tend to show that the situation is about the same as that outlined for liquid droplets in Section 5.1.3.1. The crystallites are believed to possess free translation and rotation, but the replacement partition function seems to be higher than originally envisaged by Lothe and Pound, being c. 10^8 instead of 10^3.

On the other hand, the general classical phase-integral approaches for liquid clusters have in some cases led to error, confusion and controversy. A main source of difficulty lies in relating the properties of a 'stationary' droplet to properties of the bulk liquid. Sometimes a dual nature is erroneously ascribed to the stationary droplet in which its configurational integral is supposed to describe both the surface free energy of an isolated droplet and the interactions of the cluster in bulk with its surroundings. The two requirements are, of course, mutually exclusive. In general, the errors in these treatments have led to either inclusion of free translation and rotation of the liquid cluster in the macroscopic surface free energy* of the bulk liquid, or a model for the bulk liquid in which the mathematical clusters of arbitrary size have free rotation and translation in a free volume as large as in the vapour. These operations lead to results in which the free translation and rotation of the isolated cluster in the vapour are nearly cancelled, returning thereby to the predictions of the 'classical' capillarity approximation.

The more important of the general classical phase-integral approaches will be discussed in Section 5.2. The lattice-model derivations and normal-mode computer calculations will be described in Section 5.3.

5.1.4 Recent experimental results on homogeneous nucleation from the vapour

The recognition that statistical-mechanical corrections to the capillarity approximation could be important stimulated some excellent experimental work in which supersonic nozzle and diffusion cloud chamber techniques were employed to measure critical supersaturations for condensation in a variety of systems. It would now appear that there are two classes of experimental liquids, those with unsymmetrical molecules and low surface entropies that follow the 'classical' prescription and those with symmetrical molecules and high surface entropies that follow the prescription of Lothe and Pound. This new work will not be described because of limitations of space.

5.1.5 Some important areas in homogeneous nucleation from the vapour that are omitted in this review due to limitations of space

5.1.5.1 Theoretical work on homogeneous nucleation from the vapour near the critical point

Considerations of nucleation are in this review, confined to regions near the triple point. Here the collision time is very short compared with the time between successive collisions, and hence the cluster can be represented by an ensemble of systems which are free from all external forces. This may not

---

*However, there may be *some* such release of correlation in the macroscopic surface free energy. See Section 5.2.5 for further details.

be the case for a system whose temperature is close to the critical, because the total period of interactions between a cluster and other clusters and monomers becomes appreciable. Accordingly the classical states corresponding to this time period will contribute appreciably to the partition function of the entire system of supersaturated vapour. Therefore the present authors feel that one should view with caution extrapolations of results for the region of the critical point to the region of the triple point.

Although studies of nucleation near the critical point are arbitrarily excluded from this review, it may be helpful to point out a few recent studies in this interesting field. Accordingly we cite the works of Fisher[23] and Kiang et al.[24, 25].

5.2 GENERAL CLASSICAL PHASE-INTEGRAL APPROACH TO STATISTICAL MECHANICAL CONSIDERATIONS AND THE CAPILLARITY APPROXIMATION IN HOMOGENEOUS NUCLEATION FROM THE VAPOUR

5.2.1 General

In the usual experimental situation the temperature of the system is above the Debye temperature of the condensate. Accordingly the classical phase integral may be used to investigate the statistical mechanics of nucleation. Also, the momentum integral in the classical phase integral is separable from the configurational integral when the Cartesian coordinates are used as generalised coordinates and the momentum integral always contributes $(2\pi mkT)^{1/2}/h$ per degree of freedom. Therefore only the configurational integral need be considered in detail.

In this section, some of the classical phase-integral approaches will be reviewed. The present authors believe that some of these treatments are in error. However, all of these approaches are thought to be important enough to warrant discussion, and an attempt will be made to point out the principal weaknesses in the developments.

5.2.2 The treatment by Reiss and Katz for the equilibrium concentration of nuclei

Reiss and Katz[26] applied the classical phase integral to the calculation of the system partition function of clusters and the equilibrium concentration of clusters containing n molecules. The system partition function for N molecules in a volume V of vapour is given in terms of the classical phase integral by

$$Q = \frac{[\lambda_1 q_R^{(1)} q_v^{(1)}]^N}{N!} \int_V \cdots \int_V \exp[-U(r_1 \ldots r_N)/kT] \, dr_1 \ldots dr_N \quad (5.23)$$

where

$$\lambda_1 = (2\pi mkT)^{3/2}/h^3 \quad (5.24)$$

$q_R^{(1)}$ and $q_v^{(1)}$ are the molecular rotational and vibrational partition functions, respectively. $U(r_1 \ldots r_N)$ is the potential energy of the entire assembly and depends upon the coordinates of position r of all the molecules. They first evaluate the internal part of the configuration integral of each droplet in equation (5.23) by an integration in which the clusters or droplets are fixed in position and the same molecules are kept within the same drops. The droplet is defined by a sphere constructed about the centre of mass of the molecules assigned to it. For a droplet containing n molecules, the volume of the sphere is taken to be nv_1, in which v_1 is the molecular volume of the bulk liquid. The contribution of each cluster to the configuration integral of equation (5.23) is

$$z_n = \int \ldots \int \exp[-u(n)/kT] \, dr'_2 \ldots dr'_n \qquad (5.25)$$

in which $u(n)$ is the potential energy of interaction of the n molecules within the droplet and the elements $dr'_2 \ldots dr'_n$ refer to internal coordinates. Since the centre of mass remains fixed in the internal partition function z_n, only $n-1$ elements appear. They then integrate the centre of mass of each cluster over the free volume available to it. Following standard methods, the equilibrium concentration of clusters becomes

$$c(n) = [\lambda_1 q_R^{(1)} q_v^{(1)}]^n q_n \exp(n\mu/kT) \qquad (5.26)$$

where μ is the equilibrium chemical potential of monomer in the system and q_n is related to the contribution of a single cluster to the configuration integral by

$$q_n = \frac{z_n V}{n!} \qquad (5.27)$$

They next calculate the internal Helmholtz free energy $F_{int}^{(n)}$ of a single stationary droplet containing n molecules. Now a stationary droplet is not easy to define. In fact the key to understanding the difficulty in the whole series of analyses by Reiss and co-workers, including those which will be discussed in the following sections, lies in the concept of a stationary droplet. Reiss and Katz[26] introduced the concept of the stationary droplet of size n in which the free energy is supposedly given by $n\mu_1 + \sigma S(n)$, where μ_1 is the chemical potential of the bulk liquid and σ is the surface tension of the macroscopic surface. Their model of the stationary droplet of size n is the cluster of n molecules contained within a spherical boundary that is fixed in space. The whole reasoning of their analyses is based on the assumption that this model is appropriate to represent $n\mu_1 + \sigma S(n)$. This can of course be interpreted as a definition of σ, which implies that the free rotation and the translation with fairly large free volume are included in the macroscopic surface free energy σ. However, they have not attempted to justify this model. In fact, one of the essential points of the Lothe–Pound theory[18] is the consideration of what partition function should represent $n\mu_1 + \sigma S(n)$. Lothe and Pound[18, 22] proposed that $n\mu_1 + \sigma S(n)$ is to be associated with the partition function $q_{rep} \cdot q_{int}$, in which q_{rep} is the replacement partition function as defined in Section 5.1.3.1 and q_{int} is the partition function for the $3n-6$ internal degrees of freedom of a cluster in vapour. Inasmuch as Reiss and co-workers are concerned with the internal consistency of the Lothe–Pound theory, their stationary droplet model should supposedly lead to the partition function $q_{rep} \cdot q_{int}$. An obvious and important difficulty in constructing a model for the stationary droplet is the ascription of dual properties to it, because the same partition function of a stationary droplet must describe both the surface free energy of a cluster in vapour due to the factor q_{int} and the interaction of the cluster imbedded in bulk with its surroundings due to another factor q_{rep}. These requirements are of course mutually exclusive. Other difficulties which arise include the question of how much of each molecule must remain within the spherical boundary and the problem of exchange of molecules with the bulk. However, there are even more immediate troubles with the Reiss–Katz treatment. Accordingly, we will proceed under the assumption of the simple definition of a stationary droplet given above.

In terms of internal coordinates, the canonical partition function for the internal degrees of freedom of the stationary droplet is presumed to be

$$q_n^s = \frac{\lambda_1^{n-1}[q_R^{(1)}q_v^{(1)}]^n}{n!n^{\frac{3}{2}}} \int \cdots \int_{nv_1} \exp[-u(n)/kT] \, dr_2' \ldots dr_n' \quad (5.28)$$

$$= \frac{\lambda_1^{n-1}[q_R^{(1)}q_v^{(1)}]^n z_n}{n!n^{\frac{3}{2}}} = \frac{\lambda_1^{n-1}[q_R^{(1)}q_v^{(1)}]^n q_n}{n^{\frac{3}{2}}V}$$

The exponent $n-1$ and the $n^{\frac{3}{2}}$ result from having integrated over momenta corresponding to the internal system. In other words, the translational partition function of the drop as a whole has been properly removed. (We note here that the rotational partition function of the droplet as a whole in the vapour should also have been removed.) In this way, the internal Helmholtz free energy of the droplet becomes

$$F_{int}^{(n)} = -kT \ln \frac{\lambda_1^{n-1}[q_R^{(1)}q_v^{(1)}]^n q_n}{n^{\frac{3}{2}}V} \quad (5.29)$$

Reiss and Katz then make the crucial assumption that the free volume of a stationary droplet is the volume nv_1 of the drop itself and express the total Helmholtz free energy of the stationary droplet as

$$F^s(n) = F_{int}^{(n)} - kT \ln \frac{(2\pi nmkT)^{\frac{3}{2}} nv_1}{h^3}$$

$$= -kT \ln \frac{[\lambda_1 q_R^{(1)} q_v^{(1)}]^n q_n nv_1}{V} \quad (5.30)$$

Upon applying the standard capillarity approximation of nucleation theory, in which the free energy of a cluster is given in terms of the chemical potential of the bulk condensed phase and a quantity, A, proportional to the specific surface free energy of this bulk phase, one obtains

$$F^s(n) = \mu_1 n + An^{\frac{2}{3}} - npv_1 \quad (5.31)$$

Combining equations (5.30) and (5.31) they find

$$[\lambda_1 q_R^{(1)} q_v^{(1)}]^n q_n = \frac{V}{nv_1} \exp\left[-\frac{(\mu_1 n + An^{\frac{2}{3}})}{kT}\right] \quad (5.32)$$

Reiss and Katz then substitute (5.32) into (5.26) to obtain

$$c(n) = \frac{V}{nv_1} \exp(-\Delta F^0(n)/kT)$$

$$= N\left(\frac{kT}{pnv_1}\right) \exp(-\Delta F^0(n)/kT) \quad (5.33)$$

in which

$$\Delta F^0(n) = An^{2/3} + (\mu_1 - \mu)n \quad (5.34)$$

is the free energy of formation of a cluster in terms of the capillarity approximation. Inasmuch as (kT/pnv_1) is typically of the order of unity, they recover the 'classical' expression for equilibrium concentration of clusters. See for example equation (5.5).

The present authors[27, 28] believe that the result, equation (5.33), is in error, and that the immediate sources of error are:

(a) The assumption in equation (5.30) that the 'stationary' droplet is free to translate in the volume of the droplet itself. In the terminology of Section 5.1.3.1, the second term on the right-hand side of equation (5.30) is the replacement free energy arising from the three degrees of freedom for translation of the mathematical cluster in bulk liquid. Inasmuch as the assumed free volume nv_1 for translation of the cluster in bulk is numerically about the same as that for translation of the isolated cluster in the vapour, the two quantities then cancel and one returns to the 'classical' result, with respect to translation. One notes that their assumption requires an unusual model for the bulk liquid, i.e. a rarified gas of clusters of uniform but arbitrary size.

(b) The rotation of the isolated droplet has been completely ignored.

When these two difficulties are corrected, one formally returns[3, 27, 28] to the result of Lothe and Pound (equations (5.21) or (5.22)), in which the replacement partition function q_{rep} is for six degrees of freedom and still remains to be accurately evaluated for the case of liquids. Of course, the other difficulties of the model still remain, in particular the ascription of the dual properties of an isolated cluster and a cluster in bulk to the same configurational integral, e.g. in equations (5.28) or (5.30).

5.2.3 The treatment by Reiss, Katz and Cohen for the equilibrium concentration of nuclei

5.2.3.1 Outline of the derivation

Reiss, Katz and Cohen[29] defined a droplet in the vapour as a spherical body containing n molecules of volume v_n in which the geometrical centre of the imaginary spherical boundary and the centre of mass of the droplet are constrained to coincide.

One notes at the outset that such a constraint leads to a very serious undercounting in the configurational integral. In fact, much of the entropy of the droplet is thereby overlooked[30].

The canonical partition function of the droplet in vapour thus becomes

$$q_n = \frac{\lambda_1}{n!} \int_V d\mathbf{R} \times n^3 \int \ldots \int \exp[-u(\mathbf{r}'_1 \ldots \mathbf{r}'_{n-1})/kT] d\mathbf{r}'_1 \ldots d\mathbf{r}'_{n-1} \quad (5.35)$$

in terms of the centre-of-mass coordinates $\mathbf{r}'$ where $\mathbf{R}$ is the position vector of the centre of mass in the laboratory frame. $\lambda_1 = (2\pi mkT)^{3/2}/h^3$ and n^3 is the Jacobian of the transformation to centre-of-mass coordinates. Integration over $\mathbf{R}$ yields

$$q_n = \left[\frac{(2\pi nmkT)^{3/2} V}{h^3}\right] \frac{\lambda_1^{n-1} n^{3/2}}{n!}$$

$$\times \int_{v_n} \ldots \int \exp[-u(\mathbf{r}'_1 \ldots \mathbf{r}'_{n-1})/kT] d\mathbf{r}'_1 \ldots d\mathbf{r}'_{n-1} \quad (5.36)$$

in which the translational partition function of the free cluster has been factorised, but not the partition function for free rotation of the cluster.

They then define a fixed droplet as one in which the imaginary spherical boundary with volume v_n is fixed in space *but without the constraint that the geometric centre and centre of mass must coincide.*

As mentioned in Section 5.2.2, it is difficult to define a stationary droplet, especially in relation to the bulk liquid, and some important considerations will be described in the following. However, these are not necessary for the point of the present discussion. As can be seen from the preceding comments, the point of the present discussion is that an improper constraint is being applied to the droplet in vapour but not to the stationary droplet.

Under this definition of the stationary droplet, the canonical partition function becomes

$$q_n^s = \frac{\lambda_1^n n^3}{n!} \int_{v_n} \ldots \int \exp[-u(\mathbf{r}'_1 \ldots \mathbf{r}'_{n-1})/kT] d\mathbf{r}'_1 \ldots d\mathbf{r}'_{n-1} d\mathbf{R} \quad (5.37)$$

where the integrals over $\mathbf{r}'$ are of course coupled with the integral over $\mathbf{R}$. Equation (5.37) may be rewritten as

$$q_n^s = \frac{\lambda_1^n}{n!} \int_{v_n} z(\mathbf{R}) d\mathbf{R} \quad (5.38)$$

where the configurational integral with the centre of mass fixed at $\mathbf{R}$ is denoted by $(\mathbf{R})$. Also, equation (5.36) for the droplet in vapour may be rewritten as

$$q_n = \frac{\lambda_1^n}{n!} \int_V z(0) d\mathbf{R} = \frac{\lambda_1^n}{n!} V z(0) \quad (5.39)$$

in which $z(0)$ is the configurational integral with the centre of mass constrained to remain at the geometrical centre of the sphere. Accordingly, from (5.38) and (5.39) one finds

$$q_n = q_n^s V [z(0) / \int_{v_n} z(\mathbf{R}) d\mathbf{R}] \quad (5.40)$$

$$= q_n^s V P(0)$$

where $P(0)$ is the probability density measuring the chance that the centre of mass will be found at the origin of the stationary droplet. Substitution into the standard equilibrium relationship

$$c(n) = q_n \exp(n\mu/kT) \tag{5.41}$$

gives

$$c(n) = VP(0)q_n^s \exp(n\mu/kT) \tag{5.42}$$

In relation to the capillarity approximation, the Helmholtz free energy of the stationary droplet is

$$F^s(n) = -kT \ln q_n^s = n\mu_1 + An^{2/3} - npv_1$$
$$\approx n\mu_1 + An^{2/3} \tag{5.43}$$

Substitution of equation (5.43) into (5.42) yields

$$c(n) = [VP(0)/N][N \exp(-\Delta F^0/kT)] \tag{5.44}$$

where ΔF^0 is the Helmholtz free energy of formation of a droplet in the 'classical' capillarity approximation, e.g. equation (5.1).

$$P(0) \equiv z(0) / \int_{v_n} z(\mathbf{R}) \, d\mathbf{R} \tag{5.45}$$

is the reciprocal of the volume swept out by the centre of mass of the stationary droplet and can be estimated from the standard deviation of the fluctuation in the three Cartesian coordinates of the centre of mass. It is found that this volume $P(0)^{-1}$ is an appreciable fraction of the free volume of a molecule in the vapour V/N. Thus one returns, approximately, to the 'classical' relationship (equation (5.5)).

5.2.3.2 Outline of an alternative derivation

The present authors believe that the above result is incorrect because of three principal difficulties:

(a) An improper constraint in which the centre of mass is forced to coincide with the geometric centre is applied to the droplet in vapour but not to the stationary droplet. This gives rise to a large free volume $P(0)^{-1}$ which nearly cancels the translational free volume of the droplet in the vapour.

(b) Rotation is entirely ignored in that it is tacitly assumed to be the same for the free droplet as for the stationary droplet, equation (5.40). Alternatively, the rotation is assumed to be a contribution to the macroscopic surface tension, equation (5.43).

(c) Also the ascription of dual properties to the stationary cluster remains as a large problem in the present approach. Thus the configuration integral of (5.37) is falsely presumed to describe the surface free energy of an isolated cluster as well as the interactions of an imbedded cluster with its surroundings in bulk liquid. As noted above, these requirements are mutually exclusive.

It is difficult to revise the present model for precise consideration of the effect of rotation. However, it is instructive to re-derive the effect of translation without applying improper constraints. It will emerge that a much smaller volume v_{cell} replaces $P(0)^{-1}$ and the effect of free translation of the droplet in vapour is much larger than the result of Reiss, Katz and Cohen[29].

Nishioka et al.[30] take the same model for the droplet as Reiss, Katz and Cohen, i.e. a spherical body containing n molecules, but without the false constraint that the centre of mass and the geometrical centre must coincide for the droplet in the vapour. Further, Nishioka et al. attempt to relate the model for the stationary droplet to a cluster in bulk, a relationship which is absolutely necessary for investigation of the capillarity approximation.

The canonical partition function of this *droplet in bulk liquid*, which is *artificially and incorrectly* given the property of macroscopic surface free energy in accord with the previous treatment*, is

$$q_n^s = \frac{\lambda_1^n}{n!} \int_{nv_1} \cdots \int \exp[-u(\mathbf{r}_1 \ldots \mathbf{r}_n)/kT] \, d\mathbf{r}_1 \ldots d\mathbf{r}_n$$

$$= \frac{\lambda_1^n n^3}{n!} \int_{nv_1} dS \int_{(nv_1)S} \cdots \int \exp[-u(\mathbf{r}'_1 \ldots \mathbf{r}'_{n-1})/kT] \, d\mathbf{r}'_1 \ldots d\mathbf{r}'_{n-1} \quad (5.46)$$

where S is the vector connecting the geometric centre of the sphere to the centre of mass and $\mathbf{r}'$ are position vectors relative to the centre of mass. As before, the free energy of the stationary droplet is given in terms of the capillarity approximation by

$$F^s(n) = -kT \ln q_n^s = n\mu_1 + \sigma A \quad (5.47)$$

In order to relate q_n to μ_1 and σ, the relationship between q_n and q_n^s must be obtained. Now

$$q_n = \frac{\lambda_1^n n^3}{n!} \int_V d\mathbf{R} \int_{nv_1} \cdots \int \exp[-u(\mathbf{r}'_1 \ldots \mathbf{r}'_{n-1})/kT] \, d\mathbf{r}'_1 \ldots d\mathbf{r}'_{n-1}$$

$$= \frac{\lambda_1^n n^3 V}{n!} \int_{nv_1} \cdots \int \exp[-u(\mathbf{r}'_1 \ldots \mathbf{r}'_{n-1})/kT] \, d\mathbf{r}'_1 \ldots d\mathbf{r}'_{n-1} \quad (5.48)$$

and

$$\frac{q_n}{q_n^s} = \frac{V \int_{nv_1} \cdots \int \exp[-u(\mathbf{r}'_1 \ldots \mathbf{r}'_{n-1})/kT] \, d\mathbf{r}'_1 \ldots d\mathbf{r}'_{n-1}}{\int_{nv_1} dS \int_{(nv_1)S} \cdots \int \exp[-u(\mathbf{r}'_1 \ldots \mathbf{r}'_{n-1})/kT] \, d\mathbf{r}'_1 \ldots d\mathbf{r}'_{n-1}} \quad (5.49)$$

Next consider the configurational integral in the numerator of equation (5.49). We denote the vector which connects the position O of the geometric centre of the sphere to the centre of mass P by $\mathbf{L}$ as shown in Figure 5.2. The centre of mass P is fixed in space, but the geometric centre of the integration volume O may be located at any position as long as the centre of mass is inside the sphere. Let us denote by $y(n, \mathbf{L})$ the configurational integral which is obtained under the constraint that the centre of mass is at $\mathbf{R}$ and the

---

*However, see Section 5.2.5.

geometric centre of the integration volume is represented by L

$$y(n, L) = \int_{(nv_1)_L} \cdots \int \exp[-u(r'_1 \ldots r'_{n-1})/kT]\, dr'_1 \ldots dr'_{n-1} \quad (5.50)$$

The configurational integral of the numerator in equation (5.49) is the sum of $y(n, L)$ for all the allowed values of L. It must be noted, however, that exactly the same configurational state may be included in many $y(n, L)$ with different L and of course all the configurational states have to be counted once and only once.

In order to make the argument clear, let us consider the molecules to be hard spheres of diameter d and the boundary of the integration volume to be a rigid wall. The hard spheres are considered to possess attractive forces

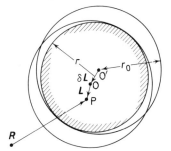

Figure 5.2 A cluster in the vapour. P is the centre of mass and O is the geometric centre of the configuration volume

between them to form liquid. If configurations within the integration sphere with the centre at O are such that the outer envelope of the configurations is given by the shaded spherical region in Figure 5.2, then those same configurations can also be accommodated within spheres with the centre shifted less than δL. L has to be shifted by $|\delta L|$ before the first type of configurations becomes impossible and new configurations must be involved. Obviously, this minimum shift of L depends on the type of configuration. However, let us consider only the spherical envelope and use the average δL for all the configurations as an approximation. This suggests a method for approximating δL in which δL is taken to be the difference between the diameter of the integration sphere and the mean diameter of the sphere which encloses n molecules as determined from averaging only the negative deviation. Then the configurational integral in the numerator of equation (5.49) is given approximately by

$$\sum_L y(n, L) = \sum_L \frac{1}{v_{\text{cell}}(L)} y(n, L) v_{\text{cell}}(L)$$

$$\approx \frac{1}{v_{\text{cell}}} \sum_L y(n, L) v_{\text{cell}}(L)$$

$$\approx \frac{1}{v_{\text{cell}}} \int_{nv_1} y(n, L)\, dL \quad (5.51)$$

in which

$$v_{\text{cell}} = \frac{4\pi}{3} (\overline{\delta L})^3 \quad (5.52)$$

Now $\int_{nv_1} y(n, L) \, dL$ and the denominator of equation (5.49) have exactly the same value and hence equation (5.49) may be rewritten as

$$\frac{q_n}{q_n^\hat{}} \approx \frac{V}{v_{\text{cell}}} \tag{5.53}$$

This result contains an important physical implication. It shows that the translational motion of a stationary droplet, which is replaced by translational motion in volume V in the vapour, is the motion within a free volume v_{cell} instead of a fluctuational motion of the centre of mass of the droplet. This is a result of the fact that all the configurational states with a fixed spherical boundary can also be realised in a cluster in vapour with fixed centre of mass because both of these involve all the same relative positions of the centre of mass to the geometric centre of the sphere.

Thus from equations (5.47) and (5.53)

$$q_n = \frac{V}{v_{\text{cell}}} \exp\left[-(n\mu_1 + \sigma A)/kT\right] \tag{5.54}$$

and from equation (5.41)

$$c(n) = \frac{V}{v_{\text{cell}} c(1)} \cdot c(1) \exp\left\{-[n(\mu_1 - \mu) + \sigma A]/kT\right\} \tag{5.55}$$

Hence the translational correction to the 'classical' nucleation theory, as given by equation (5.5), becomes

$$\frac{V}{v_{\text{cell}} c(1)} = \frac{v_g}{v_{\text{cell}}} \tag{5.56}$$

which, from a crude estimate using an ideal gas model, is $c.\ 5 \times 10^8$. This is much higher than the Reiss–Katz–Cohen estimate of 10^3–10^6. The difference is mainly due to the fact that they undercounted the configuration integral in q_n by restricting the centre of mass of a cluster in the vapour to coincide with the geometric centre of the bounding sphere. One notes that the effect of rotation has still been completely ignored in this treatment. Also the inconsistency arising from the ascription of a dual nature to equation (5.46) remains unresolved, except that a more realistic model for the translational part of q_{rep} is employed in the present analysis. As mentioned above, the single configurational integral of equation (5.46) purports to describe both the surface free energy of an isolated cluster and the interactions of a mathematical cluster in bulk liquid with its surroundings. These requirements are, of course, incompatible.

5.2.4 The Reiss–Kikuchi treatment for the equilibrium concentration of nuclei

5.2.4.1 The Reiss treatment

In a later study, Reiss[31] used essentially the same model discussed above for the stationary droplet. He considers a stationary spherical 'mathematical container' which possesses the property that the potential energy of a molecule becomes infinite outside the container. Its radius is supposed to be about the radius of the Gibbs dividing surface which is chosen to be located at the position where the superficial density becomes that of the vapour.

Again, this model has all the difficulties noted in connection with the above treatments. Namely:

(a) There is a problem in choosing the position of the boundary in order to properly describe the interactions of the droplet in bulk with the surrounding molecules.

(b) There is the problem of the force on the 'mathematical container' which arises from preventing exchange of molecules in the imbedded cluster with the surrounding bulk liquid phase. One notes that exactly the same n molecules are assumed to remain in the 'mathematical container' and comprise the cluster. Actually, there is an exchange of molecules. Hence to prevent this exchange, a force must be exerted on the 'mathematical container', whereupon it ceases to be only mathematical.

(c) More importantly, the stationary droplet so defined, if it is to be related to bulk liquid, possesses two mutually contradictory characteristics. It possesses an internal partition function which has the characteristic of the free surface for the droplet and an external partition function which has the characteristic that the drop is surrounded by molecules to make the droplet a part of the homogeneous bulk liquid. A model which satisfies both of these mutually exclusive requirements is impossible, unless of course the correct model for bulk liquid is an attenuated gas of droplets. However, if the aim is only to relate the Reiss model for a stationary droplet to the properties of a droplet in vapour, ignoring its relation to the bulk liquid, then the model does not have this difficulty[32].

Difficulty (a) has been discussed in Section 5.2.3.2 and difficulty (b) can be considered, but there is no remedy for difficulty (c) within the model. Nevertheless, there are even more immediate troubles with the present treatment, and we proceed under the present model in order to illustrate these[32].

Again, let us denote the canonical partition function of the stationary droplet by q_n^s and the configurational integral by

$$z_n^s = \int_{v_n} \ldots \int \exp\left[-u(\mathbf{r}_1 \ldots \mathbf{r}_n)/kT\right] d\mathbf{r}_1 \ldots d\mathbf{r}_n \tag{5.57}$$

where

$$v_n = \tfrac{4}{3}\pi r_c(n)^3 \tag{5.58}$$

Reiss then relates z_n^s to the configurational integral, which is a part of the canonical partition function q_n for the droplet in vapour, as follows. When the 'mathematical container' is shifted by an infinitesimal distance ds, the new contribution to the configurational integral which is not included in z_n is given by

$$z_n^s - z_n^{s'} \tag{5.59}$$

in which $z_n^{s'}$ is the configurational integral obtained by confining all the molecules within the volume of overlap of the shifted and unshifted containers.

From the relationship between Helmholtz free energy f_n and the canonical partition function,

$$\exp\left[-(f_n' - f_n)/kT\right] = z_n^{s'}/z_n^s \tag{5.60}$$

$(f_n' - f_n)$ can be obtained from

$$f_n' - f_n = \frac{\partial f_n}{\partial v_n}(-dv) = p_n\, dv \tag{5.61}$$

where dv is the difference between the volume of the container and the volume of the overlapped region and p_n is the pressure required to compress the droplet from the volume v_n to $v_n - dv$. Combining equations (5.60) and (5.61), one finds

$$z_n^{s'} = z_n^s \exp(-p_n\, dv/kT) \tag{5.62}$$

Hence one finally evaluates equation 5.59 as

$$\begin{aligned} z_n^s - z_n^{s'} &= z_n^s[1 - \exp(-p_n\, dv/kT)] \\ &= (z_n^s/kT)p_n\, dv \end{aligned} \tag{5.63}$$

z_n can now be related to z_n^s by

$$z_n = z_n^s + \frac{z_n^s}{kT}\int_V p_n\, dv \tag{5.64}$$

where V is the volume of the supersaturated vapour.

There is no problem in this method of relating z_n to z_n^s so far, except the point concerning the role of the surface tension to be discussed in this section in relation to the Kikuchi paper[33].

However, a serious problem emerges when p_n is equated to the vapour pressure in unstable equilibrium with the cluster of size n. Let us consider the nature of p_n in the following. p_n is defined as the pressure exerted by the mathematical wall of the container, as seen in equation (5.61). Since the centre of mass of the droplet is not fixed relative to the container, as emphasised by Reiss[31], the wall must sense the pressure due to the fluctuation of the centre of mass of the droplet as well as the pressure which is taken into account by Reiss. This can also be clearly seen in the expression for z_n^s given by equation (5.57), in which the states corresponding to the 'collisions' of the droplet against the wall are of course included. Thus, if one sets $p_n = p$, this implies that the volume v_n of the container should be at least as large as the volume v_g per molecule in the vapour whose vapour pressure is p. A typical numerical value of v_g for water vapour under experimental conditions is 4×10^{-19} cm^3 and this is about two orders of magnitude larger than the volume of the cluster for the usual case. Hence it is apparent that the assertion of $p_n = p$ is inconsistent with the volume of the container which is defined by equation (5.58). Now if one is to use the device of a 'mathematical container' or wall, one must duly consider all forces which the presence of this wall creates. Therefore, in order to proceed further under the present model, one must calculate p_n. It seems to be a difficult problem to evaluate the correct p_n as a function of the volume of the con-

HOMOGENEOUS NUCLEATION FROM THE VAPOUR

tainer when the volume is as small as v_n given by equation (5.58). However, it will be many orders of magnitude larger than p, as discussed in the following.

5.2.4.2 Kikuchi's extension of the Reiss treatment

Kikuchi[33] used the same model to extend the Reiss treatment by taking into consideration the role of surface tension in evaluating $(f_n' - f_n)$. In so doing, Kikuchi points out an interesting difference between one- and three-dimensional shifting in the counting process for relating z_n^s to z_n. When the isolated stationary droplet is constructed in such a way that a small deformation of the mathematical wall accompanies the same deformation of the droplet inside the wall, Kikuchi demonstrated that the three-dimensional aspect of the shifting of the mathematical wall requires consideration of the work required to push in the spherical surface against the surface tension in addition to the work to compress the container which was considered by Reiss. Denoting the pressure required to compress the 'mathematical wall' by p and the surface tension by σ, the total work is given by

$$\left(p_n + \frac{2\sigma}{r_c(n)}\right)\Delta v_n \tag{5.65}$$

where $r_c(n)$ is the radius of the 'mathematical container' and Δv_n is the volume change of the container which arises in consideration of the three-dimensional shift. The present authors agree to Kikuchi's counting process which leads to equation (5.65). However, following Reiss[31], Kikuchi incorrectly equates p_n to the vapour pressure p in unstable equilibrium with the cluster of size n. As discussed above, p_n must contain the pressure due to the collisions of the droplet against the 'mathematical wall' and as a result of this p_n will be many orders of magnitude higher than p. Nishioka et al.[32] estimate a pressure p_n of c. 10^4 atm, which is higher than $2\sigma/r_c(n) \approx 10^3$ atm.

Rigorously speaking, the present theory relates z_n^s to z_n correctly only when the volume v_n of the isolated stationary droplet is taken to be as large as v_g. When $v_n \ll v_g$, the pressure inside the droplet within the isolated stationary container is higher than the pressure inside the cluster in the vapour due to collisions of the droplet against the 'mathematical wall'. Therefore the present theory does not relate z_n^s to z_n correctly even if the correct p_n is used. When $v_n \approx v_g$, the result is correct but trivial.

5.2.4.3 Summary

From the above discussion, the Reiss–Kikuchi model and method is fraught with seemingly insuperable difficulties, the more important of which are:

(a) The dual nature of the stationary droplet model, which must contain the mutually contradictory characteristics of having a free surface while still being a part of the bulk liquid if the statistical mechanics of the capillarity approximation are to be investigated. If the statistical mechanics of the capillarity approximation are not to be investigated, then the treatment becomes merely an exercise in relating the model of the stationary droplet to the properties of a droplet in the vapour.

(b) Also, the model completely overlooks the effect of free rotation of the droplet in the vapour. In other words the 'stationary' droplet is allowed free rotation, which the cluster imbedded in bulk does not have.

(c) The pressure of the 'stationary' droplet on its 'mathematical container' has been totally neglected, and the necessary correction is difficult to estimate.

(d) Even if this huge pressure could be properly estimated, the theory would still not describe the properties of a droplet in the vapour, which of course is not under this high and artificial pressure.

It would appear that no repair of this model or procedure is possible, short of doing away with the model itself for an isolated stationary droplet in its 'mathematical container'. In so doing one must return to the concept of a mathematical cluster in the bulk liquid. In other words, one eliminates the artificial and unnecessary intermediate state of a stationary droplet in a container and considers only the actual initial and final states. Those are the cluster in bulk and the isolated droplet in the vapour, respectively. In this way, one formally returns to the Lothe–Pound result (equation (5.21)). It is perhaps worth mentioning that the high pressure due to the collisions of the droplet against the 'mathematical wall' arises from the fictitious nature of the 'mathematical wall', which is perfectly repelling and non-attracting. This difficulty does not arise in the Lothe–Pound theory, because they consider a part of the bulk liquid to obtain their replacement partition function and in this case a cohesive force exists between the cluster in bulk and the surroundings.

5.2.5 The Kikuchi treatment for the equilibrium concentration of nuclei

Kikuchi[34] uses the model for the cluster imbedded in bulk liquid, which was originally introduced by Dunning[20, 21], to relate the properties of an isolated cluster in the vapour to the bulk liquid. He thus avoids the almost insuperable difficulties with the model for the stationary droplet discussed above. Namely, it is no longer necessary to describe in one configurational integral the mutually contradictory characteristics of an isolated droplet with a free surface and a droplet imbedded in bulk liquid. Also the problem of the artificial high pressure in the stationary container is avoided.

Specifically, Kikuchi[34] defines the cluster in bulk as being comprised of n molecules contained in a fixed mathematical boundary of volume nv_1. The centre of mass, but not the entirety, of each of the n molecules is constrained to lie within the mathematical boundary. Similarly, the centre of mass but not the entirety of each of the surrounding large number of molecules N is constrained to remain outside of the mathematical boundary. As Kikuchi correctly points out, the shape of the imbedded droplet is unimportant, but a spherical shape is chosen for simplicity.

The thermodynamic properties of the whole bulk system comprised of

$n+N$ molecules is related to the canonical partition function by

$$\exp\{[-(n+N)\mu_1-\sigma A_{n+N}]/kT\} =$$

$$\frac{\lambda_1^{n+N}}{(n+N)!}\int_V\cdots\int\exp[-u(n+N)/kT]\,d\mathbf{r}_1\ldots d\mathbf{r}_{n+N} \quad (5.66)$$

where A_{n+N} is the outer area of the bulk system and the $\mathbf{r}$ are Cartesian coordinates of the molecules. The integration is of course independent of the choice of molecule number one and thus may be taken over the entire volume

$$V = (n+N)v_1 \quad (5.67)$$

to give

$$\exp\{[-(n+N)\mu_1-\sigma A_{n+N}]/kT\} =$$

$$\frac{\lambda_1^{n+N}v_1}{(n+N-1)!}\int_V\cdots\int\exp[-u(n+N)/kT]\,d\mathbf{r}_2\ldots d\mathbf{r}_{n+N} \quad (5.68)$$

where the $\mathbf{r}$ are now referred to the position of molecule number one. This operation is called the molecular volume theorem[34] and is obviously correct. Now the centre of mass of molecule number one is considered to be contained within the mathematical boundary, and the partition functions of the two regions containing n and N molecules, respectively, are separated to yield

$$\exp\{[-(n+N)\mu_1-\sigma A_{n+N}]/kT\} = \lambda_1 v_1$$

$$\frac{\lambda_1^{n-1}}{(n-1)!}\int_{nv_1}\cdots\int\exp[-u(n)/kT]\,d\mathbf{r}_2\ldots d\mathbf{r}_n$$

$$\frac{\lambda_1^N}{N!}\int_{V-nv_1}\cdots\int\exp\{-[u(n,N)+u(N)]/kT\}\,d\mathbf{r}_{n+1}\ldots d\mathbf{r}_N \quad (5.69)$$

in which $u(n,N)$ is the potential energy of interaction between the imbedded cluster and its surroundings.

Next the crucial assumption is made that the surface free energy $\sigma A_n \equiv \hat{\sigma}(n)$ of the droplet can be defined as

$$\exp[\hat{\sigma}(n)/kT] = \frac{\lambda_1^N}{N!}\int_{V-nv_1}\cdots\int\exp\{-[u(n,N)+u(N)]/kT\}\,d\mathbf{r}_{n+1}\ldots d\mathbf{r}_N/$$

$$\exp[(-N\mu_1-\sigma A_{n+N})/kT] \quad (5.70)$$

The thought process to which this definition corresponds is the removal of the surrounding liquid from the imbedded droplet whose mathematical boundary is fixed in the laboratory reference frame. One notes that no artificial high pressure arises in the resulting isolated droplet by this operation, because all configurational integrals are written for the imbedded droplet.

Substitution of equation (5.70) into (5.69) yields

$$\exp(-n\mu_1/kT) = \lambda_1 v_1\frac{\lambda_1^{n-1}}{(n-1)!}\int_{nv_1}\cdots\int\exp[-u(n)+\hat{\sigma}(n)]/kT\,d\mathbf{r}_2\ldots d\mathbf{r}_n$$

$$(5.71)$$

Making the reasonable assumption that $\hat{\sigma}(n)$ depends but little on the coordinates r, one obtains

$$\exp\left[-(n\mu_1+\sigma A_n)/kT\right] = \lambda_1 v_1 \frac{\lambda_1^{n-1}}{(n-1)!} \int_{nv_1}\cdots\int \exp\left[-u(n)/kT\right] d\mathbf{r}_2 \ldots d\mathbf{r}_n. \tag{5.72}$$

The canonical partition function of the droplet in the vapour may be expressed as

$$q_n = \lambda_1 V \frac{\lambda_1^{n-1}}{(n-1)!} \int_{nv_1}\cdots\int \exp\left[-u(n)/kT\right] d\mathbf{r}_2 \ldots d\mathbf{r}_n \tag{5.73}$$

where both the integration volume and molecule 1 are fixed in space. Thus the integral in equation (5.73) is different from that in equation (5.57). Hence by comparison with equation (5.72)

$$\exp\left[-(n\mu_1+\sigma A_n)/kT\right] = \frac{v_1}{V} q_n \tag{5.74}$$

Substitution into the standard equilibrium expression (equation (5.41)) gives for the equilibrium concentration of clusters of size n in the vapour

$$c(n) = \frac{V}{v_1} \exp\left\{-[n(\mu_1-\mu)+\sigma A_n]/kT\right\} \tag{5.75}$$

or

$$c(n) = \left(\frac{v_g}{v_1}\right) c(1) \exp\left(-\Delta F^0/kT\right) \tag{5.76}$$

where $v_g = V/c(1)$ is the molecular volume of monomer in the supersaturated vapour and v_1 is the molecular volume of bulk liquid. Thus by comparison with equation (5.5), Kikuchi's factor of correction to 'classical' nucleation theory is v_g/v_1, which is only c. 10^4 in typical experimental cases.

Kikuchi's paper[34] suggests an interesting contribution to the macroscopic surface free energy of liquids that did not receive much attention in the original Lothe–Pound treatment. This is the contribution from the release of correlation. In the Lothe–Pound theory, the entire translation and rotation of the cluster in the vapour is counted in the pre-exponential and none in the macroscopic surface free energy. It may be that some of this should be counted in the macroscopic surface free energy to simulate the release of correlation upon formation of macroscopic surface. In the Kikuchi theory[34], the entire translation and rotation of the cluster in the vapour is counted as release of correlation and thus included in the macroscopic surface free energy. The present authors believe that Kikuchi[34] overestimates this effect, in particular with regard to translation where the original Lothe–Pound estimate seems more realistic. The rotational part is more difficult to decide, and more work on just this point is needed for a definite conclusion, at least in the case of liquids. Now the main difference between the liquid and crystal cases is not so much in the bulk term as in the surface term. In the crystalline case there is obviously no pronounced release of correlation upon formation of macroscopic surface, and in this case the Lothe–Pound treatment is expected to be quite good without modification. What modifications may be needed for the liquid case still remain to be settled conclusively.

5.2.6 Summary

Most of the classical phase integral treatments of the capillarity approximation for liquids (or crystals) are seriously in error. The treatments of Reiss and Katz[26], Reiss, Katz and Cohen[29], Reiss[31] and Kikuchi[33] have as a primary difficulty the model of the stationary droplet, in which a single configurational integral must describe the mutually contradictory properties of the surface of an isolated cluster and the interactions with the surroundings of a mathematical cluster imbedded in the bulk liquid phase.

Another treatment by Kikuchi[34] does not use this model or have this difficulty. Rather it uses the model of an imbedded mathematical cluster in the bulk liquid and avoids the complications which arise from the false and unnecessary constraint of a stationary droplet. This latter approach has great promise, but in its original form[34] it probably overestimates the release of correlation contained in the macroscopic surface free energy. In fact, the entire release of correlation for both translation and rotation is included in the description of the macroscopic surface free energy in his theory. The present authors believe that this is an overestimate, particularly in regard to translation. Nevertheless, Kikuchi[34] has developed a theory in which the correlation problem in the macroscopic surface free energy is considered, and this may be a good start on a real analysis of how much of the release of correlation on forming clusters is counted in the surface tension of liquids. More work is needed to conclusively decide these important points.

In general, excluding the above paper by Kikuchi[34], these treatments have led to either inclusion of free translation and rotation of the cluster in the vapour in the macroscopic surface free energy of the bulk liquid or a model for the bulk liquid in which the liquid is comprised of an attenuated gas of clusters of arbitrary size, with no justification as to why this should be a reasonable model. The results of these treatments give essentially the 'classical' relationship, equation (5.5).

Of course, there are a number of other sources of difficulty. One of these is the lack of a good (simple) model for liquids. Inasmuch as the results should be, in principle, the same for liquids or crystals, the results for calculations on crystal models will be described in Section 5.3. However, the problem of the release of correlation upon forming a free surface is not fully illuminated by crystal calculations.

5.3 CLASSICAL PHASE-INTEGRAL APPROACHES USING CRYSTAL MODELS AND NORMAL-MODE COMPUTER CALCULATIONS

5.3.1 General

We saw that the general classical phase-integral approaches of Section 5.2 should in principle be valid for both liquids and crystals but that various difficulties may have precluded correct results. Prominent among these difficulties is the absence of a specific model for the cluster in these general approaches. There would appear to be as yet no sufficiently simple model for the liquid, but there are suitable models for the solid. In the present section, the classical phase integral is applied to lattice models in an effort to establish the nature of the required statistical mechanical corrections to the capillarity approximation for homogeneous nucleation from the vapour of either crystallites or droplets. Then the numerical results of normal-mode and Einstein computer calculations of the relevant quantities for solids will be discussed. The results of the present section are thought to be less controversial than those of Section 5.2. It is believed, as briefly discussed in Section 5.1 and also to be described in the following, that the qualitative features of the replacement partition function are the same for both liquid and crystal, i.e. the replacement partition function is due to the six translational and rotational vibrations of the imbedded cluster in bulk for which the relative positions of the molecules remain fixed. However, one notes that the actual *numerical* values of the replacement partition function may prove to be different for liquids. Also, the macroscopic surface entropy of the liquid may include additional contributions, as suggested in Section 5.2.5.

The replacement partition function was briefly described in the Introduction. It is the partition function for the free energy that the isolated cluster in the vapour does not have because it is not a part of the bulk phase. Lothe and Pound[22] pointed out that all *internal* configurations accessible to the mathematical cluster imbedded in the bulk are also accessible to the isolated cluster in the vapour. Thus the only modes of motion accessible to the cluster in bulk that are not also accessible to the isolated cluster in the vapour are the translational and rotational vibrations of the cluster in bulk *for which the relative internal coordinates remain fixed*. The motions in these six degrees of freedom are of course replaced by free translation and rotation of the isolated cluster in the vapour. The replacement partition function thus appears to be well defined and is a property of the bulk phase. Its important aspect in relation to nucleation theory is that it is an absolutely essential contribution to statistical mechanical corrections of the capillarity approximation. Lattice calculations are desirable in order to establish the nature of the replacement partition function, and some of these calculations will be described in the following.

5.3.2 The replacement partition function in a linear chain

It seemed desirable to check the above definition and properties of the replacement partition function by a precise calculation using the model of a linear chain. This was done by Abraham and Canosa[42] and independently by Lothe and Pound[43]. In these treatments, the infinite chain (bulk) is mathematically divided into equal segments (clusters in bulk), and orthogonal transformations are introduced that separate the partition function for the system into centre-of-mass parts (replacement terms) and internal parts (free cluster parts). The results are finally tested by combining the replacement term and the internal term to give the known free energy per segment in the infinite chain.

The potential energy of a long linear chain of N spring-connected masses with free ends may be expressed in terms of the harmonic approximation by the position coordinates and the spring constant β. By means of harmonic orthogonal transformations, this potential energy may be diagonalised, and the configurational integral is readily evaluated to give

$$Z = \prod_{j=1}^{N-1} (2\pi kT/m\omega_j^2)^{\frac{1}{2}} \tag{5.77}$$

where

$$\omega_j = 2(\beta/m)^{\frac{1}{2}} \sin(j\pi/2N)$$

are the normal frequencies of the chain. The vibrational free energy per atom in the infinite chain is readily calculated to obtain the standard result

$$f_\infty = -kT \ln(kT/\hbar\omega_D) \tag{5.78}$$

in which

$$\omega_D = (\beta/m)^{\frac{1}{2}} \tag{5.79}$$

is the Debye frequency.

Next, the free energy of the long chain is alternatively expressed by mathematically subdividing it into N/n equal segments of n atoms each and writing the configurational integral as the product of centre-of-mass parts and internal parts for each of the segments

$$Z = Z_Y Z_\eta^{N/n} \tag{5.80}$$

in which the centre-of-mass part for all the N/n segments is given by

$$Z_Y = n^{N/2n} \int_{-\infty}^{+\infty} \ldots \int \exp[-u(Y)/kT] \, dY_1 \ldots dY_{N/n} \tag{5.81}$$

where the Y are centre-of-mass coordinates. The internal part for each segment is

$$Z_\eta = \int_{-\infty}^{+\infty} \ldots \int \exp[-u(\eta)/kT] \, d\eta_1 \ldots d\eta_{n-1} \tag{5.82}$$

Harmonic orthogonal transformation and integration of both parts yields[42, 43]

$$F_n^{\text{int}} = -(n-1)kT \ln \frac{kT}{\hbar\omega_D} + kT \ln \sqrt{n} \tag{5.83}$$

for the internal free energy of each segment of n atoms in the linear chain and

$$F_n^{\text{ext}} = F_n^{\text{rep}} = -kT \ln(\sqrt{n}kT/\hbar\omega_D) \tag{5.84}$$

for the external free energy of each segment of n atoms in the linear chain. This latter quantity is the free energy a segment does not have when it is not a part of the infinite chain. Therefore it is the replacement free energy F_n^{rep} and $\sqrt{n}kT/\hbar\omega_D$ is the replacement partition function.

The correctness of these results may be checked by noting that

$$nf_\infty = F_n^{\text{int}} + F_n^{\text{rep}} \tag{5.85}$$

In other words, the internal parts and replacement parts combine to give the bulk free energy exactly. The importance of these results is that they tend to confirm the view that the replacement partition function describes the motions of the imbedded cluster *for which the relative internal coordinates remain fixed.*

However, there are ambiguities relating to application of the present results to a three-dimensional system. Namely, the 'surface' entropy of the isolated segment is contained in F_n^{int}, because equation (5.83) is the result for a segment with free ends. Of course, this effect is negligible for the longer segments because the 'surface' (the two end atoms) does not increase with segment length. Nevertheless, it would be desirable to repeat the above calculation for the more realistic but more complicated case of a three-dimensional crystal, and this is done in the following.

5.3.3 The replacement partition function in crystals

Nishioka et al.[44] have derived the replacement partition function and defined the surface free energy for crystallites by means of a classical phase-integral approach. The method involves the harmonic approximation and transformation to normal coordinates and is, in principle, straightforward. However, the mathematical details are somewhat tedious and therefore only the important steps in the development and the salient results will be outlined here.

Consider a monatomic bulk crystal and divide it into small clusters of the same size, containing n atoms, and shape by mathematical boundaries as shown in Figure 5.3. The lattice sites $\{a_i\}$ of the bulk crystal are defined

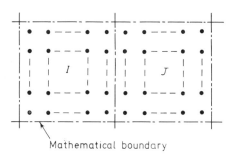

Figure 5.3 Clusters in the bulk crystal

by minimising the potential energy of interaction. Attention is now focused on the Ith cluster and the translating and rotating local Cartesian frame Σ_I is introduced such that there is no linear momentum and approximately no angular momentum of the Ith cluster relative to Σ_I. This provides a basis for discussion of the motion of the atoms at the lattice sites of the Ith cluster with respect to the lattice sites of the surrounding crystal. It is also necessary to express the kinetic contributions to the Hamiltonian and not just the potential contributions in order to separate the rotational partition function[41]. Thus the kinetic terms are written, and the potential contributions are expressed in terms of the harmonic approximation considering only nearest-neighbour interactions of the atoms *in* the Ith cluster. An orthogonal transformation is then performed to obtain the Hamiltonian of the Ith cluster in the form

$$H_I = \frac{1}{2nm}(p_{IX}^2 + p_{IY}^2 + p_{IZ}^2) + f_{I1}p_{I\theta}^2 + f_{I2}p_{I\phi}^2 + f_{I3}p_{I\psi}^2 + f_{I4}p_{I\theta}p_{I\phi} +$$

$$f_{I5}p_{I\phi}p_{I\psi} + f_{I6}p_{I\psi}p_{I\theta} + \tfrac{1}{2}\sum_i P_{Ii}^2 + \tfrac{1}{2}\sum_i \lambda_{Ii}Q_{Ii}^2 + U_{I0} \qquad (5.86)$$

where p_{IX} is the momentum conjugate to X_I and the other terms have similar meanings. The first term expresses the translational kinetic energy of the Ith cluster in the laboratory frame with the atoms at the lattice site positions, and the terms containing the Eulerian angles express its rotational kinetic energy. The term in P_{Ii} is defined by

$$\tfrac{1}{2}\sum_i P_{Ii}^2 = \tfrac{1}{2}m \sum_{j=7}^{3n} \dot{q}_{Ij}^2 \qquad (5.87)$$

which gives the kinetic energy of the cluster atoms with respect to Σ_I. The term in the $3n-6$ normal coordinates $\{Q_{Ii}\}$ describes the vibrational potential energy of the cluster atoms, where the λ_{Ii} are eigenvalues of the matrix (f_{Iij}) in

$$U_I = \tfrac{1}{2} \sum_{i=7}^{3n} \sum_{j=7}^{3n} f_{Iij} q_{Ii} q_{Ij} + U_{I0} \qquad (5.88)$$

in which q_{Ii} etc. are generalised coordinates that were introduced as linear combinations of Cartesian coordinates through an orthogonal transformation. U_{I0} is the potential energy with all the n atoms at their lattice points, excluding the potential energy of interaction of the cluster with the surrounding lattice.

Next consider the interaction energy between the clusters of the crystal. Under the approximation of nearest-neighbour interactions only, the Ith cluster interacts only with its neighbour clusters by interaction between the atoms of both sides of the mathematical boundary. Again, this potential energy is expressed by the harmonic approximation. The result for interaction with one neighbouring cluster may be given in terms of external, internal and coupling contributions by

$$U_{IJ} = U_{IJ}^{\text{EXT}}(X_I, X_J, \ldots, \Psi_I \Psi_J) + U_{IJ}^{\text{INT}}(\{Q_{Ii}\}, \{Q_{Jj}\}) + U_{IJ}^{\text{COUP}}(X_I, X_J, \ldots,$$

$$\Psi_I, \Psi_J, \{Q_{Ii}\} \{Q_{Jj}\}) + U_{IJ0} \qquad (5.89)$$

where U_{IJ0} is the interaction potential between the Ith cluster and the Jth nearest neighbouring cluster with all atoms at their lattice point positions. The total interaction energy between the Ith cluster and its surrounding clusters is

$$\sum_J U_{IJ} \qquad (5.90)$$

The Hamiltonian of the bulk crystal is given by

$$H = \sum_I \sum_J (H_I + \tfrac{1}{2} U_{IJ}) \qquad (5.91)$$

in which the summation over I spans all the clusters and the summation over J spans all the nearest neighbour clusters of the Ith cluster.

Now an argument may be presented[44] to show that

$$U_{IJ}^{COUP} \ll U_{IJ}^{INT} \tag{5.92}$$

whereupon the external and internal canonical partition functions of the imbedded cluster can be separated according to the prescription

$$Q_I = Q_I^{EXT} \cdot Q_I^{INT} \tag{5.93}$$

in which

$$Q_I^{EXT} = \frac{1}{h^6} \int \cdots \int \exp\left[-\sum_J \left\{\frac{1}{2nm}(p_{IX}^2 + p_{IY}^2 + p_{IZ}^2) + f_{11}p_{I\theta}^2 + \ldots + f_{16}p_{I\psi}p_{I\theta}\right.\right.$$
$$\left.\left. + \tfrac{1}{2}U_{IJ}^{EXT}(X_I, \ldots \Psi_I, X_J^*, \ldots \Psi_J^*)\right\}/kT\right] dX_I \ldots d\Psi_I \tag{5.94}$$

where, according to the partition function decomposition theorem[44], $X_J^*, \ldots \Psi_J^*$ are functions of $X_I, \ldots \Psi_I$, and

$$Q_I^{INT} = \frac{1}{h^{3n-6}} \int \cdots \int \exp\left[-\sum_J\left\{\tfrac{1}{2}\sum_i P_{Ii}^2\right.\right.$$
$$\left.\left. + \tfrac{1}{2}\sum_i \lambda_{Ii}Q_{Ii}^2 + U_{I0} + \tfrac{1}{2}U_{IJ}^{INT}(\{Q_{Ii}\}, \{Q_{Jj}^*\}) + \tfrac{1}{2}U_{IJ0}\right\}/kt\right] \prod_i dQ_{Ii}\, dP_{Ii} \tag{5.95}$$

where, again, $\{Q_{Jj}^*\}$ are functions of $\{Q_{Ii}\}$. The integration by the orientational variables θ_I, ϕ_I and Ψ_I are to be performed within only one of the equivalent orientations when the local lattice points $\{a_{Ij}\}$ have a rotational symmetry. Q_I^{EXT} is the partition function corresponding to the six translational and rotational degrees of freedom per cluster and *all the atoms must be treated as fixed at their local lattice sites* $\{a_{Ii}\}$ *in considering these degrees of freedom.* Q_I^{EXT} is the replacement partition function for a cluster of n atoms.

The above results must now be related to the properties of a bulk crystal. Employing the partition function decomposition theorem again and noting that all M clusters in the bulk are equivalent, the partition function of the bulk crystal can be expressed as

$$Q = (Q_I^{EXT} \cdot Q_I^{INT})^M \tag{5.96}$$

The chemical potential of the bulk crystal is

$$\mu_c = \lim_{N\to\infty} \frac{1}{N}\{-kT \ln Q + pNv_c\} \approx \lim_{N\to\infty} \frac{1}{N}\{-kT \ln Q\} \tag{5.97}$$

or, from equation (5.96)

$$\exp(-n\mu_c/kT) = Q_I^{EXT} \cdot Q_I^{INT} \tag{5.98}$$

The above results can also be demonstrated[44] for imbedded clusters of arbitrary shape.

Next we must relate the bulk properties to the properties of an isolated crystallite in the vapour. The canonical partition function of the isolated crystallite is

$$Q_n = Q_n^{EXT} \cdot Q_n^{INT} \tag{5.99}$$

in which

$$Q_n^{EXT} = \frac{1}{\gamma h^6} \int \cdots \int \exp\left[-\left\{\frac{1}{2nm}(p_{IX}^2 + p_{IY}^2 + p_{IZ}^2)\right.\right.$$

$$\left.\left. + f_{11} p_{I\theta}^2 + \cdots + f_{16} p_{I\psi} p_{I\theta}\right\}/kT\right] dX_I \, dp_{IX} \cdots d\Psi_I \, dp_{I\psi} \tag{5.100}$$

and

$$Q_n^{INT} = \frac{1}{h^{3n-6}} \int \cdots \int \exp\left[-\left\{\tfrac{1}{2}\sum_i P_{Ii}^2 + \tfrac{1}{2}\sum_i \lambda_{Ii} Q_{Ii}^2 + U_{I0}\right\}/kT\right] \prod_i dQ_{Ii} \, dP_{Ii} \tag{5.101}$$

where γ is the rotational symmetry number. Also,

$$Q_n^{EXT} = Q_n^T \cdot Q_n^R \tag{5.102}$$

in which

$$Q_n^T = \frac{(2\pi nmkT)^{\frac{3}{2}} V}{h^3} \tag{5.103}$$

and

$$Q_n^R = \frac{\sqrt{\pi}(8\pi^2 kT)^{\frac{3}{2}}(I_{xx} I_{yy} I_{zz})^{\frac{1}{2}}}{\gamma h^3} \tag{5.104}$$

where V is the volume of the container of the vapour and I_{xx}, I_{yy} and I_{zz} are the three principal moments of inertia with the atoms at their lattice sites.

Consider now the free energy difference between the isolated and imbedded crystallites

$$-kT \ln \frac{Q_n}{Q_I} = -kT \ln (Q_n^{EXT}/Q_I^{EXT}) - kT \ln (Q_n^{INT}/Q_I^{INT}) \tag{5.105}$$

In relation to the capillarity approximation, a very important and difficult question is how much of equation (5.105) should be associated with the concept of surface free energy or, more specifically, with the concept of the surface free energy of bulk surface. The first term in equation (5.105) corresponds to the translational and rotational degrees of freedom of a cluster as a rigid body with all the atoms fixed at their local lattice sites. Therefore it is conceptually difficult to ascribe it to the surface free energy, because the surface free energy is a property of the surface and should be independent of the translational and rotational motions of the cluster as a rigid body. In the second term of equation (5.105), Q_n^{INT} and Q_I^{INT} are given by equations (5.101) and (5.95), respectively. Note that U_{I0} cancels and the contribution due to the binding potential energy is

$$\tfrac{1}{2} \sum_J U_{IJ0} \tag{5.106}$$

where U_{IJ0} is the binding potential energy of the Ith cluster to its J nearest-neighbour clusters. This is a positive quantity and it is proportional to the number of surface atoms. Hence it can be ascribed to the binding potential energy contribution to the surface free energy. Consider the dynamical contributions to Q_n^{INT} and Q_I^{INT}. Note that there is zero linear and angular momentum of the cluster in these internal degrees of freedom. The other difference between Q_n^{INT} and Q_I^{INT} is the existence of $\frac{1}{2}\sum_J U_{IJ}^{INT}$ in Q_I^{INT}, which is a property of the bulk phase. When this interaction term disappears, as in creating the isolated crystallite to give Q_n^{INT}, the frequency distribution changes and surface modes appear in compensation for the disappearance of some of the bulk modes. Although it is difficult to make a more complete quantitative analysis, it seems fairly clear that the second term of equation (5.105) can be ascribed to the surface free energy $\sigma(n)$ due to the above reasoning. Also, under the Einstein approximation, the second term in equation (5.105) is obviously proportional to the number of surface atoms. For all of these reasons, one may write

$$\exp[-\hat{\sigma}(n)/kT] = \frac{Q_n^{INT}}{Q_I^{INT}} \qquad (5.107)$$

Equation (5.107) is consistent with the macroscopic concept of the surface free energy, which can be expressed as

$$\sigma = \lim_{n \to \infty} \frac{\hat{\sigma}(n)}{S(n)} = \lim_{n \to \infty} \frac{-kT \ln[Q_n/Q_I^{EXT}Q_I^{INT}]}{S(n)} \qquad (5.108)$$

where $S(n)$ is the surface area. One notes that the first term in equation (5.105) has only a logarithmic dependence on the number of surface atoms and thus does not contribute appreciably to equation (5.108).

We may now summarise the results of our work by combining equations (5.99), (5.102), (5.107), (5.96) and (5.97) to yield

$$Q_n = \frac{Q_n^T \cdot Q_n^R}{Q_I^{EXT}} \exp[-\{n\mu_c + \hat{\sigma}(n)\}/kT] \qquad (5.109)$$

or, since $Q_I^{EXT} \equiv Q_n^{REP}$ (the replacement partition function),

$$Q_n = \frac{Q_n^T \cdot Q_n^R}{Q_n^{REP}} \exp[-\{n\mu_c + \hat{\sigma}(n)\}/kT] \qquad (5.110)$$

From the standard equilibrium relationship, the equilibrium concentration of crystallites of size class n becomes

$$c(n) = \frac{Q_n^T Q_n^R}{Q_n^{REP}} \exp[-\{n(\mu_c - \mu) + \hat{\sigma}(n)\}/kT] \qquad (5.111)$$

In slightly different notation, this is the Lothe–Pound result, equation (5.21). Equation (5.107) defines the surface free energy of the isolated crystallite. $\hat{\sigma}(n)$ will not, in general, equal the macroscopic surface free energy because the surface modes which replace the bulk modes will be a function of n, as will the potential energy in equation (5.106). The replacement partition function is defined by equation (5.94). It is the partition function for the motions in six degrees of freedom that the isolated cluster does not have because it is not a part of the bulk. One notes that, in the present case of crystals, *it corresponds to the partition function for rigid-body translational and rotational vibrations of the imbedded cluster with the atoms in both cluster and surrounding bulk fixed in their own lattice site positions.* Here the lattice site positions are defined by minimising the internal potential energy of the imbedded cluster or the surrounding bulk, respectively. It should be mentioned that $X_j^*, \ldots, \Psi_j^*$ in equation (5.94) are functions of $X_I, \ldots, \Psi_I$, and these functions have not yet been explicitly found. However, by analogy with the result for the linear chain discussed in Section 5.3.2, one expects that $X_j^*, \ldots, \Psi_j^*$ may be approximated by constants which correspond to the atoms in the surrounding bulk being at their laboratory lattice sites. Further investigation of the nature of $X_j^*, \ldots, \Psi_j^*$ is desirable.

One expects that the situation for liquids will be different only in that the rigid-body translations and rotations will not be entirely vibrational. In the case of crystals, this result for the replacement partition function suggests use of the Einstein model to calculate the replacement partition function, and this will be discussed later in the present section.

5.3.4 Normal-mode computer calculations

5.3.4.1 General

The object of the normal-mode computer calculations to be described in the following is to evaluate the free energy of formation of microcrystallites and compare this 'exact' free energy with the results obtained from the capillarity approximation. It will be found, as might be expected, that the agreement is poor in the cases studied thus far. However, some interesting insights arise as to how to correct the capillarity approximation to make it practically useful. It will emerge that this may be done in terms of the replacement partition function.

5.3.4.2 The work of Burton

Burton[45-48] was apparently the first to apply normal-mode computer calculations to nucleation problems. He used f.c.c. crystallite models for argon and a pairwise additive 6-12 Lennard-Jones potential to calculate the potential energy, frequency spectrum and vibrational free energy for the $3n-6$ internal degrees of freedom of microcrystallites with complete co-ordination shells containing 13, 19, 43, 55, 79 and 87 atoms. The minimum potential energy configuration of a cluster was found by allowing the atoms to relax. Thus all-neighbour interactions were considered and complete relaxation of the crystallite was allowed. The vibrational frequencies were obtained by numerically diagonalising the force constant matrix for the cluster.

Similar calculations were performed using the same model to calculate the volume and surface free energy of bulk solid argon. These latter quantities were then used to calculate the free energy of formation by the alternative method involving the capillarity approximation (droplet model).

It was found that:

(a) The positive excess entropy per atom in the cluster, with respect to the bulk phase, is an irregular function of the number of atoms. Some of this behaviour can be explained by considering the elimination of six long-wavelength bulk modes, which comprise the replacement partition function, and the introduction of other long-wavelength surface modes. However, as will be discussed in Section 5.3.4.4, the jagged appearance of the S_{xs}/n v. n curves is probably a result of the closed-shell geometry of the model for the clusters.

(b) The positive excess potential energy per atom in the cluster with respect to the bulk phase is also an irregular function of the number of atoms. Again, some of this behaviour is probably a result of the closed-shell geometry of the model.

(c) The above qualitative results are thought to be independent of crystallographic structure or potential function, because similar calculations[48] done in the nearest-neighbour approximation for hexagonal close packing showed similar details.

(d) If the contributions from free translation and rotation of the cluster in the vapour are ignored in both the 'exact' and capillarity-approximation treatments, the free energy of formation of the critical nuclei as calculated by the 'exact' method is much higher than that found from the capillarity approximation (droplet model), leading to a predicted homogeneous nucleation rate that is lower by a factor of 10^{13}. Inclusion of the partition function for these six external degrees of freedom in the 'exact' approach suffices to give approximate agreement with the results for the uncorrected capillarity approximation[48].

(e) The 'exact' calculations suggest that the critical cluster size is nearly temperature independent, in sharp contrast to the results obtained using the capillarity approximation.

5.3.4.3 The work of Nishioka et al.

Nishioka *et al.*[49] did a normal-mode computer calculation to determine several vibrational free energies pertinent to nucleation behaviour. Their model calculations considered only a one-parameter nearest-neighbour harmonic interatomic force field with no lattice relaxation. They used an f.c.c. closed coordination shell model for the approximately spherical microcrystallites containing 13, 19, 43, 55, 79, 87, 135, 141, 177, 201, 255 and 249 atoms. The spring constant matrix was diagonalised by means of the negative eigenvalue theorem to obtain the frequency spectra. The following quantities were calculated:

(a) The vibrational free energy per atom in the bulk crystal was obtained by applying periodic boundary conditions to a cubic cluster containing 1000 atoms (and, as a check, fewer atoms).

(b) The vibrational surface free energies of the bulk crystal were obtained by considering one cube surface to be free and applying periodic boundary conditions to the other surfaces. This was done for the (111), (100) and (110) planar surfaces.

(c) The vibrational free energy for the $3n-6$ internal degrees of freedom of the isolated cluster containing n atoms was calculated.

(d) From the above quantities another quantity was obtained by difference, namely, the vibrational free energy that the isolated crystallite does not have because it is not a part of the bulk phase, i.e. the replacement free energy. This is the vibrational free energy for the missing six internal degrees of freedom of the isolated cluster and is a property of the bulk phase. It is calculated according to the prescription (see Section 5.3.3, equations (5.105) to (5.110))

$$F_n^{REP} = F_n^{BULK} - (F_n^{INT} - F_n^{SURF}) \qquad (5.112)$$

or

$$-kT \ln Q_n^{REP} = -kT \ln Q_I - [-kT \ln Q_n^{INT} - \sum_j \sigma_j n_j]$$

in which the surface free energy F_n^{SURF} is subtracted from the internal free energy of the isolated crystallite in vapour F_n^{INT} to remove the effect of the surface. This difference is then subtracted from the free energy of n atoms in bulk F_n^{BULK} to yield the replacement free energy F_n^{REP} and hence the replacement partition function Q_n^{REP}.

There are some difficulties with this definition in connection with the surface term $\sum_j \sigma_j n_j$. Here n_j is the number of surface atoms with a given number of nearest-neighbour bonds j and σ_j is the vibrational surface free energy per surface atom of type j. Such a representation is of course necessitated by the fact that many of the surface atoms of microcrystallites do not correspond to (111), (100) or (110) atoms in a macroscopic planar surface; instead they are edge or corner atoms. Also, even in the (110) surface plane of a macroscopic f.c.c. crystal there are two kinds of atoms. In order to handle this difficulty, the specific surface free energies of the (111), (100) and (110) macroscopic surfaces were also calculated by an Einstein approximation, a constant ratio was found between these results and the normal mode results, and this ratio was applied to the Einstein results for the edge and corner atoms to obtain the surface free energies per atom σ_j. Now this ratio of normal mode to Einstein surface free energies may well be different for small crystallites. In fact, there is apparently no way in the present approach to separate the effect of curvature on surface entropy from the replacement partition function.

Nevertheless, with these restrictions in mind, the replacement partition function can be defined by equation (5.112). The numerical result is c. 10^8 for $n \approx 100$ and $T/\theta = 2$, and it increases slowly with n and T as would be expected. An approximate numerical result from the theoretical prescription of Section 5.3.3, in which the replacement partition function is evaluated for the Einstein translational and rotational vibrations of the imbedded cluster for which the internal coordinates remain fixed, yields a value of 10^4 for the replacement partition function at $n \approx 100$ and $T/\theta = 2$. The discrepancy between the two results is as yet unexplained, but it may have to do with the approximations for the surface free energy mentioned above or it may arise from an error in the prescription of Section 5.3.3 by which the free volume of the replacement partition function is somewhat too small.

Finally, one notes that the results of the present study are difficult to compare with those of Burton. This is because Burton, in his capillarity-approximation calculations, did not consider the actual number and configuration of surface atoms in the microcrystallites. As will be discussed in Section 5.3.4.4, the standard capillarity-approximation estimate of the surface free energy as $4\pi r^2 \sigma$ seriously overestimates the number of surface atoms.

5.3.4.4 The work of Abraham and Dave

Dave and Abraham[50] investigated the effect of cluster shape, size and atomic interaction range on the excess potential energy of the isolated cluster over that of the same number of atoms in the bulk phase. They used a Lennard-

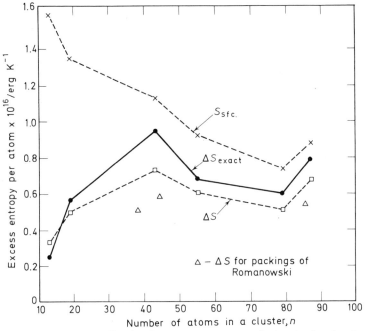

Figure 5.4 Excess vibrational entropy per atom v. cluster size for closed-coordination-shell f.c.c. argon microcrystallites according to the Einstein calculation of Abraham and Dave[52]

Jones 6-12 potential and calculated the relevant potential energies with the aid of a computer. Five regular polyhedral shapes of Romanowski and the closed coordination shell shapes mentioned above were considered. The procedure of Romanowski[51] is to construct a regular polyhedron for a given total number of atoms. It was found that

(a) For the nearest-neighbour interactions, the cubo-octahedron shapes are the configurations of lowest potential energy.

(b) For the infinite-range interactions, the shapes can be arranged in the following order of increasing potential energy: cubo-octahedron, closed coordination shell, dodecahedron, octahedron, cube and tetrahedron.

The cubo-octahedron represents the closest atomic packing.

(c) The ratio of infinite-range to nearest-neighbour potential energies is about two and increases with increase in cluster size.

Abraham and Dave[52, 53] developed an Einstein calculation to permit more convenient description of the surface and volume free energies of macroscopic crystals and the vibrational free energy of isolated crystalline clusters. They used a Lennard-Jones 6-12 potential and considered some of the shapes mentioned above. Lattice relaxation is neglected in their work. They find that their vibrational surface free energy for the (111) plane of an f.c.c. crystal is c. 8% lower than that obtained from the molecular dynamics calculations of Allen and de Wette[54].

Also, they calculated[52] the replacement partition function of an imbedded crystallite in bulk f.c.c. argon using the same model for argon as that of Burton[46] and the prescription based on rigid-body translational and rotational Einstein vibrations as described in Section 5.3.4.3.

The results for the excess entropy per atom in the cluster over that in the bulk phase v. number of atoms are given in Figure 5.4. The upper curve represents the Einstein surface entropy, considering each atom in the cluster surface. The lower dashed curve was obtained by subtracting the replacement entropy, which is due to the six vibrational motions the isolated crystallite does not have because it is not a part of a bulk crystal, from the upper curve for S_{sfc}. The solid curve gives the results of Burton[46] for the internal excess entropy of isolated crystallites. It is seen that the Einstein approximation of Abraham and Dave duplicates the results of Burton reasonably well. Of special importance is the fact that the 'down-turn' of excess entropy for small clusters is plainly due to the replacement partition function. Also, it is clear that the jagged appearance of the curve for the excess entropy of the isolated crystallite as a function of the number of atoms is occasioned by cluster shape and surface packing. Also of interest is the sensitivity of the results to cluster shape as evidenced by the triangular points for Romanowski polyhedra; again, this indicates that the jagged form of the curves is due to cluster shape. This interpretation is consistent with the fact that the replacement entropy is a property of the bulk phase and exhibits no irregularities as a function of number of atoms.

Another important result of these papers by Abraham and Dave[53] is the recognition that the original form of the capillarity approximation, in which the surface free energy is taken as $4\pi r^2\sigma$, seriously over-estimates the number of atoms in the cluster surface. This is shown in Figure 5.5.

In another paper, Abraham and Dave[55] apply a modified Einstein calculation to determine the surface and volume free energies of macroscopic crystals and the vibrational free energy of isolated crystalline clusters. They duplicate the 'exact' results of Nishioka et al.[49] for the (100) and (111) surface free energies of macroscopic f.c.c. crystals within 5%. Also, they calculate the replacement partition function by a procedure which involves taking a somewhat larger free volume for the vibrational motions of the imbedded cluster than that given by the rigid-body prescription of Section 5.3.4.3. Typical results for closed-coordination-shell f.c.c. clusters at $T/\theta = 2$ are shown in Figure 5.6. The solid $(F_n^{SURF} + F_n^{BULK})/n$ curve represents the vibrational free energy per atom of the cluster as obtained simply from the sum of volume and surface free energies. The replacement free energy is subtracted from this surface free energy to give the dashed F_n^{INT}/n curve*. The open circles are the 'exact' results of Nishioka et al.[49]. Again, one concludes that the Einstein approximation of Abraham and Dave[52, 53] can be

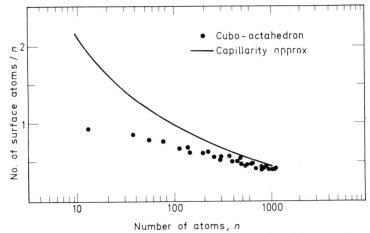

Figure 5.5 Number of surface atoms as a function of cluster size[53]

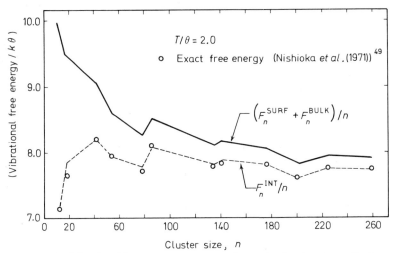

Figure 5.6 Vibrational free energy per atom v. cluster size for closed-coordination-shell f.c.c. microcrystallites according to the modified Einstein calculation of Abraham and Dave[55]

used to reproduce the normal-mode results of Nishioka et al.[49], and the 'down-turn' of the vibrational free energy at small cluster sizes is seen to be due to the replacement partition function. Also, the jagged appearance of the curve of cluster free energy v. number of atoms is clearly a property of cluster shape and surface structure.

---

*Actually, a small correction, important only at very small sizes, was introduced to account for the long-wavelength cut-off.

5.3.4.5 The work of Daee et al. and Kassner et al.

Daee et al.[56] constructed clathrate-like structure models for clusters containing 16–57 water molecules. Some were closed-cage configurations, and, in general, these roughly spherical structures were selected on the basis that the number of hydrogen bonds was maximised and that approximately tetrahedral bond angles were maintained. The clusters, while not having an ice structure, approach the crystalline structure much more closely than they resemble bulk liquid water. The hydrogen-bond potential energy is c. $10\,kT$ for room temperatures, and thus it was assumed that these tightly bound clusters of molecules are capable of supporting normal-mode vibrations. For a cluster containing g molecules of water there are $9g-6$ frequencies corresponding to the $3g$ intramolecular, $3g$ librational (hindered rotational) and $3g-6$ intermolecular vibrations. In view of the separation between these three kinds of frequencies, the vibrational partition function $Z_v(g)$ could be expressed as a product

$$Z_v(g) = Z_{\text{intra}}(g) \cdot Z_{\text{libr}}(g) \cdot Z_{\text{inter}}(g) \tag{5.113}$$

The respective normal modes were approximated with the help of some spectroscopic data. The binding energy was evaluated to force agreement of the internal free energy of the cluster with the capillarity approximation. The complete partition function of the cluster in the vapour was then formed by including the partition functions for free translation and rotation. The resulting curve of free energy of formation v. number of molecules exhibited a jagged appearance. This is apparently an effect of the shape of the cluster model, as discussed in Section 5.3.4.4. This effect may be very important in explaining certain experimental results, namely, the so-called 'heterogeneous component' observed by Allen and Kassner[57].

In another paper, Kassner et al.[58] refined and extended the methods of the above paper to consider additional water-cluster structures. Also they performed an independent calculation of the binding energy. Again the free energy of formation $v.g$ curve exhibited a jagged appearance. Furthermore, the mean locus of this curve showed approximate agreement with the original capillarity approximation, in which the statistical mechanical contributions are ignored. This agreement is probably fortuitous, particularly in view of the fact that the capillarity approximation seriously undercounts the number of surface molecules as discussed in Section 5.3.4.4.

5.3.5 A few conclusions

(a) The replacement partition function must be considered and evaluated *if one is to revise the capillarity approximation for reliable prediction of nucleation behaviour.*

(b) The replacement partition function is a well defined property of the bulk phase. It is the partition function for the six motions that an isolated cluster in vapour does not have because it is not a part of the bulk phase. In the case of a crystal, it corresponds approximately to the six rigid-body translational and rotational vibrations of a mathematical cluster imbedded in the bulk crystal for which the atoms of the cluster and the atoms of the surrounding bulk remain fixed at their respective lattice sites.

(c) The above prescription leads to a replacement partition function of $c.\ 10^4$ for a crystalline cluster of 100 atoms at $T/\theta = 2$.

(d) Normal-mode computer calculations for the same case give a replacement partition function of $c.\ 10^8$. The discrepancy may be due to difficulties and errors in evaluating the surface free energy of an isolated cluster, or it may be due to an error in the prescription of Section 5.3.3 by which the free volume of the replacement partition function is somewhat too small.

(e) The qualitative aspects of the replacement partition function are thought to be the same for liquids as for crystals. However, the numerical value may prove to be higher for liquids.

(f) Molecular dynamics and/or Monte Carlo calculations will be required to evaluate the replacement partition function for liquids.

5.4 SOME GENERAL CONCLUSIONS

(a) The question of the correct statistical mechanical prescription for describing the free energy of formation of clusters in terms of the capillarity approximation is not yet settled. Specifically, one does not know how much of the external partition functions for free translation and rotation of the cluster in the vapour are cancelled by the replacement partition function and other factors.

(b) Of the most recent classical phase integral treatments for liquid clusters, the approach by Kikuchi[34], which compares the isolated cluster in the vapour with the cluster imbedded in liquid, appears to be the most promising. The result of his treatment is that one returns almost to the 'classical' Becker–Doering prescription, which ignores statistical mechanical contributions. In other words, the free translation and rotation are nearly cancelled by other considerations. However, this result is open to question, because in his treatment the entire releases of translational and rotational correlation of the imbedded cluster are assumed to be contained in the surface tension of the bulk liquid. Further work is necessary to decide this point.

(c) For the case of crystals, the classical phase integral treatment of Nishioka, Lothe and Pound[44] appears to be correct. They compared an isolated crystallite in vapour with an imbedded crystallite in the bulk solid and concluded that the partition function for free translation and rotation of the crystallite in vapour is not cancelled by the replacement partition function. This result corresponds fairly closely to the original Lothe–Pound prescription for liquids or crystals.

(d) Normal-mode computer calculations of the replacement partition function for crystals tend to support conclusion (c). Further normal-mode and molecular-dynamics computer calculations for crystals would be desirable.

(e) Molecular-dynamics and Monte Carlo computer calculations of the replacement partition function for liquids should be useful.

(f) Experiments have not yet been of much help in settling the issue. Some liquids follow the Becker–Doering prescription, while others apparently follow the Lothe–Pound formulation. Actually, the capillarity approximation is so crude that one would not expect either prescription to predict the observations. Nevertheless, more experimental work would be desirable.

Acknowledgement

The authors gratefully acknowledge many helpful discussions with Drs. Farid Abraham and Ryoichi Kikuchi. One of us (G. M. P.) expresses appreciation to the National Science Foundation for financial aid from Grant No. GK 2159, which enabled him to participate in this work. Another (K. N.) gratefully acknowledges support from an NTNF fellowship in Norway.

References

1. Dunning, W. J. (1969). *Nucleation*, 1 (A. C. Zettlemoyer, editor) (New York: Marcel Dekker)
2. Andres, R. P. (1969). *Nucleation*, 69 (A. C. Zettlemoyer, editor) (New York: Marcel Dekker)
3. Lothe, Jens and Pound, G. M. (1969). *Nucleation*, 109 (A. C. Zettlemoyer, editor). (New York: Marcel Dekker)
4. Wegener, P. P. and Parlange, Jean-Yves (1970). *Naturwissenschaften*, **57**, 525
5. Gibbs, J. W. (1928). *The Collected Works of J. W. Gibbs, Vol. I.* (London: Longman Green)
6. Feder, J., Russell, K. C., Lothe, J. and Pound, G. M. (1966). *Advances in Physics*, **15** (57), 111
7. Farkas, L. (1927). *Z. Physik. Chem. (Leipzig)*, **125** 236
8. Volmer, M., and Weber, A. (1926). *Z. Physik, Chem. (Leipzig)*, **119**, 277
9. Becker, R., and Doering, W. (1935). *Ann. Physik.*, **24**, 719
10. Zeldovitch, J. B. (1942), *J. Exp. Theoret. Phys.*, **12**, 525
11. Zeldovitch, J. B. (1943). *Acta Phys. Chem. URSS*, **18**, 1
12. Frenkel, J. (1946). *Kinetic Theory of Liquids*, (Oxford: Oxford University Press)
13. Volmer, M. and Flood, H. (1934). *Z. Physik. Chem. (Leipzig)*, **170**, 273.
14. Volmer, M. (1939). *Kinetik der Phasenbildung* (Dresden and Leipzig: Steinkopff)
15. Rodebush, W. H. (1949). *Chem. Rev.*, **44**, 269
16. Rodebush, W. H. (1952). *Ind. Eng. Chem.*, **44**, 1289
17. Kuhrt, F. (1952). *Z. Physik*, **131**, 185, 205
18. Lothe, Jens and Pound, G. M. (1962). *J. Chem. Phys.*, **36**, 2080
19. Hirth, J. P. (1963). *Ann. N.Y. Acad. Sci.*, **101**, 805.
20. Dunning, W. J. (1965). *Proceedings of the Case Institute of Technology Symposium on Nucleation*, (Cleveland: Case).
21. Dunning, W. J. (1965). *Colloq. Intern. CNRS (Paris)*. **152**, 369
22. Lothe, Jens and Pound, G. M. (1966), *J. Chem. Phys.*, **45**, 630
23. Fisher, M. E., (1967). *Physics*, **3**, 255
24. Eggington, A., Kiang, C. S., Stauffer, D. and Walker, G. H. (1971). *Physical Review Letters*. Submitted for publication
25. Kiang, C. S., Stauffer, D., Walker, G. H., Puri, O. P., Wise, J. D. and Patterson, E.M. (1971). *J. Atmos. Sci.*, Submitted for publication
26. Reiss, H. and Katz, J. L. (1967). *J. Chem. Phys.*, **46**, 2496
27. Lothe, J. and Pound, G. M. (1968). *J. Chem. Phys.*, **48**, 1849
28. Abraham, F. F. and Pound, G. M. (1968), *J. Chem. Phys.*, **48**, 732
29. Reiss, H., Katz, J. L. and Cohen, E. R. (1968). *J. Chem. Phys*, **48**, 5553
30. Nishioka, K., Pound, G. M., Lothe, J. and Hirth, J. P. (1971). *J. Appl. Phys.*, **42**, 3900
31. Reiss, H. (1970). *J. Statistical Phys.*, **2**, 83
32. Nishioka, K., Lothe, J. and Pound, G. M. (1971). Unpublished work
33. Kikuchi, R. (1971). *J. Statistical Phys.*, **3**, 331

34. Kikuchi, R. (1969). *J. Statistical Phys.*, **1,** 351
35. Nishioka, K., Lothe, J. and Pound, G. M. (1971). Unpublished work
36. Herzberg, G. (1945). *Molecular Spectra and Molecular Structure, Vol. II* (New York: D. Van Nostrand)
37. Wilson, E. B., Jr., Decius, J. C. and Cross, P. C. (1955). *Molecular Vibrations*, 273–274. (New York: McGraw-Hill)
38. Wollrab, J. E. (1967). *Rotational Spectra and Molecular Structure*, 399. (New York: Academic Press)
39. Nishioka, Kazumi and Pound, G. M. (1970). *Amer. J. Physics*, **38,** 1211
40. Meyer, R. and Günthard, H. (1968). *J. Chem. Phys.*, **49,** 1510
41. Nishioka, K., Pound, G. M. and Abraham, F. F. (1970). *Phys. Rev. A*, **1,** 1542
42. Abraham, F. F. and Canosa, J. (1969). *J. Chem. Phys.*, **50,** 1303
43. Lothe, J. and Pound, G. M. (1969). *Phys. Rev.*, **182,** 339
44. Nishioka, K., Lothe, J. and Pound, G. M. (1971). *J. Applied Phys.*, Submitted for publication
45. Burton, J. J. (1969). *J. Chem. Phys. Letters*, **3,** 594
46. Burton, J. J. (1970). *J. Chem. Phys. Letters*, **7,** 567
47. Burton, J. J. (1970). *J. Chem. Phys.*, **52,** 345
48. Burton, J. J. (1971). *Surface Science*, **26,** 1
49. Nishioka, K., Shawyer, R., Bienenstock, A. I. and Pound, G. M. (1971). *J. Chem. Phys.*, **55,** 5082
50. Abraham, F. F. and Dave, J. V. (1971). *Surface Science*, in the press
51. Romanowski, W. (1969). *Surface Science*, **18,** 373
52. Abraham, F. F. and Dave, J. V. (1971). *J. Chem. Phys. Letters*, **8,** 351
53. Abraham, F. F. and Dave, J. V. (1971) *J. Chem. Phys.*, **55,** 1587
54. Allen, R. E. and DeWette, F. W. (1969). *J. Chem. Phys.*, **51,** 4820
55. Abraham, F. F. and Dave, J. V. (1971). *J. Chem. Phys.*, in the press
56. Daee, M., Lund, L. H., Plummer, P.L.M., Kassner, J. L., Jr., and Hale, B. N. (1971). *J. Colloid and Interface Science*, in the press
57. Allen, L. B. and Kassner, J. L., Jr. (1968). *J. Colloid and Interface Science*, **30,** 81
58. Kassner, J. L., Jr., Plummer, P. L. M., Hale, B. N., and Bierman, A. H. (1971) Preprint of a Talk at the International Weather Modification Conference, Canberra, Australia.

6
The Physical Adsorption of Gases
S. J. GREGG
Brunel University, Uxbridge

6.1	INTRODUCTION	190
6.2	DETERMINATION OF SPECIFIC SURFACE	191
	6.2.1 *Factors causing a variation in molecular area* Ω_m	192
	6.2.2 *Specific and non-specific adsorption*	194
	6.2.3 *Some further values of molecular area*	195
	6.2.4 *Adsorption on modified surfaces*	195
6.3	THE STANDARD ('UNIVERSAL') ISOTHERM	196
	6.3.1 *t-Plots*	196
	6.3.2 α_s-*Plots*	198
6.4	ADSORPTION ON MESOPOROUS SOLIDS: PORE SIZE DISTRIBUTION	199
	6.4.1 *Modes of calculation*	200
	6.4.2 *The 'model-less' method*	202
	6.4.3 *The 'tensile-strength' hypothesis*	203
6.5	ADSORPTION ON MICROPOROUS SOLIDS	205
	6.5.1 *The detection of micropores*	205
	6.5.2 *The Dubinin–Radushkevich (DR) equation*	206
	6.5.3 *Activated entry*	207
6.6	THEORETICAL DEVELOPMENTS	208
	6.6.1 *Calculations of interaction energy*	208
	6.6.2 *Statistical-mechanical treatments*	209
	6.6.3 *Equations to the adsorption isotherm*	209
	6.6.4 *Thermodynamics*	211
	6.6.5 *The Henry's law region*	211
6.7	THERMAL QUANTITIES	212
	6.7.1 *Heat of adsorption*	212
	6.7.2 *Heat of immersion*	
6.8	ADSORPTION OF WATER VAPOUR: TYPE (III) ISOTHERMS	214
	6.8.1 *Hydrophobic surfaces*	214
	6.8.2 *Hydrophilic surfaces*	216
	6.8.3 *Type (III) isotherms—other examples*	217

6.9	SOME ADDITIONAL TOPICS	218
	6.9.1 Infrared spectroscopy	218
	6.9.2 Dielectric studies	219
	6.9.3 Ellipsometry	219
	6.9.4 An experimental note	219

6.1 INTRODUCTION

Although the last few years have not seen any spectacular breakthrough either on the theoretical or the practical side of this subject, there has been a steady progress all round. Before proceeding to review this in detail, reference will be made to relevant symposia, books and general articles.

In the symposium[1] entitled *Surface Area Determination* held in 1969, the majority of papers, some 20 in all, dealt with gas–solid adsorption on both porous and non-porous adsorbents.

Books of interest include *The Gas–Solid Interface* edited by Flood[2], a collective work in two volumes, several of the authors being well-known authorities, and a second edition of de Boer's well-known work[3] *The Dynamic Character of Adsorption*, with its emphasis on the lifetime of adsorbed molecules. Also associated with the name of de Boer is the volume[4] *Physical and Chemical Aspects of Adsorbents and Catalysts* compiled by his colleagues as a tribute to him on his retirement; it is a fitting memorial to the founder and inspirer of the Dutch school, who died so tragically soon after its publication. The larger part of Gregg and Sing's[5] *Adsorption, Surface Area and Porosity* is devoted to physical adsorption, though in the restricted context of area and porosity determination; and much of the book *Gas-adsorption Chromatography* by Kiselev and Yashen[6] relates to gas–solid adsorption *per se*, as do certain chapters of *Porous Carbon Solids* edited by Bond[7].

For the period up to 1967 Steele's very comprehensive 74-page review[8] on the *Physical Adsorption of Gases on Solids*, may be profitably consulted. Other useful reviews, more restricted in scope, are those on the Elovich Equation by Aharoni and Tompkins[9] which exhaustively examines the validity of this equation in relation to the rate of both physical and chemical adsorption, and the Kendall Award Symposium on Hydrophobic Surfaces[10] which embraces a useful collection of papers on the interaction of water vapour with solid surfaces. Finally, a warm welcome must be accorded to the *Tentative Manual of Definitions, Terminology, and Symbols in Colloid and Surface Chemistry*[11], issued in 1970 under the auspices of the International Union of Pure and Applied Chemistry, which includes the Gas–Solid Interface as one of its sections. Soon to be officially published, it will form a landmark in the development of the subject. Whilst the present review is not the place for a detailed list of its recommendations, the distinction between the species actually adsorbed – the *adsorbate* – and the material in the fluid phase capable of being adsorbed – the *adsorptive* – is worthy of special mention as meeting a long-felt need. The classification of pores according to width follows current practice, namely (a) *macropores*, pores of width in excess of c. 1000 Å; (b) *micropores*, pores of width less than c. 15 Å; and (c) pores of width intermediate between (a) and (b) which are termed *mesopores*, a felicitous innovation which replaces the earlier terms 'intermediate' and 'transitional'.

For the purposes of the review it will be convenient to deal with topics as follows: specific surface and related matter; the standard isotherm; adsorption in mesopores; adsorption in micropores; theoretical aspects; thermal effects; study of the adsorbed layer by spectroscopic means; the adsorption of water as a special topic.

6.2 DETERMINATION OF SPECIFIC SURFACE

One of the main practical applications of physical adsorption is in the determination of the specific surface of solids; the monolayer capacity, n_m, is evaluated from the isotherm and the specific surface, S, calculated from it by the relation

$$S = n_m \Omega_m L \tag{6.1}$$

(Ω_m = area occupied per molecule in the completed monolayer; n_m is expressed in moles per unit mass of adsorbent; L = Avogadro's Constant.)

Summarising the present position[1,5], one can say that the standard BET method, using N_2 at 77 K, still reigns supreme for solids when the specific surface exceeds a few $m^2\ g^{-1}$; but that the value of $\Omega_m(N_2) = 16.2\ Å^2$, almost universally adopted hitherto, may need modification for certain solids. However, the claims of argon as an alternative adsorptive are being increasingly heard. For solids of surface area below c. 0.1 $m^2\ g^{-1}$ krypton is still the most popular adsorptive, but xenon[13] is also coming into favour, and areas as low as 40 cm^2 can, it is claimed, be determined to within 10%. Other vapours[5] which have been used are carbon dioxide[14], oxygen, ethane, butane[12], benzene, methanol, ethanol, and even water, but there are difficulties as to the appropriate value of Ω_m. Extension of the BET method from impermeable powders (Type (II) isotherms) to mesoporous solids (Type (IV)) introduces no complications, but if the solid contains an appreciable volume of micropores the BET method will give an erroneous value for S, and if the pores are wholly in the micropore region the method breaks down entirely.

The assignment of a value to Ω_m for each adsorbate is a recurring problem, and the painstaking review of McClellan and Harnsberger[17] in 1969 is accordingly particularly welcome. These workers collected all the values of Ω_m they could find in the literature of the previous 25 years, in which the adsorption of two or more vapours had been measured on the same solid. These values, corrected if necessary to $\Omega_m(N_2) = 16.2\ Å^2$, were compared with values of molecular area from molecular models, from liquid density, and from the van der Waals constant b. From their impressive table (compiled from 188 references) they arrived at only five recommended values of $\Omega_m/Å^2$, namely N_2 (77 K), 16.2; Ar (77 K), 13.8; Kr (77 K), 20.2; n-C_4H_{10} (273 K), 44.4; C_6H_6 (293 K), 43.0.

The inclusion of benzene is perhaps open to criticism because of the possibility of specific adsorption (Section 6.2.2).

The list quoted by Gregg and Sing[18] (though after an examination of the more commonly used adsorptives only) is likewise short (Table 6.1). The wide variation of values is obvious and merits more discussion.

Table 6.1 Values of $\Omega_m/\text{Å}^2$ (rounded to nearest Å^2)
(From Gregg and Sing[18], by courtesy of Academic Press)

	Temperature/K	$\Omega_m/\text{Å}^2$ Customary Value	Range
Nitrogen	77	16	13–20
Argon	77	14	13–17
Krypton	77	20	17–22
Xenon	77	25	18–27
Oxygen	85	14	14–18
Ethane	85	21	20–24
Benzene	298	40	30–50

6.2.1 Factors causing a variation in molecular area Ω_m

It is clear that the adsorption potential at a point Z, say, on a normal to the surface even of an ideal solid, must vary periodically as the normal moves about the surface. On the (100) plane of KCl, for example, the calculated force [19] is a maximum at a point midway between the K^+ (or Cl^-) ions and a minimum over a K^+ or a Cl^- ion. Thus the movement of an adsorbed molecule parallel to the surface will encounter energy barriers of height ΔE say, and the degree of mobility will be determined by the value of $\Delta E/kT$ (kT = average thermal energy per molecule). For $\Delta E/kT < 1$ movement will be virtually unhindered, whereas for $\Delta E/kT > 10$ it will be strongly hindered and the molecules will spend almost all their time within potential wells; adsorption will be non-localised in the first case and localised in the second. In the intermediate case where $1 < \Delta E/kT < 10$, adsorption will be part localised, part non-localised, and the time the molecules spend in the wells will increase as the value of $\Delta E/kT$ increases.

Clearly the average area Ω_m occupied by a molecule in the completed monolayer will depend on the degree of localisation. For completely non-localised adsorption, Ω_m will be determined by the size of the adsorbate molecules and the way in which they can pack together; and if this resembles that in a bulk liquid, then $\Omega_m = f(M/\rho L)^{\frac{2}{3}}$ (f is a packing factor = 1.09 for 12-coordination) which for nitrogen at 77 K gives the familiar $\Omega_m = 16.2\,\text{Å}^2$. With completely localised adsorption the molecules will reside within the potential wells, whose positions are determined by the lattice parameters of the solid; Ω_m will now be determined by the size and arrangement of the surface atoms of the *adsorbent* and will be independent of the molecular size of the adsorbate—unless this is large enough to block more than one lattice site. In the intermediate, and commonest, case where adsorption is partly localised and partly non-localised, Ω_m is a function, at present incapable of formulation, of both the lattice parameter of the adsorbent and the molecular size of the adsorbate (ρ = density of bulk liquid).

The vast majority of surfaces are far from ideal and contain point defects, dislocations and steps, which result in a considerable degree of energetic heterogeneity. The effect on Ω_m may be visualised by regarding the heterogenous surface as divided up into a number of homogenous patches. Each patch will then have its individual adsorption isotherm, with its own value of Ω_m, for a given adsorbate. Thus the relative pressure corresponding to completion of the statistical monolayer will vary from patch to patch, so that Point B[20] on the isotherm no longer corresponds to a layer exactly one molecule thick everywhere on the surface.

A quantitive formulation of this idea has been attempted by de Boer and Broekhoff[21] who followed Ross and Olivier[22] in postulating a Gaussian distribution of adsorption energies with a mean energy U and a standard deviation s. The adsorption on each uniform patch is then given by the equation

$$\ln\left(\frac{p}{k_2 p_0}\right) = \ln\left(\frac{\theta}{1-\theta}\right) + \frac{\theta}{1-\theta} - k_1 \theta \qquad (6.2)$$

where k_1 and k_2 are dimensionless constants calculable in principle from the van der Waals constants a and b, and θ ($= n/n_m$) is the surface coverage. By summation the total adsorption on the whole surface can be expressed as a function p/p_0, giving a synthetic isotherm whence Point B, the point corresponding to nominal completion of a monolayer, can be located by the BET procedure. The actual value of θ, say θ_a, at this point is already known from the synthesis and as is seen (Table 6.2) it deviates significantly from

Table 6.2   Coverage (θ_a) at Point B, calculated from synthetic isotherm
(From De Boer and Broekhoff[21], by courtesy of Kon. Ned. Akademie van Wetenschappen)

		s	
U/kcal mol^{-1}		0.200	0.400
1.8	θ_a Ω_m (app)	0.81 19.2	0.80 19.3
2.2	θ_a Ω_m (app)	0.95 16.4	0.99 15.8
2.6	θ_a Ω_m (app)	1.24 12.6	1.22 12.7

U = mean energy;   s = standard deviation of energy distribution; Ω_m(app), see text

unity; the layer is already more than one molecule thick on some patches and less than one on others and the average does not, except by chance, come to unity.

The apparent molecular area obtained by the relation b_2/θ_a (where b_2 is the two-dimensional van der Waals constant, 15.5 Å^2 for N$_2$) is clearly affected strongly by the heat of adsorption and the energy distribution. Though the analysis assumes that the adsorbed film is mobile and obeys a two-dimensional van der Waals equation, the general conclusion that the conventional value of $\Omega_m(N_2)$ corresponds to a particular energy distribution of surface sites seems inescapable.

6.2.2 Specific and non-specific adsorption

Relevant to the question of localisation is Kiselev's distinction[23] between specific and non-specific adsorption. It is based on the well-known classification of the forces operative in physical adsorption into dispersion forces and electrostatic forces. The former, which are always present, give rise to non-specific adsorption, whilst the latter which are present only when the electron distribution is sufficiently asymmetric, lead to specific adsorption. Kiselev's ideas may be summarised in the following way.

Adsorbents are divided into three types, carrying

(I) No ions or positive groups (e.g. graphitised carbon, Teflon),
(II) Concentrated positive charges (e.g. OH groups on hydroxylated oxides),
(III) Concentrated negative charges (e.g. $>O$, $>C=O$).

Adsorbates are divided into four groups, carrying

(a) Spherically symmetrical shells or σ-bonds (e.g. noble gases, saturated hydrocarbons),
(b) π-Bonds (e.g. unsaturated or aromatic hydrocarbons) or lone pairs (e.g. ethers, tertiary amines),
(c) Positive charges concentrated on peripheries of links,
(d) Functional groups with both electron density and positive charges concentrated as above (e.g. molecules with OH or NH groups).

The resultant interactions are as follows:

Adsorbate group	Type of adsorbent		
	(I)	(II)	(III)
(a)	n	n	n
(b)	n	n+s	n+s
(c)	n	n+s	n+s
(d)	n	n+s	n+s

n = non-specific, s = specific, adsorption

(From Kiselev[23], by courtesy of The Faraday Society)

The energy of specific interaction (H), can be assessed by finding the difference in adsorption heats, q^{st}, between a suitable pair of adsorbates, one of them being of group (a), which display close values of q^{st} on a non-specific adsorbant. Thus on graphitised carbon black q^{st} (for $\theta \to 0$) for Ar is 2.4 and for N_2 is 2.3 kcal mol^{-1}, but on hydroxylated silica the values are 2.1 and 2.8, so that $H \simeq 0.7$ kcal mol^{-1}. For n-pentane and ether the corresponding values are 9.2 and 8.9 for the carbon and 7.3 and 15.0 for the silica, giving an H value of c. 8 kcal mol^{-1}.

Barrer[24] on the other hand, plots a curve of q^{st} against polarisability for a number of non-polar molecules, to serve as a reference graph for evaluating the non-specific contribution $q^{st}(n-s)$ of any other adsorbate from its polarisability. The required energy of specific interaction $q^{st}(s)$ is then read off as the difference between the values of q^{st} (experimental) and $q^{st}(n-s)$; for nitrogen on chabasite, for example, Barrer found q^{st} (expt) = 9.0; $q^{st}(n-s)$ = 6.4, giving $q^{st}(s)$ = 2.6 kcal mol^{-1}.

6.2.3 Some further values of molecular area

The very elegant work of Deitz and Turner[25] in which N_2 isotherms were measured on pristine glass fibre drawn in a standard manner so as to have a known area, gave $\Omega_m(N_2)$ = 16.4 Å^2 close to the usual value of 16.2 Å^2. The value of 19.3 Å^2 reported by Pierce and Ewing[26] for a graphitised carbon black is readily accounted for by the localisation of the N_2 on the centres of the carbon hexagons in the basal plane, but Voet's value[27] of 13.8 Å^2 is more difficult to explain. For benzene on graphite, incidentally, Pierce and Ewing find that Ω_m = 47.2 Å^2, independent of temperature between 253 and 293 K, and this they explain in terms of the close fit of the hexagon of benzene to that of graphite.

The increasing use of argon[28] as an adsorbate in specific surface determination has led to deepened interest in its molecular area, particularly in comparison with that of nitrogen. Sing and his collaborators[29] in their study on a series of non-porous oxides of silica and alumina, both fully and partially hydroxylated, found that self-consistent values of specific surface could be obtained if $\Omega_m(Ar)$ were taken as 18.0 Å^2 throughout; but that $\Omega_m(N_2)$ needed to be changed from 16.2 Å^2 on the hydroxylated, to 18.0 Å^2 on the dehydroxylated aluminas, and this they attributed to the specific interaction of the Al^{3+} ions exposed on the surface, with the quadrupole of the nitrogen molecules. (In silica the Si^{4+} ions in the surface layer are almost completely screened by the O^{2-} ions.)

Since the α_s-plot (Section 6.3.2) for nitrogen, but not for argon, was distorted in the multilayer region and failed to pass through the origin when extrapolated, they inferred that a vestige of the specific (localisation) effect persisted into the higher layers. Even so, they concluded that, except when strongly localised, nitrogen is to be preferred to argon for specific-surface determination, because the heat of adsorption of argon tends to be low and Point B of the isotherm accordingly not very well defined. Again because of the specific adsorption of nitrogen, Bassett, Boucher and Zettlemoyer[30] prefer argon for hydroxylated surfaces; they found that internal consistency could be obtained with $\Omega_m(Ar)$ = 16.6 Å^2 throughout, with $\Omega_m(N_2)$ = 16.2 Å^2 on fully hydroxylated, and $\Omega_m(N_2)$ = 15.7 Å^2 on less hydroxylated silicas.

The molecular area of carbon dioxide has been considered by Clough and Harris[14], who measured isotherms on three aluminas at various temperatures in the range 193–293 K and at pressures up to several atmospheres, again with $\Omega_m(N_2)$ = 16.2 Å^2 as standard. $\Omega_m(CO_2)$ came out as 19.3 Å^2 at 191 K and rose to 26.3 Å^2 at 293 K; compared with the value 16.3 Å^2 from liquid density, this bespeaks a very considerable degree of localisation.

6.2.4 Adsorption on modified surfaces

Adsorption on surfaces which have undergone either physical or chemical modification is being increasingly studied. Prenzlow and his co-workers[31], for example, have measured adsorption isotherms of argon on the surface of graphitised carbon black modified by pre-adsorption of ethylene. The isotherm still retained the stepped form it possessed on the bare surface of the carbon, but with progressively lower adsorption in the multilayer region as the pre-adsorption increased from 2 to 26 layers of ethylene.

Chemical modification of the surface of silica gel by replacement of hydroxyl with fluoride ions has been studied by Dubinin[32]; with $\Omega_m(N_2) = 16.2$ Å^2 as standard, Dubinin found that $\Omega_m(Ar)$ changed from 17.6 Å^2 for the unmodified gel to 18.4 Å^2 and then to 18.8 Å^2 as the surface became more and

Table 6.3
(From Dubinin[32], by courtesy of the author)

Solid	Ω_m/Å^2		
	Ar (77 K)	C$_6$H$_6$ (293 K)	H$_2$O (293 K)
(I)	17.6	53	31
(II)	18.4	136	72
(III)	18.8	292	93

(Based on $\Omega_m(N_2) = 16.2$ Å^2)
(II) and (III) were obtained from silica gel; (I) by increasing replacement of OH by F

more fluoridised (Table 6.3). (If alternatively $\Omega_m(Ar)$ were taken as the standard, however, $\Omega_m(N_2)$ would have diminished to 15.6 and to 15.3 Å^2 respectively, and this would be in accord with the lesser specific adsorption of nitrogen on F as compared with OH.) The very high values for $\Omega_m(C_6H_6)$ and $\Omega_m(H_2O)$ on the fluoridised samples imply that the adsorption of water and of benzene is mostly restricted to the few OH groups still remaining on the surfaces.

Replacement of hydroxyl by methyl groups is a very effective way of reducing the polarising power of a solid and hence its capability of specific interaction (Section 6.2.2). This is well illustrated in the results of Tulbovich[33] for the adsorption of benzene and certain substituted benzenes on silica ('Aerosil') before and after its surface had been methylated by reaction with trimethylchlorosilane. The level of adsorption on the methylated silica was consistently lower than on the unmodified silica.

6.3 THE STANDARD ('UNIVERSAL') ISOTHERM

6.3.1 *t*-Plots

More than 20 years ago it was noted by Shull[34], and has since been confirmed by others[35-38], that the isotherms of nitrogen at 77 K on many different solids could be reduced within a few per cent to a common curve when plotted as $n/n_m (= N = $ the number of statistical monolayers) v. p/p_0. The reduced isotherm was used to calculate the thickness, t, of the adsorbed layer at any relative pressure, by assuming a value for the thickness, σ, of a mon-layer. A knowledge of t was required in allowing for the thickness of the adsorbed layer left on the walls during capillary evaporation when the pore size distribution was calculated from the adsorption isotherm (cf. Section 6.4). It was pointed out by Linsen and de Boer[6] that the plot of t (based on $\sigma = 3.45$ Å) against p/p_0 (the 't-curve') offers a very neat means for detection of deviations from the standard, or 'universal', isotherm caused by the onset of capillary condensation in mesopores, or by the presence of microporosity. If the experimental isotherm follows the standard exactly, the corresponding plot of n v. t (the t-plot) would be a straight line passing through the origin and having the slope $n_m/3.54$ if t is expressed in Å (Curve (I) Figure 6.1).

An upward deviation (Curve (II)) denotes the occurrence of capillary condensation in mesopores, slope QR giving the total surface area, internal plus external. If micropores are present in addition, a curve such as (IV) appears, PQ representing the micropore contribution to the adsorption and slope QR the total area excluding that of the micropore walls. If the solid

Figure 6.1 t-Plots. Graphs of amount adsorbed, n, against statistical thickness, t, of the adsorbed layer for: (I) a non-porous powder; (II) a solid containing mesopores but no micropores; (III) a powder containing micropores but no mesopores; (IV) a solid containing both micropores and mesopores

consists of discrete particles containing micropores but no mesopores, a curve such as (III) will result; slope QR now gives the external area of the particles.

The degree of accuracy called for in the t-curve is much higher when it is used for the detection of deviations than when it merely provides a correction, relatively small, on the Kelvin radius (Section 6.4); increasing attention is accordingly being paid to the refinement of the standard isotherm. In 1968 Pierce[40] critically compared the four current standard isotherms by reference to published experimental isotherms on non-porous solids; the criterion he used was that the experimental data should give a straight line passing through the origin when plotted as n against t (or N) taken from the standard isotherm under examination. He concludes that the N-curves (plots of N rather than t, v. p/p_0) of Pierce[36], of Shull[34], and of Cranston and Inkley[35] are superior to the curve of Lippens, Linsen and de Boer[37], the curve of Pierce giving the best overall fit.

Even so, the concept of a universal isotherm, or a standard t (or N) curve, for a given adsorbate such as nitrogen adsorbed on a range of adsorbents, can be no more than an idealisation: the degree of localisation is bound to differ from adsorbent to adsorbent. With oxides and hydroxides the differences may be relatively small because of the screening of the cations by the larger anions[41], so that to a first approximation the solid surface is an array of oxide or hydroxyl ions; but the degree of screening varies according to the size and valency of the cations, so that second-order differences are bound to exist, and each solid must, in fact, have its own standard isotherm. This point has been stressed by Sing and his co-workers who have published standard isotherms for hydroxylated silica[42] and γ-alumina[43]. Standard isotherms for nitrogen adsorbed on anatase and rutile samples have been established by Parfitt, Urwin and Wiseman[48]; and standard t-curves for four organic vapours (benzene, cyclohexane, isopropanol and carbon tetrachloride) derived from isotherms on seven solids including silica and alumina, have been published by Mikhail and Shebl[44]; and for carbon tetrachloride on hydroxylated silica, by Cutting and Sing[49].

6.3.2 α_s-Plots

The t-plot is essentially a neat device for comparing the shape of an experimental with that of a standard isotherm. The α_s-plot of Sing[45] retains this basic idea of the t-plot but frees it from dependence on a knowledge of the monolayer capacity and any assumption as to monolayer thickness σ. Sing plots the standard isotherm with n/n_s ($=\alpha_s$) rather than n/n_m as ordinates to give an 'α_s-curve'; n_s is the adsorption at a given value s of relative pressure, in practice taken as $s = 0.4$. The α_s-plot is then formed in exactly the same way as the t-plot except that the experimental adsorption, n(expt), is plotted against α_s instead of t. Deviations have exactly the same interpretation as before; but the slope of the linear portion is equal not to n_m but to $n_{0.4}$. Calculation of specific surface involves a knowledge of the ratio $n_m/n_{0.4}$:

$$S = n_m L \Omega_m \quad \text{(cf. equation (6.1))}$$

$$= n_{0.4}\left(\frac{n_m}{n_{0.4}}\right) L \Omega_m \quad (6.3)$$

The value of the ratio $n_m/n_{0.4}$ is determined once for all from the standard isotherm. For N_2 at 77 K on standard silica Sing[46] found $n_m/n_{0.4} = 1/1.5$, so that from equation (6.3) with $\Omega_m = 16.2 \text{ Å}^2$,

$$S = 6.5 \times 10^4 n_{0.4} \text{ m}^2 \text{ g}^{-1}$$

or

$$S = 6.5 \times 10^4 b \quad \text{m}^2 \text{ g}^{-1}$$

where b is the slope of the α_s-plot.

An important advantage of the α_s-method is that it can still be applied even where the knee of the isotherm is not sharp, i.e. c is small, whereas all methods such as the t-plot or the N-curve, which involve n_m as a normalising factor necessitate that c shall be large ($c > 50$) in order that the point corresponding to monolayer completion can be determined with reasonable certainty[50]. Thus, in principle, the α_s method is immediately applicable to adsorptives other than nitrogen. This is true even if they give rise to isotherms of Type (III). The best method of calculating the specific surface from the slope, $b(\text{expt})$, say, of the α_s-plot, is by direct comparison with the slope $b(\text{standard})$ of the α_s-plot for the same adsorptive on a standard sample of known area (determined, say, from the nitrogen isotherm).

Then
$$\frac{S(\text{expt})}{S(\text{standard})} = \frac{n_{0.4}(\text{expt})}{n_{0.4}(\text{standard})} = \frac{b(\text{expt})}{b(\text{standard})}$$

whence
$$S(\text{expt}) = \frac{b(\text{expt})}{b(\text{standard})} \times S(\text{standard})$$

The applicability of the method is exemplified by reference to the isotherms of carbon tetrachloride on three samples of non-porous silica[47], which had nitrogen-BET areas of 36, 154 and 194 m^2 g^{-1}. The isotherms of carbon tetrachloride were almost of Type (III) ($c \approx 3$) and the corresponding BET plots gave the obviously erroneous area of 27, 69 and 116 m^2 g^{-1}; the α_s-plots, however (using a flame-produced silica as standard, its nitrogen-BET area being known), gave 36, 153 and 190 m^2 g^{-1} in good agreement with the 'nitrogen' values.

The importance of reference solids, both as standards in surface area determination and for testing models of the adsorption process, has received official recognition. The Colloid and Surface Chemistry Group of the Society of Chemical Industry, with the support of the International Union of Pure and Applied Chemistry, set up in 1970 a committee to collect standard samples, and to organise the determination of their isotherms of nitrogen and krypton adsorption in a number of surface-chemical laboratories. The intention is to set up a bank of standard samples which will be in the custody of the National Physical Laboratory.

A task for the future is to extend the scope to other adsorbates such as the inert gases Ar and Xe, the simple hydrocarbons and perhaps simpler inorganic gases such as nitrous oxide.

6.4 ADSORPTION ON MESOPOROUS SOLIDS: PORE SIZE DISTRIBUTION

Analysis of the isotherm of physical adsorption offers virtually the only means of estimating the pore size distribution of solids in the mesopore range, c. 20 to c. 500 Å in width. In practice all calculations are based on the assumption that the adsorbate condenses as a liquid within the pores ('capillary condensation'), a process governed by the thermodynamic equation

$$\gamma \, ds = \Delta\mu \, dn \tag{6.4}$$

where $\Delta\mu$ is the change in chemical potential between the adsorbed and the standard state (the free liquid), γ the surface tension of the liquid adsorptive, and ds is the incremental area covered when dn moles condense. The form of the equation most frequently used is the Kelvin equation

$$\ln p/p_0 = -2\gamma V_1 \cos \phi / RT.r_k \qquad (6.5)$$

where V_1 is the molar volume, and p_0 the saturated vapour pressure of the liquid adsorptive; p is the pressure of vapour in equilibrium with the capillary-condensed liquid and ϕ is the angle of contact between the capillary-condensed liquid and the walls of the pore, which carry an adsorbed film (of thickness t, cf. Section 6.3.1). The pores are assumed to be cylindrical, but r_k is the radius, not of the pore itself, but of the 'core', i.e. the part of the pore in which capillary condensation occurs, so that pore radius r is

$$r = r_k + t \qquad (6.6)$$

In applications of equation (6.5) the simplifying assumption is invariably (and usually tacitly) made that $\phi = 0$, solely because we lack a satisfactory means for direct determination of ϕ; and though it is widely recognised that in very narrow pores, at the lower end of the mesopore range, both γ and V_1 may differ appreciably from their bulk values, it is always assumed for simplicity that the difference is negligible.

Spencer and Feredy[51] have stressed the need for caution in this respect, which they illustrate by assigning a range of possible values to the questionable factor $(\gamma/p) \cos \phi = F$, say (cf. equation (6.5)), and calculating the pore size distribution from a given adsorption isotherm of nitrogen, using standard procedure (p = density of liquid adsorptive). They found that the value of the radius corresponding to the maximum in the pore-size distribution curve changed from 14 Å for $F = 0.5$ to 34 Å for $F = 2.0$, a far from negligible variation. They concluded that the very validity of the Kelvin equation as currently used 'should be exhaustively examined'.

6.4.1 Modes of calculation

The calculation of pore size distribution is complicated by the fact that when capillary evaporation occurs within a pore of radius, r, an adsorbed film remains on the walls having a thickness, t, which is a function of p/p_0. As the equilibrium pressure is reduced from p_{i+1} to p_i, say, the amount desorbed includes not only the liquid evaporated from the core of radius $r_{k,i}$, but also that adsorbate which is desorbed from the walls of all pores having radii in excess of r_i, as their adsorbed film diminishes in thickness from t_{i+1} to t_i. Various devices have been adopted for coping with this complication, all of them somewhat tedious and demanding close attention to detail since errors are cumulative. The method recently proposed by Roberts goes a considerable way towards minimising these drawbacks.

Novel features of the Roberts method[52] are that the pore radius rather than relative pressure is taken as the independent variable, and that adsorption values are converted into the equivalent volume of liquid at the outset rather than at a later stage as in other methods. The pores are divided into groups according to their radius, each group being assigned the same average radius, $\bar{r}$. The range from 100 to 20 Å is divided into eight steps of 10 Å each, and that from 20 to 10 Å into two giving ten groups in all, numbered 1 to 10.

For each group the pore volume is related to the core volume by reference to a model, usually either cylinders or parallel-sided slits. Calculations are carried out – another innovation – in terms of pore volume rather than core volume.

In general, if W is the core volume of a group of pores of radius $\bar{r}$, the corresponding pore volume V (for cylindrical pores) is given by $V/W = \{\bar{r}/(\bar{r}-t)\}^2$; thus for a particular group having pore radius $\bar{r}_i$, film thickness t_j and core volume W_{ij} we have

$$\frac{V_i}{W_{ij}} = \left(\frac{\bar{r}_i}{\bar{r}_i - t_j}\right)^2$$

i.e.
$$V_i = Q_{ij} W_{ij} \tag{6.8}$$

where
$$Q_{ij} = \left(\frac{\bar{r}_i}{\bar{r}_i - t_j}\right)^2 \tag{6.9}$$

It will be noted that the first suffix refers to pore radius and the second to film thickness.

Consider now the first three groups of pores, which initially are all full. When p/p_0 falls from unity to $(p/p_0)_1$ (which by the Kelvin equation corresponds to core radius $(r_1 - t_1)$), an amount w_{11} expressed as liquid volume is desorbed, so that the volume V_1 of this group of pores is by equation (6.8)

$$V_1 = Q_{11} W_{11} = Q_{11} w_{11} \tag{6.10}$$

When the pressure falls to $(p/p_0)_2$, the cores of group 2 give up a volume W_{22}; also the thickness has fallen to t_2 so that the total desorption from both groups, w_2, is

$$w_2 = W_{22} + W_{12}$$

Again by equation (6.8) the volume V_2 of the second group is

$$V_2 = Q_{22} W_{22} = Q_{22}(w_2 - W_{12})$$
$$= Q_{22}(w_2 - V_1/Q_{12})$$

By an extension of the argument, the volume V_3 of the third group is

$$V_3 = Q_{33} W_{33}$$
$$= Q_{33} \{w_3 - V_2/Q_{23} - V_1/Q_{13}\} \tag{6.11}$$

and so on up to group 10.

It is now only necessary to draw up a table of the values of Q_{ij}, calculated from equation (6.9) for all values of i and j ranging from 1 to 10; $\bar{r}_1$ is, of course, 95 Å, $\bar{r}_2$ is 85 Å etc. The values of t_1, t_2 etc. are read off from a t-curve (Section 6.3.1) for values of $(p/p_0)_1$, $(p/p_0)_2$, etc. which through equation (6.5) (with $\phi = 0$) and equation (6.6) correspond to $\bar{r}_1$, $\bar{r}_2$ etc.

The great advantage of the method is that once drawn up the table is applicable to all experimental isotherms. (Separate tables are required for a parallel plate model and for interstices between spheres.)

A full table for the cylindrical model appears in Roberts' paper, and can be made up into a standard worksheet into which the experimental data can be inserted so that the arithmetic operations can be carried out rapidly and easily. Examples of its application to experimental isotherms of nitrogen are also given. Though the procedure can readily be converted into a computer program, there is nevertheless a signal advantage in having at hand, in the laboratory, a means of immediate calculation of the results of a pore size distribution experiment.

All cylindrical pore models hitherto involve a squaring of the pore radius in the course of the calculation. The modified method of John and Bohra[56] is an attempt to circumvent this stage with its attendant multiplication of errors. In effect, however, their procedure ignores the substantial change in the area of the film on the walls as evaporation proceeds and is therefore invalid.

Hitherto, virtually all calculations have been based on the isotherm of nitrogen at 77 K; there would, however, be much advantage in extension to other adsorptives and a start in this direction has been made, with organic vapours on silica gel, by Mikhail and Shebl[44].

6.4.2 The 'model-less' method[53]

Brunauer and his colleagues[54] have pointed out that virtually all the existing methods of pore size analysis assume, at an early stage, a model of pore shape in order to convert core volumes into pore volumes; yet most of the pores of most actual solids have irregular shapes for which no model is possible, particularly since the actual shapes themselves are usually unknown. They suggest therefore that it is better in the first instance to calculate the *core size* distribution rather than the pore size distribution itself. For this purpose they recommend the use of the hydraulic radius r_h (volume:surface ratio) in preference to equivalent cylindrical radius; and they calculate the surface area of the cores (i.e. the area of the film left on the walls after capillary evaporation) by use of an integrated form of equation (6.4), namely

$$s = (RT/\gamma) \int \ln(p/p_0)\, dn \qquad (6.12)$$

When p/p_0 is lowered from 1.0 to 0.95, and a quantity δn_1, say, is desorbed, a group of cores of volume $\delta v_1 = \delta n_1 \cdot V_l$ is emptied of capillary condensed liquid (V_l = molar volume of liquid adsorptive). The area δs_1 of the adsorbed film is obtained by integration of equation (6.12) between the limits of n corresponding to $p/p_0 = 1.0$ and $p/p_0 = 0.95$. The average hydraulic radius of this group of cores is $r_{h,1} = \delta v_1/\delta s_1$. Similarly, when the relative pressure is lowered to 0.92 and a further δn_2 is desorbed the hydraulic radius of the second group of cores is $r_{h,2} = \delta v_2/\delta n_2$. Proceeding thus down the desorption branch of the isotherm one arrives at a core size distribution, a curve of $\delta v/\delta r_h$ v. r_h. So far no correction for the thinning of the adsorbed film on the walls has been made; Brunauer maintains, however, that the *core size* distribution graph is well suited to most practical purposes.

To correct for the thinning effect it is necessary to assume a model of pore shape and to use a t-curve. If δn_2 is the amount desorbed in the second stage (say, from 0.95 to 0.92 p_0) and $\delta n'_2$ the amount desorbed from the walls of the first group, then the amount desorbed from the cores of the second group is $(\delta n_2 - \delta n'_2)$, and the volume is $(\delta n_2 - \delta n'_2)V_l$. The corrected surface area S_2 (corr) of the cores of this group is obtained by integrating equation (6.12) with $(\delta n_2 - \delta n'_2)$ as upper limit and $n_0 (=$ saturation adsorption) as the lower. Thus the corrected hydraulic radius of the second group of cores is $(\delta n_2 - \delta n'_2)V_l/S_2$(corr). Calculation of $\delta n'_2$, if the walls are parallel plates is given by $\delta n'_2 V_l = (t_1 - t_2)S_1$. (If they are cylinders this is still true within 1 or 2%.)

Finally to convert core parameters into pore parameters one must use a model; if the shape is unknown Brunauer recommends the cylindrical model, for which $v(\text{pore})/v(\text{core}) = (2r_h + t)^2/(2r_h)^2$.

The problem of the calculation of *surface area* from the isotherm in the hysteresis region has been considered at length by Broekhoff and de Boer[57, 58]. Again, they make use of an integrated form of equation (6.4), in which $\Delta \mu$ has been set equal to $\mu_l - \mu_g - F(t)$; here μ_l and μ_g are the chemical potentials of the liquid and vapour respectively and $F(t)$ is a function of t which can be evaluated from the t-curve. Applicability to various shapes of pore and types of hysteresis loop is discussed in detail. As the procedure is concerned only with the region of the isotherm beyond Point B, it offers the possibility (at any rate when the mesopores are slit shaped) of evaluating specific surface area even if micropores are present.

6.4.3 The 'tensile-strength' hypothesis

In 1965 Harris[59] drew attention to the fact that the isotherms of nitrogen at 77 K very frequently close at a relative pressure ~ 0.45, but never below. Interpreted in terms of the Kelvin equation this would mean that in a large number of adsorbents the pore size distribution shows a sudden cut-off at exactly the same radius, $r = 18$ Å, corresponding to $p/p_0 = 0.45$, a highly improbable state of affairs. Harris accordingly suggested that a change in the mechanism of adsorption occurred at this point but gave no details.

Actually, in 1948 Schofield[60] had already attributed the closure point to limitations on the capillary condensation mechanism imposed by the tensile strength of the liquid adsorbate, an idea which has been elaborated by Flood[61], by Melrose[62] and by Everett[63].

The pressure difference, Δp, across a concave meniscus of liquid, of mean radius r_k, in equilibrium with its vapour, is given by the Laplace equation

$$\Delta p = p_l - p = -2\gamma/r_k \qquad (6.13)$$

where p_l and p are the pressures in the liquid and vapour phases. Thus the liquid will be subject to a negative pressure, or tension $\tau(= -p_l)$.

Since by the Kelvin equation (with $\phi = 0$), cf. equation (6.5),

$$\ln p/p_0 = -2\gamma V_l/RT.r_k \qquad (6.14)$$

elimination of r_k between equations (6.13) and (6.14) shows that the tension set up in the liquid condensed in a pore of radius r^k will be given by

$$\tau + p = -(RT/V_1) \ln p/p_0 \quad (6.15)$$
$$\text{or} \quad \tau \simeq -(RT/V_1) \ln p/p_0 \quad (6.16)$$

since $p \ll \tau$.

As soon as τ exceeds the tensile strength, τ_0, of the liquid adsorptive the capillary-condensed liquid will become unstable and will tend to evaporate. This will occur as soon as p/p_0 falls to the value $(p/p_0)_c$ given by insertion of τ_0 in equation (6.16), i.e. will take place in pores of radii equal to or less than the value $r_{k,c}$, in turn given by putting $(p/p_0) = (p/p_0)_c$ in equation (6.14). The hysteresis loop will therefore close at the relative pressure determined by the tensile strength of the liquid adsorptive, no matter whether the pore system extends to radii below $r_{k,c}$ or not.

A direct test of the hypothesis would involve a comparison of the value of τ calculated from the closure point of the loop by equation (6.16), with an independently determined value of τ_0. Unfortunately, the direct determination of the tensile strength of a liquid is experimentally difficult, and very few recorded values are extant; consequently the tests that have been carried out so far have had to be somewhat indirect.

Thus Kadlec and Dubinin[64], from calculations of molecular forces, conclude that $\tau_0 = 2.06/d_0$, where d_0 is the mean separation of molecules, calculated from the bulk liquid density; and since (cf. equation (6.13)) $\tau = 2\gamma/r_{k,c}$ one obtains $d_0/r_{k,c} = 1.03$ independently of the nature of the adsorbate. In fact, for four different adsorbates on a number of adsorbents the values of $d/r_{k,c}$ were about the same, 0.34 ± 0.02, but widely divergent from the expected value; water gave 0.21–0.30. The modification of properties of liquids in fine tubes is suggested as an explanation of the discrepancies.

Everett and Burgess[65] have adopted a different approach; they point out that the temperature dependence of τ calculated by equation (6.16) should correspond with that expected for the tensile strength of a liquid obeying an equation of the van der Waals type; they calculated τ from the closure points of the isotherms of a number of adsorbates, including nitrogen and benzene, at several temperatures, and plotted T/T_k against τ/p_k (T_k and p_k are the critical temperature and pressure of the bulk adsorptive). The reasonably good fit of the data to the line predicted by a modified van der Waals equation (Guggenheim[66]) lends support to the hypothesis.

Further (indirect) support is provided by the fact that isotherms of butane at 273 K or samples of graphite ball-milled[67] for periods up to 1000 hours all showed a shoulder at the same relative pressure of 0.40 corresponding to $r_{k,c} = 19$ Å. The average particle size varied widely – the specific surface ranged from 10 to 600 m² g⁻¹ – and it is highly unlikely that all the specimens would have a system of interparticulate pores with exactly the same minimum r_k value.

Similar arguments apply to the nitrogen isotherm at 77 K obtained on seven different powders compacted[15, 16] at a number of different pressures up to 100 ton in⁻². The constituent particles included spheres (e.g. silica), plates (mica), tubes (halloysite) of various sizes, so that the interstitial pores must have been correspondingly varied; nevertheless in all cases the loop closed at a relative pressure of 0.45 or above but never lower.

6.5 ADSORPTION ON MICROPOROUS SOLIDS

The special role of micropores — pores having a width not exceeding three of four molecular diameters — first pointed out by Pierce and Smith[68] and by Dubinin[69], and emphasised by Kiselev[70], is slowly receiving its proper recognition. Most workers would now agree that when micropores are present in a solid, the isotherm is distorted in the low-pressure region, so that the surface area calculated by the BET procedure is too high — by an amount depending on the exact width and volume of the micropores. This is particularly true of Type (I) isotherms since these result from systems in which the pores are wholly in the micropore region and the external surface is negligible.

6.5.1 The detection of micropores

The use of the t-plot of Lippens and de Boer[37] and the newer α_s-plot of Sing[45], as a means of detecting the presence of micropores when the nitrogen isotherm is of Type (II) or Type (IV) has already been outlined. Since the adsorbate

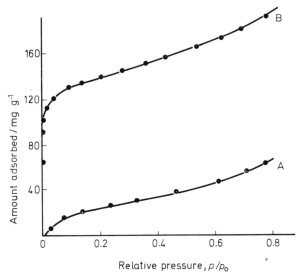

Figure 6.2 Detection of microporosity by pre-adsorption of nonane. Isotherms of nitrogen at 77 K on a microporous carbon: (A) after pre-adsorption of nonane; (B) before pre-adsorption of nonane (From Langford[16], by courtesy of the author)

within the micropores is believed to be in a state closely resembling, but not identical with, that of the liquid, it is possible to estimate the micropore volume by converting the intercept on the n-axis of the t- or α_s-plot to a liquid volume by multiplication with the molar volume of the liquid adsorptive.

Distortions of the isotherm, and therefore of the t- or α_s-plot may arise however not only from microporosity but also from adsorption on specially active sites, or from chemisorption. The pre-adsorption method described by Gregg and Langford[71] enables a distinction to be made between these possibilities. A microporous carbon black, after outgassing, was exposed at 77 K to n-nonane vapour which was subsequently removed from the external surface but not from the micropores by opening to the pumps with the carbon now at ambient temperature. The nitrogen isotherm (Type (II)) yielded a BET surface area close to the geometrical area as found by electron microscopy, and in the multilayer region it ran parallel to the nitrogen isotherm obtained before the pre-adsorption of nonane (Figure 6.2). The vertical separation of the curves in this region is equal to the micropore contribution. Nonane would not be adsorbed particularly strongly on active sites, nor, on the vast majority of solids, would it be chemisorbed.

By way of confirmation, Diano[72,73] — with hexane rather than nitrogen as adsorbate — found that the heat of adsorption in the micropores was higher than on the external surface of the carbon black.

Enhancement of the heat of adsorption, implicit in the concept of micropores, was also demonstrated by Anderson and Horlock[74] with a magnesium oxide, shown by the t-plot to be microporous. The isosteric heat of adsorption of nitrogen and of argon was consistently much higher than the theoretical value for the exposed (100) plane of magnesium oxide, as calculated from interatomic forces (Section 6.6.1); e.g. for one sample, at a coverage of $\theta = 0.2$, q^{st} was 3 kcal mol^{-1} and the theoretical value 1.9 kcal mol^{-1}.

6.5.2 The Dubinin–Radushkevich (DR) equation[75,76]

A method for the calculation of micropore volume from the adsorption isotherm, based on the Polanyi potential theory of adsorption, was put forward some years ago by Dubinin and Radushkevich (DR) who arrived at the equation

$$\log_{10} a = \log_{10}(V_0 \rho) - D(\log_{10} p_0/p)^2 \qquad (6.17)$$

Here a is the adsorption in mass units; V_0 is the micropore volume and ρ is the density of the liquid adsorptive. D is a constant related in principle to properties of the adsorbent and the adsorbate, but empirical in practice. A plot of $\log a$ v. $(\log p_0/p)^2$ should therefore yield a straight line with an intercept whence the micropore volume could be immediately calculated.

A critical examination of the applicability of the DR equation by reference to a range of adsorbents and adsorbates has been undertaken by Marsh and his co-workers[77]. They found that there are, indeed, numerous cases in which a straight line, covering the very wide range of relative pressures from 10^{-5} to near unity, is obtained when the data are plotted according to equation (6.17). Often, however, deviations appear. Sometimes the plot deviates upwards at the high-pressure end, presumably through capillary condensation in mesopores or multilayer adsorption on the external surface; occasionally *two* straight lines are obtained, and since the high pressure (low $(\log p_0/p)^2$) branch has the lesser slope extrapolation of the low-pressure branch would lead to erroneously high values of V_0. Sometimes the plot is curved and the extrapolation for V_0 breaks down entirely.

It would thus seem that the DR plot has not as yet established itself as a means for calculation of micropore volume from the low pressure region of the isotherm. If the isotherm is of Type (I) it is necessary to continue the measurements into the plateau region to confirm that the DR plot does not deviate from linearity at high pressures; the plot then becomes merely an analytical method for estimating the saturation value of adsorption, more satisfying mathematically perhaps than simple extrapolation of the isotherm itself to $p/p_0 = 1$, but still empirical. If the isotherm is of Type (II) or (IV) the method is inapplicable; the adsorption values in the region of the isotherm prior to Point B include contributions from both the filling of micropores and the formation of a monolayer on the remainder of the surface, and there are no means of deciding the proportion of each contribution.

Attempts to solve the important problem of calculation of the pore size distribution in the micropore region continue to be made. Brunauer and his associates[53, 55] have developed the method of de Boer and his co-workers[78], which was based on the principle that specific surface is proportional to the slope of the t-plot (Section 6.3.1). Tangents are drawn to the t-plot in the micropore region and the slope of each is read off. If the slopes for $t = t_1$ and $t = t_2$ give surfaces of S_1 and S_2 respectively, then if the pore walls are parallel plates the volume v_2 of the group of pores having width $(t_1 + t_2)$ is $\frac{1}{2}(S_1 - S_2)(t_1 + t_2)$. It is shown that the same value of v_2 will result if the pores are cylindrical. The analysis is continued till there is no further decrease in the slope of the t-plot, and the pore size distribution is the curve of v v. hydraulic radius ($=$ half-width, e.g. $\frac{1}{2}(t_1 + t_2)$).

The authors recognise that because of enhanced heat of adsorption the t-curve is, in the micropore region, distorted from its course for an open surface. To meet this difficulty they have at their disposal a series of t-curves, each corresponding to a particular c-value and obtained from isotherms on non-porous substances[44]; and they select that one which has the same c-value as does the isotherm on the microporous substance under test.

This device is open to criticism, however: the c-value is calculated from the linear part of the BET plot which corresponds to the limited range around Point B of the isotherm. The micropore region however extends from the origin upwards and, because of the size distribution includes a *range* of values of the heat of adsorption, so that there is no one value of c covering the whole region.

The reviewer is reluctantly forced to the conclusion that at present there exists no reliable method for estimation of pore size distribution in the micropore region.

6.5.3 Activated entry

Discussion of microporosity is incomplete without reference to the activated-entry effect[79] which is produced by presence of very fine constrictions (only slightly greater than the diameter of an adsorbate molecule) within micropores. Passage of adsorbate molecules through the constriction into the space beyond will encounter an energy of activation, and consequently the measured adsorption will *increase* with rise in temperature. An example is provided by isotherms on carbon rendered microporous by controlled oxidation in air[80]; with one typical sample the volume (calculated as liquid) of butane adsorbed at saturation was 2.4, 8 and 21 mm^3 g^{-1} at 196, 227 and 273 K respectively; the value for carbon dioxide was 21 mm^3 even at 196 K, because the carbon dioxide molecule is smaller than butane (diameters 2.8 and 4.9 Å respectively). Activated entry is frequently the explanation of a hysteresis which persists to 'zero' pressure.

6.6 THEORETICAL DEVELOPMENTS

Attempts are constantly being made to refine, and extend the scope of, calculations of adsorption parameters from known properties of the participating substances. These may start either with the laws of interatomic forces, or from a statistical model.

6.6.1 Calculations of interaction energy

The energy of interaction $u(r)$ of two isolated atoms as a function of the distance of separation is given by an expression of the general form $u(r) = -A/r^m + B/r^n$; common values of m and n are $m = 6$ and $n = 12$. A is a constant referring to attraction and is calculable in principle from the polarisability of the two atoms; B is a constant, in practice empirical, referring to repulsion. The procedure, which is well reviewed by Pitzer[81], is to sum pair-wise the interaction of each atom in the semi-infinite solid with each atom in the molecule of adsorbate, with allowance for electrostatic interactions if permanent dipoles or quadrupoles are present. One thus arrives at an expression for the adsorption energy $U(z)$ of an adsorbate molecule in terms of its distance z from the surface. Additional refinements are the inclusion of allowance for the interactions of adsorbed molecules with each other, sometimes with an additional correction for the perturbation of a pair-wise interaction by the presence of a molecule nearby (three-body effect)[82, 83]. In the majority of cases attention has perforce been restricted to adsorbates with very simple molecules. By means of a calculation of this kind Kiselev[84] and his associates showed that $U(z)$ is higher for argon adsorbed at a growth step (-2.94 kcal mol^{-1}) than on the smooth surface (-1.98 kcal mol^{-1}) of sodium chloride—as indeed would be expected intuitively. The occurrence of two-dimensional condensation at temperatures below the two-dimensional critical temperature of 80–85 K was also predicted.

The case of carbon monoxide on sodium chloride has been dealt with by Folman and co-workers[85], with allowance for induced dipole–quadrupole interaction. They were led to infer that the most stable position for the carbon monoxide molecule was in a normal orientation above a Na^+ ion. They also predicted the displacement of the infrared band to be 9 cm^{-1} in quite good agreement with the experimental value of 10–14 cm^{-1}.

Anderson and Horlock's calculation of $U(z)$ for nitrogen on magnesium oxide[74] has already been mentioned (Section 6.5.1). In Melrose's treatment[86] of the water–graphite system attention is focused primarily on the liquid–solid rather than the vapour–solid interface. Mayer[87] stresses the importance of taking into account the anisotropy of polarisability when graphite is the adsorbent. If this anisotropy is taken into account the edge planes exhibit an adsorption energy which is higher than the basal in the ratio of 1.5:1, whereas if the anisotropy is neglected it is *lower* in the ratio of 1.3:1.

A very detailed calculation by Ricca and his associates[88] for helium adsorbed on the (100) and (110) faces of krypton and xenon led to the conclusion that there is a high degree of mobility in the adsorbed layer and that sojourn times are $c.\ 10^{-10}$ and 10^{-8} seconds on solid krypton and xenon respectively.

From measurements of the change of the work function of palladium (outgassed in ultra-high vacuum) with increasing coverage of adsorbed xenon, Palmberg[89] deduced that the polarisability of xenon in the adsorbed state was 8.2×10^{-24} cm^3 molecules^{-1}, more than twice that for free atoms of xenon (4.0×10^{-24}). This points to a difficulty in calculations of $U(z)$, since these are conventionally based on normal values of polarisability.

6.6.2 Statistical-mechanical treatments

Numerous calculations, particularly of chemical potential, entropy and heat of adsorption, along statistical-mechanical lines have been made by the Kiselev school[90–93], mostly for carbon or zeolites as adsorbents and the noble gases or alkanes as adsorbates. Various models are considered, some including adsorbate–adsorbate interaction, with and without localisation of adsorption.

The properties of the adsorbent are taken into account for the first time in statistical-mechanical treatments of adsorption by Pierotti[94], whose work has been extended by Rudzinski[95].

6.6.3 Equations to the adsorption isotherm

Attempts are constantly being made to increase the range over which the course of the isotherm can be represented by an analytical expression. Amongst these is the equation of Brunauer, Skalny and Bodor[96] which is a modification of the BET equation for the case that the adsorbed film is only 5–6 molecules thick at a relative pressure of unity, i.e. when the saturated vapour pressure is reached. This equation

$$\frac{kx}{n(1-kx)} = \frac{1}{n_m c} + \frac{c-1}{n_m c} \cdot kx \qquad (6.18)$$

is identical with the conventional BET equation except that the relative pressure x is replaced throughout by kx. It is stated to fit the Shull composite isotherm[34] well, but it suffers from the drawback of containing an additional empirical constant k.

Greenleaf and Halsey[97] propose the equation

$$p = b'n \exp\left[(c'/b')n\right] + p_0 \exp\left[-a'(n/n_m)^{-3}\right] \qquad (6.19)$$

where $c' < 0$, $a' > 0$, $b' > 0$; a' is a measure of the adsorption potential; b' and c' can be expressed in terms of gas–solid virial coefficients (cf. Steele[8]). This equation has the advantage of reducing to Henry's law at low coverages and to Halsey's θ^3 law at high coverages: when $\theta(=n/n_m)$ is small, equation (6.19) becomes $p = b'n + c'n^2$ or for $\theta \ll 1$, $p = b'n$. When θ is large, it reduces to

$$\ln(p/p_0) = -a'(n/n_m)^{-3} = -a'\theta^{-3}$$

It is claimed to give good agreement with experimental isotherms.

The isotherm of Takizawa[98] is based on a hypothetical division of the surface into regions of different adsorption potential on each of which adsorption follows the BET equation which may be recast[99] into the form:

$$n = \frac{x}{1-x} \cdot \frac{n_m}{1+(c-1)x}$$

Takizawa's equation thus reads

$$n = \frac{x}{1-x} \sum \left\{ \frac{n_{m,i} \cdot c_i}{1+(c_i-1)x} \right\} \tag{6.20}$$

It is shown that, so long as $c > 10$, the apparent monolayer capacity evaluated from a conventional BET plot is almost equal to the sum of the individual monolayer values. An analytical method is proposed for obtaining the individual values of $n_{m,i}$ and c_i from the low pressure part of the isotherm but the procedure is essentially one of curve fitting with an array of disposable constants, inasmuch as there are no independent means of finding the relative values of $n_{m,1}$, $n_{m,2}$, etc. or of the corresponding values of c_1, c_2, etc.

The Langmuir mechanism of layer by layer condensation and evaporation which lies at the basis of the BET and Langmuir equations has been challenged by Jovanović[100]; he concludes that the resultant mathematical expressions are incorrect because they fail to allow for the probability that a molecule which has only just become detached from the surface may be driven back by collision with an oncoming molecule. He arrives at the isotherm equation

$$n = n_m[1-\exp(-a''x)]\exp(b''x) \tag{6.21}$$

where a'' and b'' are constants referring to adsorption in the first and higher layers respectively; they are given by

$$a'' = f\tau_1 \text{ and } b'' = f\tau_L$$

where

$$f = \sigma_1 p_0 (2\Pi m k T)^{-\frac{1}{2}}$$

τ_1 and τ_L are the 'settling times' (sojourning times) in the first and higher layers, σ_1 is the area of an adsorbed molecule and m its mass; k is the gas constant per molecule.

In equation (6.21) three disposable constants a'', b'' and n_m are involved; but if the knee of the (Type (II)) isotherm is reasonably sharp, then the value of a'' is high ($20 < a'' < 50$) so that the term $\exp(-a''x)$, for relative pressures above 0.25, is negligible compared with unity (<0.01). Consequently, in the multilayer region equation (6.21) simplifies to $n = n_m \exp(b''x)$
or
$$\ln n = \ln n_m + b''x \tag{6.22}$$

so that a plot of adsorption n against relative pressure p/p_0 should give an intercept equal to ln (monolayer capacity).

Thus the monolayer capacity is obtained from the properties of the higher layers where the effect of surface heterogeneity is much diminished (cf. Sing, Section 6.2.3).

The rather limited tests adduced show that equation (6.22) fits the experimental data closely over the range $0.2 < p/p_0 < 0.7$; and the calculated values of τ_1 and τ_L have the reasonable values $c.\ 10^{-8}$ and $c.\ 10^{-13}$ seconds respectively for butane on glass at 195–273 K.

6.6.4 Thermodynamics

In his comprehensive paper on the thermodynamics of the gas–solid phase Erikson[101] lays particular emphasis on the dimensional changes which occur on adsorption. He reiterates the distinction first made by Gibbs between the work of stretching the surface, which is related to the surface tension, γ, and the work of cleavage (work of forming new area) which is related to the excess free energy surface density, g_{ex}^s. He maintains that by dealing with these two separate quantities and using two dividing surfaces, the rather unsatisfactory concept of surface pressure can be avoided, and that a satisfactory framework for analysing the dimensional changes caused by adsorption is provided. This treatment enables him to explain the anomalous contractions which are found at coverages below $c.\ 0.5$ when polar gases such as NH_3 or CH_3Cl are adsorbed on certain solids.

The thermodynamics of physical adsorption of gases is also treated very fully by Schay[102] in a recent paper. He approaches the subject through the concept of excess quantities, and takes account of the fact that the surface area of a solid (unlike that of a liquid) cannot be regarded as an independent variable in a surface chemical experiment, because the area of a given sample cannot be varied at will during the experiment.

6.6.5 The Henry's law region

Theoretical aspects of adsorption are also dealt with in a paper by Everett[103] in his discussion of two high-temperature methods for the determination of specific surface S. The first uses the Henry's law region ($\theta < 0.01$), the theory of which will depend only on adsorbent–adsorbate interactions and will be independent both of the interactions between the adsorbate molecules and of their size. The second applies to measurements just outside the Henry's law region, so that deviations are caused by interactions between pairs of adsorbate molecules; the theory now requires a knowledge of the laws governing forces between adsorbed molecules; this method for S has the advantage that, since it is largely independent of interactions between the adsorbent and the adsorbate, it should be relatively insensitive to heterogeneity of the solid surface.

For the first method, Everett arrives at the expression

$$k_H = \frac{Sz_0}{RT} F\left(\frac{\varepsilon^*}{kT}\right)$$

where k_H is the Henry's law constant ($n = k_H p$); z_0 is the distance from the Gibbs dividing surface to the surface of the solid; ε^* is the depth of the minimum in the curve of ε (= adsorption potential) v. z. Everett evaluates $F(\varepsilon^*/kT)$ from various established expressions for interatomic force, and z_0 from, for example, the van der Waals radii, and proceeds to the calculation of S for a graphitised carbon black from the experimental isotherm. Adsorbates included the noble gases and hydrogen and although the values of S obtained varied fairly widely amongst themselves (up to 100%) they cluster around the BET-nitrogen value of 12.5 m² g^{-1}.

For the second region, it has been shown that

$$\ln n/p = \ln k_H - 2nB^*/\Omega \tag{6.23}$$

where B^* is the two-dimensional second virial coefficient[104] which describes the deviation of the spreading pressure Π from ideality:

$$\Pi S = nRT[1 + nB^*/S + \cdots]$$

B^* was again evaluated from an expression for $U(z)$, the 'three-body effect' (Section 6.6.1) being allowed for in four different ways. Applied through equation (6.23) to the experimental isotherms of krypton, xenon and methane, they led to values for the specific surface of carbon black which agreed well at 8.8 ± 1.0 m² g^{-1}. The discrepancy from the BET value of 12.5 ± 2 m² g^{-1} (from Ne, Ar and Kr isotherms) is considered not to be greatly outside the error of the two methods.

Whilst more work is clearly required, the outlook for the high-temperature methods, particularly the one using the second virial coefficient, is regarded as promising when used with the noble gases. Its use with more complex molecules is likely to be of limited value owing to complications such as hindered rotation and variation in orientation.

6.7 THERMAL QUANTITIES

6.7.1 Heat of adsorption

The role of thermal quantities in physical adsorption is largely one of testing, confirming or supporting the results or conclusions from other measurements, notably of adsorption isotherms. As such, references to the heat of adsorption will be spread throughout the review, but a brief section devoted to the subject *per se* may not be amiss.

Of the various methods available, calculation from isotherms at neighbouring temperatures to give the 'isosteric heat of adsorption', q^{st}, is by far the commonest because of the relative ease of experimentation. A high degree of experimental precision is called for, however; moreover unless complete thermodynamic reversibility is present spurious results will be obtained.

As a variant of the 'isosteric method', the adaptation of gas chromatography[6] has become increasingly popular, both because of the relative ease and rapidity of operation and on account of the extended range of temperatures which is offered by the use of high-boiling adsorptives. The slope of the curve of $\log_{10} t$ against $1/T$ (t = corrected retention time) is equal to $q^{st}/2.303R$. Unfortunately, however, since the method is restricted to low coverages, the results apply only to the more active parts of the surface. Theoretical aspects have been discussed by Amariglio[105].

The chromatographic method has been used by Elkington and Curthoys[106] to study the effect of treating carbon black with concentrated nitric acid. As expected from Kiselev's arguments on specific adsorption (Section 6.2.2) q^{st} increased markedly on the oxidised carbon if the adsorbate was polar or unsaturated (e.g. for diethyl ether, from 8.7 to 16.1, for benzene, from 8.9 to 12.4 kcal mol^{-1}) whereas with saturated non-polar compounds the increase was only slight (e.g. for neopentane, from 8.3 to 9.3 kcal mol^{-1}).

By using column temperatures up to 500 °C, Kiselev and his co-workers[107] were able to study a wide range of n-alkanes (C_5–C_{25}) on a variety of adsorbents, including silica, carbon and zeolites; they found that q^{st} was a linear function of the number of carbon atoms.

Calorimeters for the direct determination of the heat of adsorption q^c are usually diathermal rather than adiabatic or isothermal in type. The paper of Černy, Ponec and Hladec[108] on theoretical aspects of this type of calorimeter is, therefore, of special interest. By use of an analog computer it is shown that for slow adsorption processes, calorimeters of high time constant τ_c are needed. For fast processes τ_c may be either high or low; the latter is preferable ($\tau_c = C/k$; C = thermal capacity, k = Newton's cooling constant).

A particularly neat device is the twin adsorption calorimeter[109], the adsorbent being in one bulb and the liquid adsorptive in another; the amount adsorbed is calculated from the latent heat of evaporation of the adsorptive when the two bulbs are put in connection. This calorimeter was used to measure the heat of adsorption of carbon tetrachloride on carbon black. From the curve of q^c v. θ the heat of two-dimensional condensation was 1.3 ± 0.2 kcal mol^{-1} at 234 K.

A twin adiabatic calorimeter has been used, also by Kiselev and his co-workers, to carry out the very difficult measurements of the heat capacity C_m of an adsorbed film[110]. Each bulb contained adsorbent and was provided with a heater and resistance thermometer. Adsorptive was added to one bulb only and the difference in temperature rise between the bulbs was measured. For benzene adsorbed on carbon black C_m was lower than in the bulk liquid, implying less degrees of freedom. Measurements were also carried out with alkanols. Theoretical aspects of the subject with reference to models were also studied[111].

6.7.2 Heat of immersion

The heat of immersion q^i of a solid in a liquid adsorptive is of interest in relation to the energetics of the interface, as an indirect means of determination of the heat of adsorption, and for the determination of surface area.

In their 'absolute' method for surface area determination Harkins and Jura[112] measured the heat of immersion in a liquid adsorptive of the solid which had been exposed to the saturated vapour of the adsorptive. It was postulated that the adsorbed film would be liquid-like in character and would have a surface enthalpy, h^L, per unit area equal to that of the bulk liquid ($h^L = \gamma - T d\gamma/dT$). Some recent work suggests that pre-charging to *saturation* — with its attendant risk of capillary condensation in the interstices between the solid particles, and thus of a reduction in the effective area of the film — is not necessary. Goodman and his co-workers[113] measured the heat of immersion in the hydrocarbons n-hexane to n-hexadecane of graphitised carbon black having a known area, pre-coated from the vapour phase with various known amounts of the hydrocarbon. At a coverage of only 1.5 monolayer the heat of immersion per unit area of solid, h^i, had already fallen from its high initial value to a figure close to the surface enthalpy, h^L, of the liquid hydrocarbon. The adsorption isotherms were of Type (II). For carbon tetrachloride on non-porous silica, which yields a *Type (III)* isotherm (Section 6.3.1), Gregg and Diano[72, 73] found that h^i became equal to h^L at a coverage of *c*. 0.1 monolayer, i.e. a relative pressure of *c*. 0.4.

The heat of immersion in polar liquids has been used by Wade[144] to monitor the surface changes produced by thermal degradation of carbon black.

A critical review of heats of immersion and of adsorption, in relation both to their calorimetric determination and their applicability in physical adsorption we owe to Lafitte and Rocquerol[114]. These workers conclude that the heat of immersion is more appropriate where a single calorimetric measurement is sufficient, as in the comparison of a series of surfaces which differ only in one parameter such as chemical nature or porosity, or for the determination of specific surface by the Harkins–Jura method; but that direct determination of heat of adsorption is more suitable for a study of the energetics of progressively covering the surface of a solid with adsorbate.

6.8 ADSORPTION OF WATER VAPOUR: TYPE (III) ISOTHERMS

6.8.1 Hydrophobic surfaces

The unique properties of water, which stem from the ability of a water molecule to form hydrogen bonds with its four nearest neighbours, are well known. Because of this hydrogen bonding water is a far less volatile liquid than it otherwise would be. By simple extrapolation from the other hydrides of the Periodic Group (H_2S, H_2Se, H_2Te) it is readily found that the boiling point and latent heat of non-hydrogen-bonded water would be *c*. 100 K and *c*. 5 kcal mol^{-1} respectively; and by insertion of these values into the Clausius–Clapeyron equation one finds that the vapour pressure of this hypothetical water would be *c*. 120 bar at 298 K, *c*. 3800 times higher than the experimental value of 25.4 Torr.

In the context of adsorption this implies that if the adsorbent is such that it cannot promote the formation of the tetrahedral arrangement, then water will behave as an adsorptive 3800 times more volatile than normal water. The relative pressure corresponding to, say, a humidity of 50% would then be only 1.3×10^{-4}, and so would correspond to a region close to the origin of the isotherm, where the adsorption is extremely small: the surface should be 'hydrophobic'. Once a few tetrahedral clusters have been formed however – usually around the few hydrophilic sites still present on an otherwise hydrophobic surface – they can grow by a co-operative process and an isotherm convex to the pressure axis – a Type (III) – results. Hydrophobic surfaces are much less common than hydrophilic ones, where the surface is largely covered with polar groups, especially hydroxyl, capable of nucleating the tetrahedral structure of water.

The hydrophobic nature of strongly outgassed graphite is well known[115] and Boehm[116] has found that, not unexpectedly, diamond powder likewise gives an isotherm with water vapour which is close to Type (III) in form, if it has been treated with hydrogen at 800 °C to remove chemisorbed oxygen. Many polymeric materials of hydrocarbon type also tend to be hydrophobic. Thus Jellinek and co-workers[117] found that with polystyrene no adsorption of water vapour could be detected at pressures below $0.85p_0$, but that at $0.91p_0$ the adsorption was equivalent to a monolayer; this corresponds to an extreme form of Type (III) isotherm. Polymethylmethacrylate gave a clear Type (III) isotherm with water vapour at 0 °C; q^{st} was close to the latent heat of water, whereas polyacrylonitrile, consistent with its greater concentration of polar centres, yielded an almost linear isotherm (Henry's law type).

If a polar solid is coated with an organic material so as to turn its surface into an array of hydrocarbon groups, its isotherm with water vapour will change from Type (II) to Type (III) (from (IV) to (V) if it is mesoporous). Pope and Howe[118] modified the surface of titania by immersing it in a solution of aqueous sodium oleate of different pH values and determined the isotherm of water vapour on the dried solid. The sample from pH 12 gave a Type (II) isotherm, whereas that from pH 8 gave a Type (III); that from pH 10 was an intermediate case, an isotherm of Type (II) with a very small c-value. The product from pH 8 had its surface completely covered with a layer of chemisorbed oleate groups.

Curiously, calcium carbonate, in the experiments of Hall and co-workers[119], did not become completely hydrophobic even when it was covered with a multilayer of oleate, the isotherm remaining of Type (II); at thick coverages (c. 6 layers) a steep rise in the water isotherm was observed at c. $0.3p_0$, and was explained by penetration of the water molecules between layers of oleate ions which were lying parallel to the surface of the solid. In these experiments q^{st} was found to be about equal to the latent heat of condensation. The degree of hydrophobicity depends markedly on the orientation of the pre-adsorbed molecules.

The converse of the adsorption of water vapour on a solid is the adsorption of another vapour (A) on ice or on liquid water. If A is non-polar and therefore incapable of undergoing hydrogen bonding with the ice or water surface, then between water and vapour A only dispersion forces will operate and these are of course relatively feeble. As coverage increases the mutual interaction of the adsorbate molecules becomes significant, so that a Type (III) isotherm should again result. Interesting work along these lines has been done by Adamson and his co-workers[120]. n-Alkanes adsorbed on ice gave

Type (III) isotherms with an isosteric heat of adsorption close to the latent heat of condensation of the alkane. Nitrogen adsorbed on non-annealed ice on the other hand yielded an isotherm which was still of Type (II) (the value of c lying between 31 and 66) no doubt because of the enhancement of the interaction by the quadrupole of nitrogen. The adsorption of a hydrocarbon, toluene, on *liquid* water, calculated from changes in the surface tension of the water, was also found by Ottewill and Hauxwell[121] to give a Type (III) isotherm.

6.8.2 Hydrophilic surfaces

Hydrophilic behaviour is promoted, above all, by the presence of hydroxyl groups on the surface. An important group of hydroxylated surfaces are those derived from metal oxides[133] and in the last few years considerable effort has been devoted in this context more especially to silica, alumina, titania[134, 135], magnesia and ferric oxide.

The surface hydroxylation of silica is the subject of a comprehensive paper by Hockey and his co-workers[122] who examined the stoichiometry of reactions of the surface hydroxyl with $Si(CH_3)_2Cl_2$, BCl_3 and $TiCl_4$. By analysing the adsorbed phase for chloride (mere estimation of the evolved HCl is inadequate) and using infrared examination along with data for loss of weight at definite temperatures in a supporting role, they were led to the conclusion that fully hydroxylated silica carries two types of hydroxyl site: (A) isolated OH groups and (B) interacting OH groups, present on neighbouring sites. As the temperature is raised from ambient to 500 °C, A sites are progressively eliminated, but B sites remain, and these in turn disappear between 500 and 800–900 °C; the completely dehydroxylated surface is hydrophobic – it gives a Type (III) isotherm with water vapour – but can be rendered hydrophilic again by exposure to liquid water at elevated temperature.

For titania[116] Boehm proposes a model in which a molecule of water is first physisorbed on each exposed Ti^{4+} ion; one proton then moves from each molecule to a neighbouring oxygen ion of the lattice, converting it into an OH group. The surface thus becomes an array of two types of OH group, one of them acidic and one basic, as can be demonstrated by titration with, for example, ammonia and acetic acid. Manuera and Stone[123], from temperature programmed desorption and other techniques, inferred that water can be taken up by the rutile form of titania in *three* ways: (a) as molecular water weakly adsorbed on isolated O^{2-} ions; (b) as molecular water strongly adsorbed on isolated Ti^{4+} ions; and (c) dissociatively chemisorbed, i.e. in the form of OH groups.

The hydroxylation of α-ferric oxide has been studied by McCaffery and Zettlemoyer[124], according to whom water is taken up by a dissociative mechanism, similar to that described by Boehm for titania, to give a chemisorbed layer. Further water can be physisorbed on the chemisorbed film, the first physisorbed layer (as shown from data for heat of immersion and dielectric relaxation together with calculations of entropy) being immobile, and succeeding layers mobile. With thorium dioxide[126], Holmes, Fuller and Secoy[125] found three modes of binding of adsorbed water, corresponding to three successive layers, the first, chemisorbed, the second, hydrogen-bonded, and the third resembling liquid water. The values of the isosteric net heat of adsorption ($q^{st} - L$) were respectively -19.0, -8.6 and -1.8 kcal mol^{-1} (L = latent heat of condensation).

Standard isotherms (cf. Section 6.3) for the physical adsorption of water on a variety of solids, including zirconium silicate, calcium carbonate and titania, have been published by Brunauer, Mikhail and Hagymassy[127]. The isotherms are classified according to their c-values, the statistical monolayer being complete at relative pressures as follows: for $50 < c < 100$ at c. 0.1; for $c = 23$ at 0.20; for $10 < c < 14.5$ at 0.25; for $c = 5$, at 0.3.

Van Olphen[128] has reviewed the adsorption of water vapour by clays. The isotherms are usually of Type (II). With the nitrogen-BET area as standard, the value of $\Omega_m(H_2O)$ for a series of kaolinites with the cation varied by base exchange ranged from 7.9 to 16.5 Å^2, and the number N of water molecules per cation from 5 to 10 for univalent, and 14 to 20 for bivalent, cations. For expanding clays such as bentonite similar ranges of $\Omega_m(H_2O)$ and N were found. The heat of immersion in water varied very widely, from 55 to 500 erg cm^{-2}. Clearly water is not a suitable adsorptive for estimating the surface area of clays.

A novel approach has been adopted by Deitz and Turner[129] in their measurements of the dynamic adsorption of water vapour on fibres drawn from a melt of Vycor glass. Small, definite, doses were admitted into a fixed volume containing the fibres, and readings of pressure against time taken at each of four temperatures 60, 80, 100 and 120 °C. Isochrones – curves of adsorption against pressure for different fixed times between 10 and 3000 minutes – were then constructed; when analysed kinetically the results led to the conclusion that two processes were involved: (a) chemisorption of water and (b) physisorption of water on the localised products of (a).

6.8.3 Type (III) isotherms—other examples

The emergence of a Type (III) isotherm is to be expected wherever the adsorbent–adsorbate interaction is substantially the same in strength as the adsorbate–adsorbate interaction, i.e. when specific interactions (Section 6.2.2) are absent. It follows therefore that one should be able to obtain a Type (III) isotherm with an adsorbate other than water by selecting a solid which has the required weak interaction with the adsorbate. Thus, using butane as adsorbate, Davis[130] obtained a Type (III) isotherm with calcium fluoride as adsorbent, as did Gregg and Gammage[131] with ball-milled calcite carrying a monolayer of water on its surface; in the latter case the state of surface must have been somewhat similar to that in the experiments of Adamson, already mentioned.

In the study carried out by Day, Peacock and Parfitt[132], the adsorbate was pentane and the adsorbent rutile whose surface had been modified in a controlled manner by pre-adsorption of various adsorbates. After pre-treatment with either ethanol or hexan-1-ol, the pentane isotherm was of Type (III), whereas on unmodified titania it was of Type (II): the alkanol molecules are anchored to the titania surface with their hydroxyl groups turned inwards, so that the outer surface of the adsorbent is now composed of methyl groups, which experience only dispersion interaction with pentane molecules. When the titania surface was modified with pre-adsorbed *water*, on the other hand, the pentane isotherm remained of Type (II); this is probably because the water had hydroxylated the titania surface producing a close array of OH^- ions which were able to polarise the pentane and thus enhance the force of interaction.

Confirmation that a Type (III) isotherm is associated with a nearly zero net heat of adsorption ($q^{st} - L$) is afforded by results on the system carbon tetrachloride–silica[46, 47]. The isotherm is almost of Type (III) ($c \approx 3$) and the value of the net heat of adsorption (except at coverages below $c.$ 0.05) is within experimental limits, zero as determined both from the isotherms at 0 and 20 °C, and from the curve of heat of immersion against coverage[72]. Close to the origin of the isotherm was a very small knee, corresponding to a heat of adsorption in this region numerically greater than the latent heat of condensation. It could well be that this is an essential feature and reflects the presence of 'active spots' which are necessary for the nucleation of clusters of adsorbate molecules.

6.9 SOME ADDITIONAL TOPICS

6.9.1 Infrared spectroscopy

Perhaps the major application of infrared spectroscopy in the adsorption field has been in the study of chemisorption, where the displacement of characteristic bands is relatively large. As the technique has become more and more refined however its use in physisorption is becoming more widespread, usually in association with the more conventional techniques of adsorption. Hockey and his co-workers applied it in this way in their study of the hydroxylation of silica[122], already mentioned. A similar investigation with germania by Low, Madison and Ramamurthy[136] led to analogous result; in addition the adsorption of carbon tetrachloride was shown to be physical in that the band shift from 3673 to 3622 cm^{-1} was reversed on pumping at 25 °C.

Kozirovski and Folman[137] also used infrared spectroscopy in their work on nitrous oxide adsorbed on evaporated films of sodium chloride; they concluded that the nitrous oxide is for spatial reasons adsorbed on the Na^+ rather than the Cl^- sites and that it assumes a perpendicular orientation with its O atom turned towards the Na^+ ion. A similar study was also carried out with carbon monoxide[85]. (Section 6.6.1).

It has been pointed out by Kiselev and co-workers that the heating of the specimen by the infrared radiation will tend to give misleading results because of thermal desorption and decomposition effects and they have devised a modified technique to minimise this drawback[138]. A further development, of considerable promise, is laser Raman spectroscopy, which is described by Hendra and Loder[139]. As with infrared spectroscopy, the bands arise from vibration of the adsorbed species, but the new technique has the advantage that bands due to the adsorbent are not prominent. Application to silica ('Aerosil') showed that the adsorption of carbon tetrachloride was non-specific and that of acetonitrile was weakly specific (Section 6.2.2).

6.9.2 Dielectric studies

Measurements of dielectric relaxation also, are commonly used alongside other techniques of adsorption. The adsorption of water vapour, ethanol and acetone by kaolin has been studied in this way by Sutton[140] and his associates; from the dielectric data together with calculations of entropy they concluded that the adsorbate was in a state different from that of the normal liquid and controlled markedly by the surface. For benzene adsorbed on silica gel Thorpe[141] and his co-workers found that the dielectric isotherm fell into three linear branches which they correlated with monolayer adsorption, capillary condensation in mesopores and bulk condensation on the external surface respectively.

Again using dielectric measurements in conjunction with conventional adsorption isotherms, Hall and Kouvarellis[142] examined the adsorption of water vapour on an alum which had first been dehydrated to a composition approximating to $6H_2O:KCr(SO_4)_2$. From the results it was inferred that at coverages below one monolayer the water is physically adsorbed but that at higher coverages water begins to dissolve irreversibly in the solid phase, i.e. it becomes water of hydration.

6.9.3 Ellipsometry

Ellipsometry is a powerful technique within its own limited field. It consists in the measurement of the change in the state of polarisation of a light beam on reflection from the surface of a thin film of a crystal and with suitable assumptions it can be used to calculate the area per adsorbed molecule in the sub-monolayer region. In this way Bootsma and Mayer[143] found, for example, a value of $\Omega_m = 20 \text{ Å}^2$ for krypton adsorbed on crystals of germanium.

6.9.4 An experimental note

Space does not permit of a general review of practical methods, but brief mention must be made of an important source of error arising from temperature inequality. Using an electronic microbalance Cutting[47, 146] found that the isotherm of nitrogen was markedly lower, in the region of higher relative pressures, than that measured by a conventional volumetric technique. Quantitative investigation showed that the temperature of the sample, even with a depth of immersion of 30 cm, was higher by 0.8 K than that of the surrounding bath. By use of a specially designed metal hangdown tube, and by keeping the sample bucket in contact with the bottom of the tube except during the few seconds whilst a reading was being taken, Cutting succeeded in virtually eliminating the difference between the two isotherms.

A related source of error, in both gravimetric and volumetric methods, has been discovered by Teichner and Nicolaon[145], who found that the coldest point of an adsorption system was the part in contact with the surface of the bath, the sample being warmer by at least 0.16 K. It is therefore impossible ever to reach a point on the isotherm which truly corresponds to $p/p_0 = 1$, so that calculations of pore size distribution based on the experimental isotherm in this region are bound to be in error; the use of a mercury penetration method, which is relatively insensitive to temperature, is indicated for the upper range of pore size. Since the magnitude of the error depends on the temperature difference between the bath and its surroundings, it is negligible for isotherms measured at temperature around ambient.

Fortunately, the values of S calculated from the nitrogen isotherm by the BET procedure are far less sensitive to the exact value of p_0, and are little affected.

References

1. (1970). *Surface Area Determination. Proc. Int. Symp. (1969)*. (London:Butterworths)
2. Flood, E. A. (editor) (1967). *The Gas–Solid Interface*. (London: Arnold)
3. De Boer, J. H. (1968). *The Dynamical Character of Adsorption*. (Oxford: Clarendon Press)
4. Linsen, B. G. (editor) (1970). *Physical and Chemical Aspects of Adsorbents and Catalysts*. (London: Academic Press)
5. Gregg, S. J. and Sing, K. S. W. (1967). *Adsorption, Surface Area and Porosity*. (London: Academic Press)
6. Kiselev, A. V. and Yashein, Ya. I. (1968). *Gas-Adsorption Chromatography*. (New York: Plenum Press)
7. Bond, R. L. (editor) (1967). *Porous Carbon Solids*. (London: Academic Press)
8. Steele, W. A. (1967). *Advan. Colloid and Interface Sci.*, **1**, 3
9. Aharoni, C. and Tompkins, F. C. (1970). *Advan. Catal.*, **21**, 1
10. (1968). *J. Colloid Interface Sci.*, **28**, 343
11. Information Bulletin No. 3 of the International Union of Pure and Applied Chemistry (1970). *Tentative Manual of Definitions, Terminology and Symbols in Colloid and Surface Chemistry*
12. Hall, P. G. and Stoekli, H. F. (1969). *Trans. Faraday Soc.*, **65**, 3335
13. Watanabe, K. and Yamashina, T. (1970). *J. Catal.*, **17**, 272
14. Clough, P. S. and Harris, M. R. (1969). *Chem. Ind. (London)*, 343
15. Gregg, S. J. (1968). *Chem. Ind. (London)*, 611
16. Langford, J. (1967). *Ph. D. Thesis*, Exeter University
17. McClellan, A. L. and Harnsberger, H. F. (1967). *J. Colloid. Interface Sci.*, **23**, 577
18. Gregg, S. J. and Sing, K. S. W. (1967). *Adsorption, Surface Area and Porosity*, 116 (London: Academic Press)
19. Orr, W. J. C. (1939). *Trans. Faraday Soc.*, **35**, 1247
20. Brunauer, S. and Emmett, P. H. (1937). *J. Amer. Chem. Soc.*, **59**, 1553
21. De Boer, J. H. and Broekhoff, J. C. P. (1967). *Proc. Koninkl. Nederl. Akad. Wetens.*, **B, 70**, 342
22. Ross, S. and Olivier, J. P. (1964). *On Physical Adsorption*. (New York: Interscience)
23. Kiselev, A. V. (1965). *Discuss. Faraday Soc.*, **40**, 205
24. Barrer, R. M. (1966). *J. Colloid Interface Sci.*, **21**, 415
25. Deitz, V. R. and Turner, N. H. (1970). *Surface Area Determination. Proc. Int. Symp. (1969)*, 43. (London: Butterworths)
26. Pierce, C. and Ewing, B. (1967). *J. Phys. Chem.*, **71**, 3408
27. Voet, A. (1969). *J. Colloid Interface Sci.*, **30**, 264
28. Gammage, R. B., Fuller, E. L. and Holmes, H. F. (1970). *Surface Area Determination. Proc. Int. Symp. (1969)*, 161. (London: Butterworths)
29. Carruthers, J. D., Payne, D. A., Sing, K. S. W. and Stryker, L. J. (1971). *J. Colloid Interface Sci.*, **36**, 205

30. Bassett, D. R., Boucher, E. A. and Zettlemoyer, A. C. (1968). *J. Colloid Interface Sci.*, **27,** 649
31. Prenzlow, C. F., Beard, H. R. and Brundage, R. S. (1969). *J. Phys. Chem.*, **73,** 969
32. Dubinin, M. M. (1970). *Surface Area Determination. Proc. Int. Symp. (1969).* 123. (London: Butterworths)
33. Tulbovich, B. L. (1969). *Russ. J. Phys. Chem.*, **43,** 901
34. Shull, C. G. (1948). *J. Amer. Chem. Soc.*, **70,** 1405
35. Cranston, R. W. and Inkley, F. A. (1957). *Advan. Catal.*, **9,** 143
36. Pierce, C. (1959). *J. Phys. Chem.*, **72,** 3673
37. Lippens, B. C., Linsen, B. G. and De Boer, P. H. (1964). *J. Catal.*, **3,** 32
38. De Boer, J. H., Linsen, B. G. and Osinga, Th. J. (1965). *J. Catal.*, **4,** 643
39. Lippens, B. C. and De Boer, J. H. (1965). *J. Catal.*, **4,** 319
40. Pierce, C. (1968). *J. Phys. Chem.*, **72,** 3673
41. Verwey, E. J. W. (1946). *Rec. Trav. Chim.*, **65,** 521
42. Carruthers, J. D., Cutting, P. A., Day, R. E., Harris, M. R., Mitchell, S. A. and Sing, K. S. W. (1968). *Chem. Ind. (London)*, 1772
43. Payne, D. A. and Sing, K. S. W. (1969). *Chem. Ind. (London)*, 918
44. Mikhail, R.Sh. and Shebl, F. A. (1970). *J. Colloid. Interface Sci.*, **32,** 505
45. Sing, K. S. W. (1968). *Chem. Ind. (London)*, 1520
46. Sing, K. S. W. (1970). *Surface Area Determination. Proc. Int. Symp. (1969)*, 25. (London: Butterworths)
47. Cutting, P. A. (1969). *Ph. D. Thesis, Brunel University*
48. Parfitt, G. D., Urwin, D. and Wiseman, T. J (1971). *J. Colloid Interface Sci.*, **36,** 217
49. Cutting, P. A. and Sing, K. S. W. (1969). *Chem. Ind. (London)*, 268
50. Gregg, S. J. and Sing, K. S. W. (1967). *Adsorption, Surface Area and Porosity*, 114. (London: Academic Press)
51. Spencer, D. H. T. and Feredy, F. (1968). *Chem. Ind. (London)*, 847
52. Roberts, R. F. (1967). *J. Colloid Interface Sci.*, **23,** 266
53. Brunauer, S. (1969). *Chem. Eng. Progr. Symp.*, No. 96, 1
54. Brunauer. S., Mikhail, R. Sh. and Bodor, E. E. (1967). *J. Colloid Interface Sci.*, **24,** 451
55. Brunauer, S., Mikhail, R. Sh. and Bodor, E. E. (1968). *J. Colloid Interface Sci.*, **26,** 45
56. John, P. T. and Bohra, J. N. (1967). *J. Phys. Chem.*, **71,** 4041
57. Broekhoff, J. C. P. and De Boer, J. H. (1970). *Surface Area Determination. Proc. Int. Symp. (1969),* 97
58. Broekhof, J. C. P. and De Boer, J. H. (1968). *J. Catal.*, **10,** 391
59. Harris, M. R. (1965). *Chem. Ind. (London)*, 269
60. Schofield, R. K. (1948). *Discuss. Faraday Soc.*, **3,** 105
61. Flood, E. A. (1967). in *The Solid–Gas Interface*, **1,** 54. (E. A. Flood, editor) (New York: Dekker)
62. Melrose, J. C. (1966). *Amer. Inst. Chem. Eng. J.*, **12,** 986
63. Everett, D. H. (1967). in *The Gas–Solid Interface*, **2,** 1086. (E. A. Flood, editor) (New York: Dekker)
64. Kadlec, O. and Dubinin, M. M. (1969). *J. Colloid Interface Sci.*, **31,** 479
65. Everett, D. H. and Burgess, C. G. V. (1970). *J. Colloid Interface Sci.*, **33,** 611
66. Guggenheim, E. A. (1967). *Thermodynamics*, 142. (Amsterdam: North-Holland)
67. Gregg, S. J. and Hickman, J. (1971). *Third Conference on Industrial Carbon and Graphite* of the Society of Chemical Industry, 1970, 145. (London: Academic Press)
68. Pierce, C. and Smith, R. N. (1953). *J. Phys. Chem.*, **57,** 64
69. Dubinin, M. M. (1955). *Quart. Rev. Chem. Soc.*, **9,** 101
70. Kiselev, A. V. (1958). *The Structure and Properties of Porous Materials*, 51. (London: Butterworths)
71. Gregg, S. J. and Langford, J. (1969). *Trans. Faraday Soc.*, **65,** 1394
72. Diano, W. and Gregg, S. J. (1971). *Colloques Internationaux du Centre National de la Recherche Scientifique*, No. 201
73. Diano, W. (1969). *Ph.D. Thesis*, Exeter University
74. Anderson, P. J. and Horlock, R. F. (1967). *Trans. Faraday Soc.*, **65,** 251
75. Dubinin, M. M. and Zaverina, E. D. (1949). *Zhur. Fiz. Khim.*, **23,** 1129; Dubinin, M. M. (1965). *Russ. J. Phys. Chem.*, **39,** 697
76. Radushkevich, L. V. (1949). *Zhur. Fiz. Khim.*, **23,** 1410
77. Freeman, E. M., Siemieniewska, T., Marsh, H. and Rand, B. (1970). *Carbon*, **8,** 7

78. De Boer, J. H., Van der Plas, Th. and Zonderman, G. J. (1965). *J. Catal.*, **4**, 649
79. Maggs, F. A. P. (1953). *Research*, **6**, S13
80. Gregg, S. J., Olds, F. M. W. and Tyson, R. F. S. (1971). *Third Conference on Industrial Carbon and Graphite* of the Society of Chemical Industry, 1970. (London: Academic Press)
81. Pitzer, K. S. (1959). *Advan. Chem. Phys.*, **2**, 59
82. Pitzer, K. S. and Sinanoglu, O. (1962). *J. Chem. Phys.*, **32**, 1279
83. McLachlan, A. D. (1964). *Mol. Phys.*, **7**, 381
84. Kiselev, A. V., Lopatkin, A. A., Laurie, B. I. and Shpigel, S. (1969). *Russ. J. Phys. Chem.*, **43**, 1498
85. Giverzman, R., Kozirovski, Y. and Folman, M. (1969). *Trans. Faraday Soc.*, **65**, 2206
86. Melrose, J. C. (1968). *J. Colloid Interface Sci.*, **28**, 403
87. Mayer, E. F. (1967). *J. Phys. Chem.*, **71**, 4416
88. Ricca, F., Pisani, C. and Garrone, E. (1969). *J. Chem. Phys.*, **51**, 4079
89. Palmberg, P. W. (1971). *Surface Sci.*, **25**, 598
90. Kiselev, A. V. and Poshkus, D. P. (1967). *Russ. J. Phys. Chem.*, **41**, 1433
91. Kiselev, A. V., Poshkus, D. P. and Afreimovich, A. Ya. (1968). *Russ. J. Phys. Chem.*, **42**, 1345, 1348
92. Kiselev, A. V., Poshkus, D. P. and Afreimovich, A. Ya. (1970). *Russ. J. Phys. Chem.*, **44**, 545
93. Poshkus, D. R. and Afreimovich, A. Ya. (1968). *Russ. J. Phys. Chem.*, **42**, 626
94. Pierotti, R. A. (1968). *Chem. Phys. Lett.*, **2**, 385
95. Rudzinski, W. (1971). *Chem. Phys. Lett.*, **10**, 183
96. Brunauer, S., Skalny, J. and Bodor, E. E. (1969). *J. Colloid Interface Sci.*, **30**, 546
97. Greenleaf, C. M. and Halsey, G. D. (1970). *J. Phys. Chem.*, **74**, 677
98. Tazikawa, A. (1968). *Kolloid Z. Z. Polymer.*, **222**, 143
99. Gregg, S. J. and Sing, K. S. W. (1967). *Adsorption, Surface Area and Porosity*, 45. (London: Academic Press)
100. Jovanović, D. S. (1969). *Kolloid. Z. Z. Polymer*, **235**, 1203
101. Erikson, J. C. (1969). *Surface Sci.*, **14**, 221
102. Schay, G. (1971). *J. Colloid Interface Sci.*, **35**, 254
103. Everett, D. H. (1970). *Surface Area Determination. Proc. Int. Symp. (1969)*, 181. (London: Butterworths)
104. Johnson, J. D. and Klein, M. L. (1967). *Trans. Faraday Soc.*, **63**, 1269
105. Amariglio, H. (1968). *Surface Sci.*, **12**, 2
106. Elkington, P. A. and Curthoys, G. (1969). *J. Phys. Chem.*, **73**, 2321
107. Keibal, V. L., Kiselev, A. V., Savinov, I. M., Khudyakov, V. L., Shcherbakova, K. D. and Yashin, Ya. I. (1967). *Russ. J. Phys. Chem.*, **41**, 1203
108. Cerny, S., Ponec, V. and Hladec, L. (1970). *J. Chem. Thermodynamics*, **2**, 391
109. Berezin, G. I., Kiselev, A. V., Sagatelyan, R. T. and Serdobov, M. V. (1969). *Russ. J. Phys. Chem.*, **43**, 118
110. Berezin, G. I., Kiselev, A. V. and Sinitsyn, V. A. (1967). *Russ. J. Phys. Chem.*, **41**, 490
111. Berezin, G. I. and Kiselev, A. V. (1969). *Russ. J. Phys. Chem.*, **43**, 683, 894
112. Harkins, W. D. and Jura, G. (1944). *J. Amer. Chem. Soc.*, **66**, 1362
113. Clint, J. K., Clunie, J. S., Goodman, J. F. and Tate, J. R. (1970). *Surface Area Determination. Proc. Int. Symp. (1969)*, 299. (London: Butterworths)
114. Lafitte, M. and Rocquerol, J. (1970). *Bull. Soc. Chim. Fr.*, 3335
115. cf. Pierce, C., Smith, R. N., Wiley, J. W. and Cardes, H. (1951). *J. Amer. Chem. Soc.*, **73**, 4551
116. Boehm, H. P. and Herrmann, M. (1967). *Zeits. Anorg. Allgem. Chem.*, **352**, 156
117. Jellinek, H. H. G., Luh, M. D. and Nagarajan, V. (1969). *Kolloid Z. Z. Polymer.*, **232**, 758
118. Pope, M. I. and Howe, T. M. (1970). *Powder Technology*, **3**, 367
119. Hall, P. G., Lovell, V. M. and Finkelstern, N. P. (1970). *Trans. Faraday Soc.*, **66**, 2629
120. Adamson, A. W., Dormant, L. M. and Orem, M. (1967). *J. Colloid Interface Sci.*, **25**, 206
121. Hauxwell, F. and Ottewill, R. H. (1968). *J. Colloid Interface Sci.*, **28**, 564
122. Armistead, C. G., Tyler, A. J., Hambleton, F. H., Mitchell, S. A. and Hockey, J. A. (1969). *J. Phys. Chem.*, **73**, 3947; cf. Peri, J. P. and Hensley, A. L. (1968). *J. Phys. Chem.*, **72**, 2926
123. Munuera, G. and Stone, F. S. (1971). *Discuss. Faraday Soc.*, **52**, 205
124. McCaffery, E. and Zettlemoyer, A. C. (1971). *Discuss. Faraday Soc.*, **52**, 239

125. Holmes, H. F., Fuller, E. L. and Secoy, C. H. (1968). *J. Phys. Chem.*, **72**, 2095
126. Gammage, R. B., Fuller, E. L. and Holmes, H. F. (1970). *J. Phys. Chem.*, **74**, 4276
127. Hagymassy, J., Brunauer, S. and Mikhail, R. Sh. (1969). *J. Colloid Interface Sci.*, **29**, 485
128. Van Olphen, H. (1970). *Surface Area Determination. Proc. Int. Symp. (1969)*, 255
129. Deitz, V. R. and Turner, N. H. (1971). *J. Phys. Chem.*, **75**, 2718
130. Davis, B. W. (1969). *J. Colloid Interface Sci.*, **31**, 353
131. Gammage, R. B. and Gregg, S. J. (1972). *J. Colloid Interface Sci.*, **38**, 118
132. Day, R. E., Parfitt, G. D. and Peacock, J. (1971). *Discuss. Faraday Soc.*, **52**, 215
133. Morimoto, T., Nagao, M. and Tokuda, F. (1969). *J. Phys. Chem.*, **73**, 243
134. Dawson, P. T. (1967). *J. Phys. Chem.*, **71**, 838
135. Morimoto, T., Nagao, M. and Omori, T. (1969). *Bull. Chem. Soc. Japan*, **42**, 943
136. Low, M. T. D., Madison, N. and Ramamurthy, P. (1969). *Surface Sci.*, **13**, 238
137. Kozirovsky, Y. and Folman, M. (1969). *Trans. Faraday Soc.*, **65**, 244
138. Galkin, G. A., Kiselev, A. V. and Lygin, V. I. (1968). *Russian J. Phys. Chem.*, **42**, 765
139. Hendra, P. J. and Loder, E. J. (1971). *Trans. Faraday Soc.*, **67**, 823
140. Nelson, S. M., Huang, H. H. and Sutton, L. E. (1969). *Trans. Faraday Soc.*, **65**, 225
141. Cutfield, S. K. and Thorpe, J. M. (1969). *Trans. Faraday Soc.*, **65**, 869
142. Hall, P. G. and Kouvarellis, G. K. (1968). *Trans. Faraday Soc.*, **64**, 1940
143. Bootsma, G. A. and Meyer, F. (1969). *Surface Sci.*, **13**, 110
144. Wade, W. H. (1969). *J. Colloid Interface Sci.*, **31**, 111
145. Nicolaon, G. and Teichner, S. J. (1968). *J. Chim. Phys.*, **65**, 871
146. Cutting, P. A. (1970). *Vacuum Microbalance Techniques*, **7**, 71. (New York: Plenum Press)

7
Chemisorption on Tungsten Crystal Planes

H. WISE
Stanford Research Institute, Menlo Park, California

7.1	INTRODUCTION	225
7.2	COVALENT MOLECULAR ADSORPTION: CARBON MONOXIDE–TUNGSTEN	226
7.3	COVALENT ATOMIC ADSORPTION: HYDROGEN–TUNGSTEN	230
7.4	COVALENT MIXED ADSORPTION: NITROGEN–TUNGSTEN	234
7.5	SOME OBSERVATIONS ON COVALENT CHEMISORPTION ON TUNGSTEN	238
	NOTE ADDED IN PROOF	239

7.1 INTRODUCTION

In an active field of scientific endeavour, such as the interaction of gases with solid surfaces, a review paper runs the risk of obsolescence before the ink is dry on the printed page. At the same time, however, the rapid accumulation of scientific data makes it invaluable to consider at frequent intervals the 'state of the art.' Thus this review is meant to be more of a signpost than a hitching post.

During the last decade, the development and application of several scientific tools for the examination of solid surfaces on the atomic level have given major impetus to the study of chemisorption on surfaces of single crystals. Prominent among the techniques employed are molecular flow sorption kinetics (ultra-high vacuum flash desorption (UHV) and constant-flow or constant-pressure adsorption)[1-3], low-energy electron diffraction (LEED)[4-8], field emission[9], field ion microscopy[10], contact-potential measurements[11, 12], and electron-stimulated desorption[13]. A detailed discussion of the underlying principles of some of these experimental methods and their limitations may be found in a recent review[14]. The results obtained with these techniques under well-defined conditions of surface structure and surface cleanliness have begun to yield some of the information needed to elucidate the equilibrium properties and the kinetics and mechanisms of various adsystems.

In this review, we shall consider the interactions of various gases with tungsten surfaces of different crystal orientations. Tungsten was selected because of its preponderance as a substrate for scientific investigation and its suitability for examination by different experimental techniques. The gases chosen, carbon monoxide, hydrogen, and nitrogen, represent three types of covalent adsorption: molecular, atomic and mixed.

7.2 COVALENT MOLECULAR ADSORPTION: CARBON MONOXIDE–TUNGSTEN

Because of its non-dissociative nature, the adsorption of CO on polycrystalline surfaces of tungsten has been studied in some detail[15, 16, 16a]. The results suggested different binding sites occupied to varying degrees and exhibiting electropositive and electronegative adsorptions. At low temperatures ($T < 170$ K), a weakly bound state was observed (virgin state v) part of which is converted to the β-states (β_1 and β_2) stable at temperatures above 500 K. More recent studies on single-crystal substrates have focused on the origin of the different binding modes, originally considered to be indigenous to substrate orientations in the polycrystalline sample. The largest body of data has been accumulated on W(110) and W(112) by means of UHV sorption, LEED studies, and a modification of the conventional flash-desorption technique, step desorption[17], in which a field emitter located in front of the solid surface under study detects the gas desorbed.

On W(110) the following processes[18] occur at a surface coverage $\theta > 0.25$:

$$CO_v \begin{array}{l} \xrightarrow{250\ K} CO_{gas} \\ \xrightarrow{500\ K} CO_{\beta_1} \xrightarrow{700\ K} CO_{\beta_2} \xrightarrow{1000\ K} CO_{gas} \end{array}$$

Surface abundance data of the different CO adspecies observed during desorption experiments exhibit the pattern summarised in Table 7.1. The adsorption process exhibits high initial sticking probabilities (S_0) over a range of temperatures (Figure 7.1). The high value of S_0 at 100 K suggests non-activated adsorption for v-CO. Of special interest is the observation[4] that the maximum surface coverage for β-CO is nearly one-half that of v-CO ($\theta_\beta/\theta_v \approx \frac{1}{2}$). LEED experiments[19] at 300 K, based on intensity measurements of the diffraction spots, confirm the rapid adsorption of CO on W(110) with a high sticking probability ($S_0 \approx 1$). However, no distinct surface structure due to the adsorbate is revealed until the adsystem is heated in excess of 1000 K at which time a C(9 × 5)–CO structure* predominates. The complex LEED pattern observed may be due to surface reconstruction at the high experimental temperatures employed ($T > 1200$ K).

---

*A discussion of two-dimensional space groups may be found in Wood, E. A. (1964). *J. Appl. Phys.*, **35**, 1306 and in Ref. 8.

Also on W(100) the adsorption of CO at 300 K takes place with a high sticking probability (Table 7.2). Although complete surface coverage $(10 \times 10^{-14} \text{ cm}^{-2})$ is attained[20], no ordered superstructure is observed unless the adsorbate is heated in excess of 1000 K. Under these conditions the surface coverage decreases to one-half its original value ($\theta = \frac{1}{2}$), and new

Table 7.1 Desorption kinetics of carbon monoxide–tungsten adsystem

Crystal Plane	Adsystem	Surface structure (LEED)*	Desorption peaks†/K	Desorption activation energies/kcal mol^{-1}	Ref.
(110)	Virgin (v)	—	375	7–10	18
	β_1	—	975	40–55	18
	β_2	—	1125	66–69.5	18
	β_3	C(9 × 5)	1200	—	19
(100)	β_2	C(2 × 2)	1150	—	20
	β_3	—	1380	—	20
(112)	β_1	C(2 × 4) P(2 × 1)	1000	—	23
	β_2	C(6 × 4)	1200	—	23
Poly	β_1	—	1000	—	‡§
	β_2	—	1200–1300	—	‡§
	β_3	—	1300–1500	—	‡§

*LEED pattern observed only after heating of adsystem to $T > 600$ K.
†Temperature corresponding to maximum desorption rate.
‡Redhead, P. A. (1961). (1961). *Trans. Faraday Soc.*, **57**, 641
§Rigby, L. J. (1964). *Can. J. Phys.*, **42**, 1256

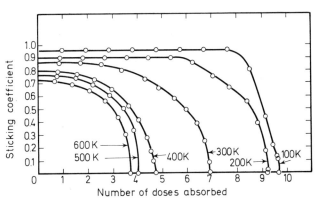

Figure 7.1 Sticking coefficient of CO on W(110)
(From Kohrt and Gomer[18], by courtesy of North-Holland)

diffraction spots appear with a C(2 × 2)–CO structure, similar to that observed during nitrogen[21] and hydrogen[22] adsorption on W(100) (*vide infra*).

On W(112) two β-states were observed by Chang[23] in addition to a weakly bound α-state, which could be depopulated at 300 K. On this crystal plane, the surface density in the β_1-state appeared to equal that of the β_2-state.

The LEED measurements exhibited a diffuse pattern that acquired ordered surface structures when the sample was heated in excess of 600 K (Table 7.1).

The relationship between different binding sites and substrate orientation has been further elucidated by determination of work function changes during CO sorption[20, 24]. For adsorption at substrate temperatures in excess of 100 K, the presence of CO on the surface of different crystal planes of tungsten results in a negative dipole, i.e. a work function increase. Semi-quantitatively the relative change in work function appears to be highest for the more

Table 7.2 Initial sticking probability of CO on tungsten

Crystal plane	S_0	Technique*	Ref.
(110)	1	AF	18
(100)	1	LEED	20
(112)	>0.9	LEED	23
	0.25	AF	23
(113)	0.62	AF	†
	0.2	AF	‡

*AF: Adsorption, Flash desorption;
LEED: Low-energy electron diffraction
†Gavrilyuk, V. M. and Medvedev, V. K. (1963). *Sov. Phys. Solid State*, **4**, 1737
‡Eisinger, J. (1957). *J. Chem. Phys.*, **27**, 1206

Table 7.3 Change in work function of tungsten due to CO sorption
(From Kohrt and Gomer[18], by courtesy of North-Holland)

Crystal plane	W-Atom surface density ($\times 10^{-14}$)	Dosage for half coverage	Work function/eV Substrate	Work function/eV Adsystem	$\Delta\phi$
(100)	10.00	4	4.90	5.6	+0.7*
(110)	14.14	12	5.80	7.0	+1.2
(111)	5.77	9	4.60	5.6	+1.0
(120)	—	5	4.46	5.4	+0.9
(211)	8.16	—	5.05	6.2	+1.1

*For CO adsorbed at 100 K on W(100) Yates and Madey report a work function change of $\Delta\phi = +0.6$ eV (*J. Chem. Phys.* (1971), **54**, 4969).

densely packed crystal planes, such as W(110). Possibly, since the change in work function due to adsorption is proportional to the number density of surface dipoles formed, the change in $\Delta\phi$ (Table 7.3) may reflect the number of CO complexes present on a given crystal plane. A measure of this quantity is to be found in the number of CO doses, N, required to reach half coverage ($\theta = \frac{1}{2}$), after which the relative change in work function per dose ($\Delta\phi/\Delta N$) decreases due to dipole–dipole interaction between the admolecules.

In our discussion of W(110) we have not mentioned the α-states of adsorption originally considered[17] to be part of v-CO but more recently recognised

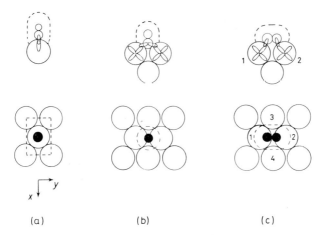

Figure 7.2 Binding modes of CO on W(110). Upper drawings represent cross-sections (cuts perpendicular to the surface), lower drawings are top views. Dashed lines represent van der Waals dimensions of CO. (a) Adsorption on a single W atom via C. (b) Adsorption in an upright position between two next-nearest neighbour W atoms. The geometry in (b) is the same for adsorption via sp^2 or sp hybrids of C. (c) Adsorption in a prostrate mode between two next-nearest neighbour W atoms. For the mode shown there is also some interaction with atoms 3 and 4.
(From Kohrt and Gomer[18], by courtesy of North-Holland)

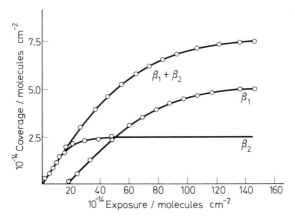

Figure 7.3 Hydrogen adsorption in different states on W(110)
(From Tamm and Schmidt[27], by courtesy of the American Institute of Physics)

to be an independent adlayer desorbing at $T < 300$ K [18]. This α-bonded CO is of interest because of its electropositive dipole in contrast to the negative dipole moment of the v- and β-states. But the detailed physical properties of this state are not established as yet.

Although incomplete in detail, the conversion of v-CO into β-CO, which occupies twice the effective area required by v-CO, has led to considerable speculation on adsorbate configurations. On the assumption that bonding is most favoured for configurations with maximum overlap between the bonding orbitals of the surface and those of the adsorbate, a molecular-orbital approach has been considered for the CO–W(110) adsystem[18]. By analogy to single metal atom bonding in carbonyls, it is suggested that three configurations are feasible as depicted in Figure 7.2: (a) adsorption on a single W atom via the C atom by involvement of the sp_z orbital of C and the d_{z^2} orbital of W; (b) adsorption via the C atom between two next-nearest neighbour W atoms involving either two sp^2 C orbitals (or the sp_z C orbital) and d_{yz} W orbitals; and (c) bonding in a prostrate configuration on next-nearest W atoms involving d_{yz} W orbitals, sp^2 C orbitals, and p (or sp^2) O orbitals. Such models may be employed in assigning a configuration of Type (a) to virgin-bonded CO and of Type (b) or (c) to the β-state (Figure 7.3). The half coverage observed in β-bonding would fit a bridge-bonded adsorbate involving two next-nearest W atoms in W(110). Some authors[20, 23] speculate in their assignment of Type (b) bonding to the $β_1$-states and Type (c) bonding to $β_2$.

7.3 COVALENT ATOMIC ADSORPTION: HYDROGEN–TUNGSTEN

A system representative of predominantly covalent atomic adsorption is that of hydrogen on tungsten. Contrary to expectations, the sorption studies on specific crystal planes have indicated a complexity suggestive of the influence of surface structure on the higher bonding orbitals of hydrogen. The electron reflection experiments of Armstrong[25] indicate that on W(100), W(211), W(110), and W(111), hydrogen is adsorbed on top of the tungsten surface. The observation that the scattering properties of the hydrogen adatom are similar to those of a free hydrogen atom makes the adatom location below the geometric surface plane rather unlikely.

In Table 7.4 we have summarised the work function changes observed by several investigators for this adsystem. Field emission and surface-potential measurements indicate (a) a linear variation of work function with surface coverage, and (b) a negative surface potential on many of the crystal faces except W(100). The observation of predominantly electronegative adsorption has elicited considerable speculation. It is considered more likely that covalent bonding occurs with electron transfer to the hydrogen atom (due to differences in the electronegativity of tungsten and hydrogen) rather than the inclusion of protons in the surface substrate layer[26].

The sorption process appears to be sensitive to the substrate structure (Table 7.5). In spite of considerable scatter in the data, there is an indication that the initial sticking probability for the hydrogen adsystem parallels that of the nitrogen adsystem. Conspicuous is the lack of reactivity of the W(110)

Table 7.4 Work function change due to adsorption of hydrogen on tungsten*

Crystal plane	$\Delta\phi$	Method†	Reference
(100)	+0.54	SP	26
	+0.88	SP	‡
	+0.9	FE	§
	+0.9	SP	22
(110)	−0.14	SP	26
	0	SP	‡
	+0.35	FE	§
	+0.38	SP	¶
(111)	+0.30	SP	26
	+0.15	SP	‡
	+0.5	FE	§
(112)	+0.72	SP	30
(113)	+0.43	SP	26
(210)	+0.55	FE	§
(211)	+0.7	FE	§
(310)	+0.75	FE	§
(320)	+0.3	FE	§
(411)	+0.7	FE	§

*At maximum surface coverage with hydrogen.
†SP: surface potential; FE: field emission.
‡Armstrong, R. A. (1966). *Can. J. Phys.*, **44**, 1753
§Becker, J. A. (1961). *2nd Int. Congr. Catalysis, Technip., Paris*, 1777
¶Stern, R. M. (1965) *J. Vac. Sci. Technol.*, **2**, 286

Table 7.5 Adsorption of hydrogen on tungsten at 300 K

Crystal plane	S_0	Reference
(100)	0.1	25
	0.65	22
	0.13 (β_2)	27, 28
	0.07 (β_1)	27, 28
	0.66	†
(110)	$<10^{-3}$	25
(111)	$\sim 10^{-2}$	25
	0.42	†
(112)	0.85*	30
(210)	0.23	†
(211)	0.3	25
(310)	0.66	†
	0.26	†
(320)	0.27	†
(411)	0.19	†

*At 600 K
†Becker, J. A. (1961). *2nd Int. Congr. Catalysis, Technip., Paris*, 1777

plane, the most densely packed plane (1.41×10^{15} W atoms cm^2), as compared to W(100), on which full surface coverage is attainable. The LEED data[22] for W(100) indicate that at $\theta \leqslant \frac{1}{2}$ no substrate reconstruction occurs during hydrogen adsorption. A surface structure is formed having the unit mesh C(2×2)—H, indicative of a square array with edges twice the lattice

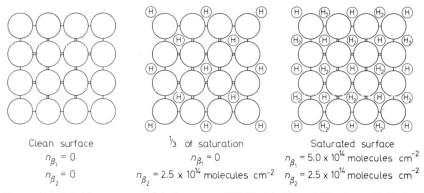

Figure 7.4 Model structures of hydrogen on W(110). (a) Clean (100) plane. (b) Surface saturated with H atoms in the β_2-state. (c) Surface saturated with hydrogen in both β_1- and β_2-states, $n_{total} = 7.5 \times 10^{14}$ molecules cm^{-2}
(From Tamm and Schmidt[27], by courtesy of the American Institute of Physics)

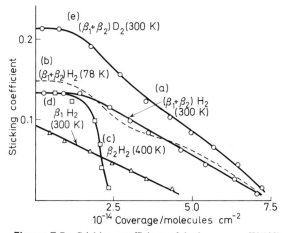

Figure 7.5 Sticking coefficient of hydrogen on W(100); initial gas temperature 300 K
(From Tamm and Schmidt[28], by courtesy of the American Institute of Physics)

spacing of tungsten and a central hydrogen species for each unit. On the same plane flash desorption experiments have disclosed the presence of two different binding sites[27]. The β_2-state exhibits second-order desorption kinetics, while the β_1-state obeys first-order kinetics. Hydrogen adsorption as a function of surface coverage[27] reveals that the β_1-state begins to fill after the β_2 sites have been saturated (Figure 7.3). A model structure has been proposed with β_2 sites occupied by hydrogen atoms and β_1 sites by hydrogen molecules (Figure 7.4). The adsorption studies have been interpreted to exhibit different sticking probabilities for the two sites[28] (Table 7.5), with a linear decrease in S_{β_1} as a function of surface coverage, but a constant value for S_{β_2} independent of surface coverage (Figure 7.5).

The desorption activation energies measured by flash-desorption studies amount to $E_d = 32.3$ kcal mol^{-1} for β_2 and $E_d = 26.3$ kcal mol^{-1} for β_1. The binding energy $D(W\text{—}H)$ for the β_2-state is found to be 69 kcal mol^{-1} from the relationship $D(W\text{—}H) = \frac{1}{2}[E_d + D(H\text{—}H)]$ where $D(H\text{—}H)$ is the dissociation energy of hydrogen. For the β_1-state, the binding energy of 26 kcal mol^{-1} appears surprisingly high for bonding of molecular hydrogen. It has been suggested that the β_1-structure involves a hydrogen molecule oriented vertically with one of the H atoms buried below the surface of the lattice. More recently[29] a non-random recombination of neighbouring hydrogen atom pairs has been advanced as the reason for pseudo-first-order desorption kinetics.

On W(112) the LEED pattern observed[30] before and after adsorption of hydrogen exhibits no change in the unit mesh (1 × 1). Since the adsorbate is in one-to-one registry with the substrate, the adatoms are located on individual tungsten sites. Again flash-desorption experiments indicate different

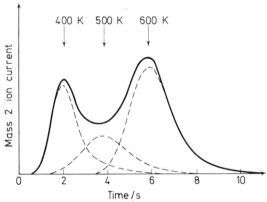

Figure 7.6 Flash-desorption kinetics of hydrogen on W(112)
(From Adams, Germer and May[30], by courtesy of North-Holland)

binding sites with desorption peaks occurring at 400, 500 and 600 K (Figure 7.6). After the β_{600}-state is fully covered, adsorption occurs in the two other states ($\beta_{400} + \beta_{500}$). Also the observed work function change appears to parallel the β_{600}-site occupation[29].

The existence of the different binding states on the various crystal planes indicates a complexity unexpected on the basis of the simple atomic structure of substrate and adsorbate. In an attempt to interpret these results, Tamm and Schmidt[28] examined qualitatively the hybridisation of the 5d and 6s orbitals of W(100) and the resulting directionality of these orbitals on the surface of the (100) plane. The β_2-state is considered to result from the directionality of the unhybridised orbitals which can participate in the bonding of a hydrogen atom (Figure 7.7). Occupation of such a β_2-site will cause the d-electrons of four neighbouring W atoms to overlap with the s-electron of the hydrogen atom forming an s–d hybrid and exhibiting an orbital configurative corresponding to a C(2 × 2) LEED structure. For the

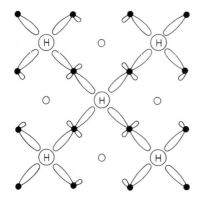

Figure 7.7 Bonding of hydrogen on W(100)
(From Tamm and Schmidt[27], by courtesy of the American Institute of Physics)

β_1-state involving molecular hydrogen, the participation of e_g orbitals of the substrate is suggested.

In comparison with chemisorption studies on polycrystalline W surfaces, the results indicate that a change in occupation of different binding sites with surface coverage is responsible for the observed variation in heat of adsorption. The qualitative similarity between the desorption spectra of hydrogen on W(100) and on polycrystalline samples is possibly due to the fact that (100) orientation predominated on the polycrystalline W foils examined.

7.4 COVALENT MIXED ADSORPTION: NITROGEN–TUNGSTEN

Before examining the data obtained for different crystallographic planes of this adsystem it may be well to summarise the results for polycrystalline surfaces[31–37]. Early flash-desorption studies demonstrated the existence of four different binding states. Two of these, the α- and γ-states, were interpreted to involve weakly bound molecular species. The more strongly bound β_1- and β_2-states were suggested to represent adatoms, with first-order desorption kinetics for the β_1-state, and second-order desorption kinetics for the β_2-state. Adsorption of nitrogen on polycrystalline surfaces was considered[36] to involve a sequence of steps with initial molecular adsorption in the low binding energy α-state, followed by diffusion over the surface, and finally dissociative chemisorption on sites representative of the β-states[38] with maximum surface coverage $\theta < 0.6$. The energetics of nitrogen sorption indicate a binding energy of 10–20 kcal mol^{-1} for the α- and γ-states. For the β-states, the desorption energies are close to the heat of adsorption[39] of c. 80 kcal mol^{-1}. For the sticking coefficient at zero coverage (S_0) on W films and ribbons at 298 K, values were reported ranging from 0.03 to 0.6 [40]. Although such variation may be associated with differences in surface roughness of the specimen, the results may be an indication of the strong dependence of adsorption on surface crystallography.

By a combination of experimental techniques, including flash desorption, LEED, work function measurements and field ion microscopy, the role of crystal orientation in this adsystem has been examined. In 1962, studies with the field ion microscope[41] indicated different rates in dissociative chemisorption of nitrogen on the various planes of tungsten at room temperature: $(100) > (111) > (211) \gg (110)$. Only at temperatures below 190 K the (110) plane was found to be occupied by an adsorbate with a heat of adsorption of c. 9 kcal mol^{-1}. The observation of second-order desorption kinetics from the (110) plane[42] indicated the presence of adatoms and correlated with the properties of the γ-state previously noted on polycrystalline samples. Subsequently it was found[43] from nitrogen adsorption studies on close-packed planes of tungsten that, at room temperature, the initial sticking coefficient of nitrogen on W(100) is c. 0.25, on W(111) < 0.04 and on W(110) nearly zero. These observations of crystal plane selective nitrogen adsorption have been greatly extended by more recent studies, as summarised in Table 7.6. The data indicate further that saturation coverage corresponds to a

Table 7.6 Sticking coefficient of nitrogen on various tungsten crystal planes at 300 K

Crystal plane	$10^{-14} \times$ W Atom surface density/cm^{-2}	S_0	Adsorbate surface structure
(100)	10.00	0.25[43], 0.5[21], 0.41[45]; 0.37[44a]; 0.16 †	$C(2 \times 2)$—N[21, 44a]
(110)	14.14	0[43]; 10^{-2} †	
(111)	5.77	<0.04[43]	
(210)	4.47	0.25[44a]	$P(2 \times 1)$—N[44a]
(211)	8.16	<10^{-2} [44a]	
(310)	6.32	0.28[44a]	$\begin{cases} P(2 \times 1)\text{—N}^{44a} \\ C(2 \times 2)\text{—N}^{44a} \end{cases}$
(311)	3.02	0.29 *	
(321)	5.35	<10^{-2} [44a]	

*Eisinger, J. T. (1958). *J. Chem. Phys.*, **28**, 165
†Madey, T. E. and Yates, J. T., Jr. (1967). *Suppl. Nuovo Cimento*, **5**, 483

value of nearly half a monolayer, an observation in line with the interpretation of the surface structures observed in the LEED patterns (Table 7.6). No substrate reconstruction is observed. On W(100) the nitrogen atoms form domains of $C(2 \times 2)$ structures, which means that they are double-spaced along the [001] and [010] directions and single-spaced along the [011] directions as shown in Figure 7.8.

Adams and Germer[44] conclude that the W(100) site is the active centre for dissociative chemisorption of nitrogen, and they depict on a stereographic triangle of the various crystal planes of a cubic structure those planes which have (100) sites and on which nitrogen adsorption would occur by a non-activated process. Such a model proves valuable in synthesising the chemisorption properties of a polycrystalline specimen on the basis of the relative contribution of the various crystal planes and their relative preponderance.

Further information on the nature of chemical bonding of the adsorbate

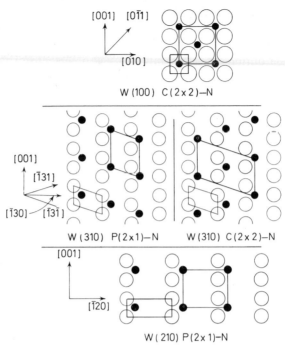

Figure 7.8 Nitrogen surface structures on W planes. Substrate primitive unit meshes are indicated by light lines, overlayer meshes by heavy lines. ○ Tungsten; ● nitrogen (From Adams and Germer[44b], by courtesy of North-Holland)

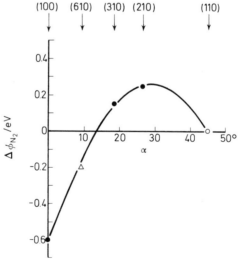

Figure 7.9 $\Delta\phi$ values for saturation nitrogen coverage on planes $(hk0)$ of the $[001]$ zone plotted against the angle α between $(hk0)$ and (100). ● From Ref. 44b; ○ from Ref. 1 of Ref. 44b; from Ref. 34 of Ref. 44b.

and the electron distribution normal to the surface can be deduced, in principle, from the measurements of changes in the work function of the adsystem. Such an approach of the changes in sign and magnitude of the dipole moment during adsorption and desorption and their dependence on the structural characteristics of the solid has been widely used for the system under discussion. The results obtained for a number of crystallographic planes by field emission and surface potential methods are summarised in Table 7.7. Although the adsorbate dipole moments differ in magnitude depending on the experimental method employed, the sign of the work function change for various crystal planes is of special interest. Strong covalent bonding of nitrogen would be expected to result in electron transfer to the adsorbate. However such an increase in work function is not observed

Table 7.7 Work function changes for tungsten–nitrogen adsystem at maximum coverage* at 300 K

Crystal Plane	$\Delta\phi$
(100)	−0.9† (FE), −0.60 [44a] (SP), −0.63 [21] (SP), −0.4 [43], (CP), −0.57¶ (TRP)
(110)	0 [43] (SP), +0.14§ (SP), +0.03¶ (TRP)
(111)	+0.15‡ (FE), +0.3 † (FE), +0.17 [43] (SP)
(210)	+0.27 [44a] (SP), +0.5 [21] (SP)
(310)	+0.20 [44a] (SP), +0.39 † (FE)
(311)	+0.26 † (FE)
(112)	+0.46 § (SP)
(332)	+0.5 ‡ (FE)
(610)	−0.2 ‡ (FE)

*SP = surface potential; FE = field emission; TRP = thermionic retarding potential.
†Holscher, A. A. (1964). *J. Chem. Phys.*, **41**, 579
‡Ozuri, T. (1964). *J. Phys. Soc. Japan*, **19**, 83
§Hopkins, B. J. and Usami, S. (1968). *J. Appl. Phys.*, **39**, 3500
¶Madey, T. E. and Yates, J. T., Jr. (1967). *Suppl. Nuovo Cimento*, **5**, 483

for W(100) or W(610) (Table 7.7). Since surface reconstruction is not associated with the sorption process on W(100), as demonstrated by the LEED results, the observed work function decrease is not simply a matter of differences in electronegativity between adsorbate and substrate. The surface dipoles observed may be indicative of weakly bonded molecular adsorption accompanied by dissociation and strong binding during heating of the specimen and possibly bond formation of partially ionic character.

Alternatively different metal orbitals may be participating in bond formation with the adsorbate. Employing the latter approach, Schmidt and co-workers[27,45] suggest that on W(100) bond formation with the adatom involves sd_{xy} hybrids from four adjoining tungsten atoms and the atomic orbitals of the nitrogen atoms ($2p_x$ and $2p_y$) located coplanar with and central to the four tungsten atoms. Symmetry of the sd hybrids about each tungsten atom will exhibit a vacant sd orbital in the $\langle 10 \rangle$ direction. As a result, the adjacent W site is inactivated for bond formation with nitrogen, but adsorption on every other site takes place in the $\langle 11 \rangle$ direction, producing the C(2 × 2)–N structure of the adsorbate lattice observed by LEED.

An interesting 'zero-order empirical model' is proposed[44] to interpret the variation in work function change with nitrogen coverage on different crystal planes. Assuming that the dipole has a positive outward orientation for a nitrogen atom above the geometric plane of W(100), and a negative outward orientation for a nitrogen atom below the plane, the work function may be related to the angle α between the ($hk0$) and (100) planes of the [001] zone. On this basis the correlation shown in Figure 7.9 is obtained.

We may conclude that in this adsystem involving molecular and atomic binding the sorption kinetics are sensitive to the substrate structure of the crystal. However, the binding energies observed are not simply related to the electron distribution around a nitrogen atom on a specific tungsten site, nor is the sign of the surface dipole formed insensitive to configurational changes that occur during heating of the specimen.

7.5 SOME OBSERVATIONS ON COVALENT CHEMISORPTION ON TUNGSTEN

Details of the chemisorption process on specific crystal planes of tungsten, as reviewed on the preceding pages, have revealed an unexpectedly complex role of surface structure. It appears that for both molecular and atomic adsorbates bonding occurs in distinct states, which exhibit specific binding energies and surface dipoles. For a given adsorbate different crystal planes show various types of binding. A single plane may have more than one binding state, and single- and multiple-bonded adsorbates may be formed on the same plane. In addition to the substrate-related sorption characteristics, the apparent long-range interactions between adsorbate species affect with progressive coverage the type of bonding exhibited by the adsystem. Yet no definitive or clearcut relationship emerges between the crystallography of the surface and the binding state. On W(110) both hydrogen and nitrogen interact most weakly, while carbon monoxide experiences strong binding modes. Could the weak interaction be due to a geometric effect or the electron distribution at the surface of this plane in the presence of the adsorbate?

The existence of distinct binding states, and the changes in work function associated with electron transfer to the adsorbate point to covalent bonding. Quantum mechanical considerations of delocalised electrons and electron redistribution at the surface due to the presence of an adsorbate[46, 47] have been juxtaposed with the molecular orbital point of view involving examination of the overlap of substrate and adsorbate orbitals[18, 27, 28, 48].

Concerning the geometric configuration of the adsorbate, the LEED results indicate in many cases ordered surface structures indicative of large heats of adsorption and long-range order unperturbed by surface diffusion, since $\Delta H_{ads} > E_{diff}$. In the case of tungsten, none of the adsystems examined in this report has caused surface reconstruction as observed, for example, with nickel[49]. Examination of the ordered surface structures resulting from chemisorption has been interpreted by Somorjai and co-workers[8] in terms of the following 'ordering rules': (a) the surface structure tends towards the smallest unit cell which can be attained by close packing arrangement and controlled by adsorbate–adsorbate and adsorbate–substrate interactions; (b) the rotational symmetry of the adsorbate reflects that of the substrate plane; and (c) the unit cell vectors of the adsorbate are in coincidence with the periodicity of the substrate.

Chemisorption studies on individual crystal planes have provided some answers to the complex pattern of solid–gas interactions. At the same time these studies have raised new and interesting questions which offer exciting challenges for the immediate future.

Note added in proof

Adams points out in a recent private communication that the peaks observed at 400 K and 500 K (Figure 7.6) may have been caused by desorption from the W-crystal support leads.

References

1. Ehrlich, G. (1963). *Advan. Catal.*, **14**, 255
2. Redhead, P. A. (1962). *Vacuum*, **12**, 203
3. Gibson, R., Bergsnov-Hansen, B., Endow, N. and Pasternak, R. (1963). *Trans. 10th Nat. Vac. Symp.*, 88
4. Lander, J. J. (1965). *Recent Progress in Solid State Chemistry*, Vol. 2, 26. (New York: Pergamon Press)
5. Estrup, P. J. and McRae, E. G. (1971). *Surface Sci.*, **25**, 1
6. McRae, E. G. and Jennings, P. J. (1969). *The Structure and Chemistry of Solid Surfaces*, G. A. Somorjai, editor), John Wiley & Sons, Paper No. 7
7. Stern, R. M., Taub, H. and Gervais, A. (1969). ibid, Paper No. 8
8. Somorjai, G. A. and Szalkowski, F. J. (1971). *J. Chem. Phys.*, **54**, 389
9. Gomer, R. (1961). *Field Emission and Field Ionization*, (Cambridge, Mass.: Harvard University Press)
10. Müller, E. W. (1960). *Advan. Electronics and Electron Physics*, **13**, 83
11. Culver, R. V. and Tompkins, F. C. (1959). *Advan. Catal.*, **11**, 67
12. Herring, C. and Nichols, M. H. (1949). *Rev. Mod. Phys.*, **21**, 185
13. Madey, T. E. and Yates, Jr., J. T. (1971). *J. Vac. Sci. Technol.*, **8**, 525
14. Swanson, L. W., et al. (1967). *Literature Review of Adsorption on Metal Surfaces*, Field Emission Corp., McMinnville, Oregon
15. Swanson, L. W. and Gomer, R. (1963). *J. Chem. Phys.*, **39**, 2813
16. Ehrlich, G. (1966). *Ann. Rev. Phys. Chem.*, **17**, 2956
16a. Ford, R. R. (1970). *Advan. Catal.*, **21**, 51
17. Bell, A. E. and Gomer, R. (1966). *J. Chem. Phys.*, **44**, 1065
18. Kohrt, C. and Gomer, R. (1971). *Surface Sci.*, **24**, 77
19. May, J. W. and Germer, R. H. (1966). *J. Chem. Phys.*, **44**, 2895
20. Anderson, J. and Estrup, P. J. (1967). *J. Chem. Phys.*, **46**, 563
21. Estrup, P. J. and Anderson, J. (1967). *J. Chem. Phys.*, **46**, 567
22. Estrup, P. J. and Anderson, J. (1966). *J. Chem. Phys.*, **45**, 2254
23. Chang, C. C. (1968). *J. Electrochem. Soc.*, **115**, 354
24. Engel, T. and Gomer, R. (1969). *J. Chem. Phys.*, **50**, 2428
25. Armstrong, R. A. (1966). *Can. J. Phys.*, **44**, 1753
26. Hopkins, B. J. and Pender, K. R. (1966). *Surface Sci.*, **5**, 316
27. Tamm, P. W. and Schmidt, L. D. (1969). *J. Chem. Phys.*, **51**, 5352
28. Tamm, P. W. and Schmidt, L. D. (1970). *J. Chem. Phys.*, **52**, 1150
29. Yates, J. T. and Madey, T. E. (1971). *J. Vac. Sci. Technol.*, **8**, 63
30. Adams, D. L., Germer, L. H. and May, J. W. (1970). *Surface Sci.*, **22**, 45
31. Ehrlich, G. (1961). *J. Appl. Phys.*, **32**, 4
32. Becker, J. A. and Hartman, C. D. (1953). *J. Phys. Chem.*, **57**, 153
33. Ehrlich, G. (1961). *J. Chem. Phys.*, **34**, 29

34. Redhead, P. A. (1962). *Proc. Symp. Electron Vac. Phys. Hungary, 1962,* 89
35. Oguri, T. (1963). *J. Phys. Soc. Japan,* **18,** 1280
36. Rigby, L. J. (1965). *Can. J. Phys.,* **43,** 532
37. Robbins, J. L., Warburton, W. K. and Rhodin, T. N. (1967). *J. Chem. Phys.,* **46,** 665
38. Smith, T. (1964). *J. Chem. Phys.,* **40,** 1805
39. Kisliuk, P. (1959). *J. Chem. Phys.,* **31,** 1605; (1959). **30,** 174
40. McKee, S. and Roberts, M. W. (1967). *Trans. Faraday Soc.,* **63,** 1418 (in this paper earlier data are reviewed.)
41. Ehrlich, G. and Hudda, F. G. (1962). *J. Chem. Phys.,* **36,** 3233
42. Ehrlich, G. and Hudda, F. G. (1963). *Phil. Mag.,* **8,** 1587
43. Delchar, T. A. and Ehrlich, G. (1965). *J. Chem. Phys.,* **42,** 2686
44a. Adams, D. L. and Germer, L. H. (1971). *Surface Sci.,* **26,** 109
44b. Adams, D. L. and Germer, L. H. (1971). *Surface Sci.,* **27,** 21
45. Clavenna, L. R. and Schmidt, L. D. (1970). *Surface Sci.,* **22,** 365
46. Grimley, T. B. (1960). *Advan. Catal,* **22,** 1
47. Rhodin, T. N., *et al,* (1969). *The Structure and Chemistry of Solid Surfaces,* (G. A. Somorjai, editor). (New York: John Wiley and Sons)
48. Bond, G. (1966). *Discuss. Faraday Soc.,* **41,** 20
49. Germer, L. H. and MacRae, A. U. (1962). *Proc. Nat. Acad. Sci. U.S.,* **48,** 997

8
Hydrosols

D. H. NAPPER and R. J. HUNTER
University of Sydney

8.1	INTRODUCTION		242
8.2	PREPARATION AND CHARACTERISATION OF DISPERSIONS		242
	8.2.1 *Preparation*		242
	8.2.1.1 *Inorganic sols*		243
	8.2.1.2 *Polymer latices*		243
	(a) *Electrostatically stabilised*		243
	(b) *Sterically stabilised*		244
	8.2.2 *Particle size distributions*		245
	8.2.3 *Electron microscopy*		246
	8.2.4 *Light scattering*		247
	8.2.5 *The Coulter counter*		249
	8.2.6 *Surface-area determinations*		250
	8.2.7 *Electrokinetics*		250
	8.2.7.1 *Streaming potential*		251
	8.2.7.2 *Measurement of streaming potential*		253
	8.2.7.3 *Electrophoresis*		254
	(a) *Relaxation and retardation corrections*		254
	(b) *Particle conductivity and surface conductivity*		256
	(c) *Concentrated suspensions*		257
	8.2.7.4 *Measurement of electrophoretic mobility*		258
	8.2.7.5 *Variation of dielectric constant and viscosity in the double layer*		259
8.3	CHARGE AND POTENTIAL DISTRIBUTION AT THE SOLID/SOLUTION INTERFACE		260
	8.3.1 *The potentials involved*		260
	8.3.2 *Comparison with mercury/solution interface*		261
	8.3.3 *Double layer at the solid/solution interface*		262
	8.3.4 *Specific adsorption and the discreteness of charge effect*		265
	8.3.5 *The oxide/solution interface*		267
	8.3.6 *The diffuse part of the double layer*		268

8.4	COLLOID STABILITY		269
	8.4.1 Attraction		269
		8.4.1.1 Calculation of the van der Waals attraction	269
		8.4.1.2 Direct measurement of the van der Waals attraction	271
	8.4.2 Electrostatic stabilisation (DLVO theory)		271
		8.4.2.1 Calculation of the repulsion	272
		8.4.2.2 Tests of the DLVO theory	273
		(a) Equilibrium studies	274
		(b) Kinetic measurements	277
	8.4.3 Steric stabilisation		279
		8.4.3.1 Theories of steric stabilisation	279
		(a) Classical thermodynamic approach	279
		(b) Statistical thermodynamic theories	280
		8.4.3.2 Experimental studies	288
		(a) Displacement flocculation	288
		(b) Thermodynamic flocculation	289
		(c) Molecular models of steric stabilisation	291
8.5	COAGULATION OF HYDROSOLS		292
	8.5.1 Coagulation by hydrolysable cations		292
	8.5.2 Mechanical and surface coagulation		294
	8.5.3 Flocculation by polymers		294

8.1 INTRODUCTION

Hydrosols are dispersions of colloidal particles in water. This review discusses some recent developments in the understanding of hydrosols, though where necessary (e.g. if unifying principles are involved) we have not hesitated to include relevant discussion of non-aqueous dispersions as well. It has not proved possible to examine all facets of the phenomenology of hydrosols; the choice of topics, especially those reviewed at greater depth, reflects both areas of high activity and significant progress, as well as those areas where our own interests lie. Significant gaps and imbalances can be redressed in subsequent reviews. Where uncertainty and controversy arise, we have not flinched from expressing a viewpoint, although we hope a reasonably balanced account of the opposing opinions is presented.

8.2 PREPARATION AND CHARACTERISATION OF DISPERSIONS

8.2.1 Preparation

The past decade has seen increasing importance placed on the preparation of monodisperse colloidal dispersions. These provide model dispersions for studies of, e.g. colloid stability, light scattering and rheology. The exploitation of monodisperse systems has crossed interdisciplinary boundaries[1]: monodisperse latices, e.g., have been used to measure the pore size of filters and biological membranes, in serologic diagnostic tests and in the studies of the reticuloendothelial system. Latices have also found application in the calibration of various instruments (e.g. Coulter counter, electron microscope, light scattering photometer).

8.2.1.1 Inorganic sols

Only a few inorganic sols have so far been prepared in a monodisperse condition. Methods for generating reasonably monodisperse gold sols have long been known[2]. La Mer[3] prepared monodisperse sulphur sols and these classical studies have provided inspiration for subsequent investigations, such as those of Ottewill[4] who prepared monodisperse silver halide sols. Dispersions of barium sulphate[5,6] and lanthanum and lead iodate[7] with sharp distributions have also been generated.

Watillon and Dauchot[8] have now produced monodisperse selenium hydrosols by reducing selenious acid with hydrazine hydrate. To obviate uncontrolled heterogeneous nucleation, very small gold particles were added as growth centres. Monodisperse sols ranging in particle diameters from 40 to 500 nm were successfully prepared. The larger diameter sols were generated by growth of smaller monodisperse selenium sols.

Another metal sol of interest that has recently been investigated is Carey Lea's colloidal silver sol. This interest arises because, unlike many hydrophobic sols, it exhibits reversible coagulation; removal of the coagulating electrolyte by simply washing the coagulum results in spontaneous redispersion of the silver particles. Apparently the positively charged silver particles are covered by specifically adsorbed citrate ions in superequivalent amounts, leading to a net negative charge[9].

Stöber[10,11] has described a technique for preparing spherical silica particles of uniform size with diameters in the range 50–2000 nm. The particles were generated by hydrolysis of alkyl silicates with subsequent condensation of the resulting silicic acid in alcoholic solutions. Radioactive tracers can be incorporated into these silica particles. Those metals (e.g. ^{51}Cr, ^{59}Fe, ^{141}Ce, ^{124}Sb, ^{7}Be and ^{58}Co) which form oxides of low solubility at pH ≥ 7 were easily incorporated into the bulk of the particles during growth, if present at concentrations less than 10 p.p.m. relative to the mass of silica in the systems[11]. Such tagged particles should find many biological applications. These silica sols will provide an alternative to the commercial Ludox sols.

Demchak and Matijevic[12] have for the first time prepared a monodisperse sol of a metal hydroxide. Chromic hydroxide sols, which exhibited higher order Tyndall spectra, were generated by the simple expedient of ageing chromium sulphate or chrome alum solutions (e.g. 10^{-3} M chrome alum at 75 °C for 18 h). Particle sizes ranged from 200 to 500 nm, depending upon the exact conditions of preparation. The particles were positively charged.

8.2.1.2 Polymer latices

(a) *Electrostatically stabilised* – Monodisperse electrostatically-stabilised polystyrene latices have been available commercially for some time now.

These latices form highly attractive model dispersions being composed of rigid, non-porous spheres whose radius and surface charge-density can be varied at will. One disadvantage of commercial latices for colloid stability studies has been that they are usually stabilised by an unspecified surfactant. It now appears[1] that the Dow polystyrene latices are prepared in the presence of one of the following surfactants: Aerosol MA (sodium bis-1,3-dimethylbutyl sulphosuccinate), Aerosol OT (sodium di-2-ethylhexyl sulphosuccinate), potassium oleate or sodium lauryl sulphate. Apparently Aerosol MA is the emulsifier commonly used[13]. Dezelic, Petres and Dezelic have described in detail procedures for preparing monodisperse polystyrene latices using Aerosol MA as stabiliser[13].

There is good evidence to suggest that the latex particles produced by such emulsion polymerisation possess both physically adsorbed molecules of surfactant and chemically bound groups derived from the initiator (e.g. sulphate half-ester groups derived from the decomposition of potassium persulphate[1]).

Ottewill and Shaw[14-19] pioneered the preparation of monodisperse latices containing a single ionogenic surface species, in this case carboxylic acid groups. Polystyrene was polymerised by emulsion polymerisation using sodium dodecanoate as the emulsifier and hydrogen peroxide as initiator. The particles were sufficiently monodisperse to exhibit higher order Tyndall spectra. Their diameter was in the range 60–300 nm. The coefficient of variation was less than 10%; with some latices this index was as small as 1–2%.

Extensively dialysed latices were stable despite the fact that all the sodium dodecanoate had been removed[19]. Electrophoretic and titrimetric studies strongly suggested that the residual charge was derived solely from carboxylic acid groups[17]. I.R. studies of the particles implied the existence of phenylacetic acid residues bound into the particles[18]. Vanderhoff[20-23] has also shown how the surfactant can be removed from Dow polystyrene latices by using mixed-bed ion-exchange resins. This results in latices that are stable, presumably due to sulphate half ester residues incorporated into the particles.

The foregoing methods demand the removal of the surfactant by either dialysis, which is slow, or ion exchange, which may contaminate the latex with polyelectrolytes unless rigid precautions are followed. To obviate these unwieldy procedures, Kotera, Furusawa and Takeda[24] have prepared highly monodisperse polystyrene latices in the total absence of added soap. Potassium persulphate was used as initiator and so oligomeric surfactant molecules were generated *in situ* by the sulphate anion free radicals. Only relatively large particles with diameters in the range 350–1400 nm were prepared in this way. Similar poly(vinyl acetate)[25,26] and poly(methyl methacrylate)[27] dispersions have been prepared. The production of monodisperse poly(vinyl acetate) latices by emulsion polymerisation has also been descirbed[28]. Finally we note the appearance of the proceedings of an Amer. Chem. Soc. symposium on polymer colloids, which we have not yet sighted[29,30].

(b) *Sterically stabilised* – The foregoing account is confined to electrostatically-stabilised latices. Techniques have now been devised for preparing monodisperse latices that are stabilised solely by a steric mechanism. The breakthrough was really achieved in non-aqueous dispersion media by Osmond et al.[31,32] but the same general principles are applicable in aqueous systems[33]. The latices are generated by the heterogeneous polymerisation of a monomer in the presence of a suitable amphipathic macromolecular surfactant. The latter is usually either a block or graft copolymer which may be added prior to the commencement of polymerisation or generated *in situ* by a precursor. The copolymer consists of nominally insoluble anchor moieties coupled with soluble stabilising moieties (e.g. for preparing hydrosols poly-(ethylene oxide-*b*-styrene) is a typical surfactant). Particles prepared in this way exhibit no observable electrophoretic mobility[33]. Latices stabilised by a combination of electrostatic plus steric mechanisms can be prepared by adsorbing suitable non-ionic surfactants onto electrostatically stabilised particles[34].

8.2.2 Particle size distributions

It is generally agreed that many, though certainly not all, dispersions exhibit size distributions that are positively skewed[35-37]. Presumably such distributions are analogous to the distribution of heights in the human population; the latter is slightly skewed in the positive direction because negative values for height have no physical meaning.

Kerker[38] has recently summarised the distribution functions that are most frequently ascribed to colloidal dispersions. These are the traditional normal and the logarithmic normal distribution functions and the more recent zeroth order logarithmic distribution function (ZOLD). Note that ZOLD uses the modal rather than the geometric mean size as a parameter in the distribution function:

$$F(x) = (\sigma_0 x_M \exp\{\sigma_0^2/2\} (2\pi)^{\frac{1}{2}})^{-1} \exp\{-(\ln x - \ln x_M)^2/2\sigma_0^2\} \quad (8.1)$$

where x_M = modal size and σ_0 = zeroth order logarithmic standard deviation or breadth parameter. The pre-exponential factor is, of course, the normalisation constant. The parameter σ_0, which is a measure of the width and skewness of the distribution, is related to the mean size ($\bar{x}$) and the modal size through

$$\ln \bar{x} = \ln x_M + 1.5\,\sigma_0^2 \quad (8.2)$$

For sufficiently narrow distributions ($\sigma_0 \ll 1$), σ_0 closely approximates the coefficient of variation ($\sigma/\bar{x}$ where σ = standard deviation). Changes in σ_0 alter both the width and skewness at constant mode but the skew is always positive. We also note that Epenschied *et al.*[39] introduced the idea that a general family of logarithmically skewed distributions exists of which the logarithmic normal and ZOLD are special members. Watterson[40] has challenged this view and asserted that only one logarithmic normal distribution function exists.

Although it has been possible to find distribution functions of different form that represent either positive or negative skew, changing the sign of the skew has usually necessitated the rather awkward procedure of changing functional form. Rowell and Levit[41] therefore devised two new functions that permit the skewness to vary continuously from negative to positive.

Note that these functions are mathematical inventions and are not derived from mechanistic considerations of how colloidal dispersions are formed.

The first function, the skewed normal distribution (SND), is defined by

$$F(x) = [(1-s^2)/\sigma(2\pi)^{\frac{1}{2}}] \exp\{(-(x-x_M)-\dot{s}|x-x_M|)^2/2\sigma^2\} \quad (8.3)$$

where s = skew parameter ($-1 < s < 1$), x_M = modal size and σ = breadth parameter. (Rowell and Levit actually give $-1 > s > 1$ as the limits for s. These limits would demand that $F(x)$ be negative. All their calculations were for $|s| < 1$.) For $s = 0$, the function is a normal distribution function. As s varies from -1 through 0 to $+1$, the skew varies continuously from negative, to no skew, to a positive skew. The SND must be employed with caution, however, for colloidal dispersions, because it admits of negative values of x which are, of course, physically inadmissible.

The skewed zeroth-order logarithmic distribution (SZOLD) is the second function discussed by Rowell and Levit. It is defined as

$$F(x) = (s/\sigma x_M(2\pi)^{\frac{1}{2}} \exp\{\sigma^2/2\}) \exp\{-\ln^2(1+s[x-x_M]/x_M)/2\sigma^2\} \quad (8.4)$$

This function becomes the positively skewed ZOLD if $s = +1$. For $s = -1$, SZOLD transforms into a negatively skewed distribution function defined originally by Wallace and Kratohvil[42]:

$$F(x) = (\sigma_0 x_M \exp\{\sigma^2/2\}(2\pi)^{\frac{1}{2}}) \exp\{-\ln^2(1+[x_M-x]/x_M)/2\sigma^2\} \quad (8.5)$$

The curves for $s = \pm 1$ are in fact mirror images of each other. The minus sign arising in $F(x)$ when s is negative is to be ignored because it simply indicates the inversion of the coordinate axis.

We also note in passing some recent developments in mechanistic approaches to particle size distributions in aqueous latices prepared by emulsion polymerisation. O'Toole[43] has discussed the stochastic broadening of latex size distributions during polymerisation while Saidel and Katz[44] have provided an alternative description of this phenomenon. Gardon[45] has considered the broadening in latex size distribution that results from distribution of free radicals over the particles.

8.2.3 Electron microscopy

It is often important to know the value of the number of particles per unit volume (N) of a colloidal dispersion. The usual method of determining particle numbers by electron microscopy is indirect; the particle size distribution must first be found and from this the mean volume per particle calculated. The particle concentration N is then obtained from the volume fraction of disperse phase present, estimated by an independent (e.g. gravimetric) technique. This method requires the measurement of the size of a large number of particles (e.g. 1000) to eliminate statistical fluctuations[46]. Moreover, any errors in size determination are magnified almost threefold in computing N. Finally, particles of irregular geometric shape are troublesome.

Alexander and Robb[47] have developed an electron microscopic technique that permits the number of particles to be counted more or less directly. Both natural and synthetic polymers, when spread at the air/water interface, form films which have sufficient mechanical strength to support colloidal particles embedded in them[48]. The method consists of spreading a known volume of the dispersion with a suitable embedding polymer on a surface of known area, e.g. on a Langmuir trough. Poly(vinyl alcohol) (88% hydrolysed from poly(vinyl acetate)) was found to be a suitable spreading agent for polystyrene latex particles. A portion of the spread film was then transferred to a nitrocellulose film on an electron-microscope grid. The average number of particles for each grid aperture of known area was found by direct counting of suitable electron micrographs. Knowledge of the average number of particles per unit area of spread film permitted N to be found by simple calculation. Care must be taken to ensure uniform spreading of the latices. Unadsorbed soap should be removed from the latex before spreading. The results appear to be accurate to $\pm 10\%$, which is quite a satisfactory precision for most purposes.

Sheppard and Tcheurekdjian[49, 50], have actually determined the sizes of latex particles using a surface balance. A known weight of polystyrene particles was spread at the air/water interface of a Langmuir trough with the aid of organic liquids (e.g. alcohols). A layer one particle thick was obtained. Extrapolation of the surface pressure v. area curves to zero surface pressure gave an area per particle. Assuming either hexagonal or cubic close packing, the particle radius could be calculated. The results were in good agreement with other methods. This technique is unlikely to be widely adopted because of the need to use non-aqueous liquids for spreading.

8.2.4 Light scattering

Light scattering is often a useful method for determining particle size and particle size distributions. Measurements can usually be performed directly on the dispersions.

Kerker's definitive book[38] on light scattering appeared in 1969. It provides a comprehensive survey of the literature relevant to the use of light scattering for size determination. Literature up to and including 1968 is covered. We will be content therefore with pointing out several of the more recent developments.

The application of laser techniques to light scattering continues. Finnigan et al.[51] have recently reviewed previous laser studies and discussed the advantages of laser light scattering photometers. The construction of one such photometer of their own design is described. An example of the versatility of laser techniques is its application by Chu and Schoenes[52] to investigate the translational motions of polystyrene latices relevant to transport properties. Accurate values of the diffusion coefficient were obtained by measurement of the power spectrum of the light scattered by the particles using a self-beating technique.

Gucker[53] has re-examined the now classical formulation of the Mie angular light-scattering phase functions for spheres in terms of the Legendre polynomials. He has derived two alternative, and in some ways more simple formulations that involve sines and cosines only. These new series should prove to be more convenient in analytical studies. Gucker[54] has also derived a simple method for checking the self-consistency of values of the amplitude functions. The method is based upon the fact that for non-absorbing spheres each value of the amplitude functions lies on a circle in the complex plane with its centre at $(1/2, 0i)$ and radius $1/2$.

Phillips, Wyatt and Berkman[55] have reported the first measurements of the intensity of polarised light scattered from a single latex particle. These studies are reminiscent of the rather less accurate measurements of Gucker on single aerosol droplets[56]. A high-speed computer was used to fit the measured intensities to theoretical curves and so to obtain the particle size (with a 1% accuracy) and the refractive index. No problems with the uniqueness of the fitting of the experimental points were encountered.

The usual dissymmetry method of determining particle sizes is commonly only applicable for spheres when the ratio of the particle diameter to the wavelength is, say $1:2$. Maron and Pierce[57] have shown how to extend the dissymmetry method to cover diameter to wavelength ratios of 2 or more, at least for polystyrene latices. The principle of the method rests in re-defining the dissymmetry in terms of the ratio of the intensities at two forward angles differing by 10 degrees (e.g. 45 and 35 degrees). Measurement of these intensities permits the average particle size to be estimated. Like the orthodox dissymmetry method, this new procedure is sensitive to polydispersity.

A second light scattering method for determining the size of latex particles has appeared recently[58]. The method merely requires the specific turbidity of the latex in the limit of infinite dilution to be measured, along with the refractive index increment. Obviously only relatively simple equipment is needed. In this context we also mention the simple method devised by Kratohvil[59] for determining the number concentration of polydisperse spheres. A single measurement of the Rayleigh ratio at a given wavelength is all that is required.

Livesey and Billmeyer[60] have discussed how the light scattering technique can be extended to evaluate sizes in the range 10^2–10^5 nm. What is needed are measurements down to very small scattering angles. Small-angle x-ray scattering may also be used to determine particle size distributions in dilute polydisperse suspensions of spheres[61]. One advantage of this method is that it is applicable to particles whose diameter is < 100 nm.

We also note that Jennings[62, 63] has apparently resolved the origin of the curious sign reversal that bentonite sols exhibit as a function of concentration when their angular scattering patterns are measured in the presence and absence of an electrical field. Apparently the particles, which are both anisotropic and anisometric, possess both permanent and induced moments in an electric field. These moments are comparable in magnitude but associated with axes which are in quadrature.

Bryant and Latimer[64] have discussed the anomalous diffraction by large particles with different geometric shapes (discs, cylinders and ellipsoids of revolution). Their results should be applicable to sols containing particles large in comparison with the wavelength and with a refractive index close to that of the medium.

8.2.5 The Coulter counter

One drawback with the light scattering method of measuring particle-size distributions is the elaborate mathematical computations needed. Further, the difficulties associated with non-spherical geometrical forms, especially those with edges and corners, place limitations on its applicability. Alternative procedures, e.g. electron microscopy often require the initial disperse state to be disturbed.

One instrument that virtually permits *in situ* estimation of the particle-size distribution of aqueous dispersions of non-conducting particles is the Coulter counter[65]. (A Particle Data counter has also been manufactured.) Distributions in the size range 700–4000 nm may be measured. The effective volume of the particle is determined by measuring the change in resistance of an electrolyte solution that occurs when a single non-conducting particle, emitted from a small aperture, passes between two electrodes. Clearly particles of non-spherical shape can be handled by this method, given suitable calibration of the counter. Although the principle of the instrument is simple, its operation and the interpretation of the results are by no means straightforward.

Several groups[66-72] have examined critically some of the problems encountered in the use of the Coulter counter. It is commonly accepted that the counter may give distributions that are both broader and more skewed than those obtained by, e.g. electron microscopy. Spielman and Goren[67] have claimed that the broadening derives from particles following different paths near and through the sensing aperture. They proposed an ingenious method of hydrodynamic focusing to obviate this variability and so to increase the size resolution of the instrument. The method consists of feeding the dispersion through a focusing aperture prior to a passage through the sensing aperture. The procession of particles as a result tends to travel on the same streamline and to follow the same approach trajectory. Focusing led to better agreement between the diameters of polystyrene latex particles as determined by the Coulter counter and by electron microscopy. Matthews and Rhodes[72] have claimed that focusing is unnecessary at larger sizes.

Cooper and Parfitt[88] have attributed the increase in positive skewness of the distribution to coincidence effects. Two types of coincidence are distinguished: first, simultaneous passage of two particles through the sensing aperture (horizontal coincidence); second, when the passage of two small particles produces a pulse large enough to be counted (vertical coincidence). The latter effect would tend to impart a positive skew to the measured distribution. Princen and Kwolek[73] have asserted that the Coulter method of coincidence correction may be inadequate if greater than 3% coincidence occurs. Together with Wales and Wilson[74] they have proposed alternative correction procedures. At $1-2\%$ coincidence all three procedures give similar results so that this is clearly the least ambiguous domain in which to operate the instrument.

The Coulter counter is being used increasingly to follow the kinetics of flocculation[28, 69, 71, 72]. The stability ratio W can be measured for any species, singlet, doublet, etc., that can be detected. Matthews and Rhodes[72] have critically evaluated this type of application and concluded that accurate measurements are possible. They showed experimentally that neither passage through the orifice nor dilution with the Coulter electrolyte influenced the measurements of W so that aggregates appear to retain their integrity on passage through the counter.

8.2.6 Surface-area determinations

Ottewill and Shaw[16] have measured the adsorption/desorption isotherm of krypton onto polystyrene latex particles at liquid nitrogen temperatures. The specific surface area obtained from the B.E.T. plot was almost identical with that derived from electron microscopy. This agreement coupled with the absence of hysteresis, implies the absence of gross porosity in the particles.

A novel method of determining surface areas of latices has been developed by Stryker and Matijevic[75]. Hafnium, at a pH of 4.5–5, exists in aqueous solutions in the form of a soluble, neutral hydrolysed complex $Hf(OH)_4$. This is a reasonably compact and symmetric species. The complex was shown to adsorb onto poly(vinyl chloride) latices forming a close-packed monolayer. The area per molecule was 23 $Å^2$. The use of the long-lived ^{181}Hf facilitates such adsorption measurements. One virtue of this method is that it permits the surface area to be determined without the necessity of drying down the latex.

8.2.7 Electrokinetics

Despite its limitation the measurement of zeta (ζ) potential by one or other of the electrokinetic techniques remains the most widely used method for obtaining information about the charge and potential distribution at the colloid/solution interface. To summarise even the most recent applications of the ζ-potential would be a formidable task since the literature appears in a wide variety of journals devoted to agriculture, medicine, biology, engineering and technology (especially mineral processing). Recent reviews of the work in medicine and biology appear in the two volumes edited by Bier[76, 77] whilst a brief summary of the theory and some applications to technology have been given by Sennett and Olivier[78]; their compilation is, however, rather uncritical.

More recently, MacKenzie[79] has reviewed the current ideas on the interpretation of ζ-potential and in particular its application to the study of the flocculation and flotation processes used in the treatment and separation of mineral ores.

Of the four principal methods of obtaining electrokinetic information (electrophoresis, streaming potential, electro-osmosis and sedimentation potential) the first two remain the most widely used. The advent of millivoltmeters with very high input impedance ($c.$ 10^{14}–10^{15} Ω) has given the streaming potential a considerable advantage over the electro-osmotic method for measurements on capillaries or porous plugs. At the same time it has not led to a great increase in the use of the sedimentation potential method. Despite the early success of Pearce and Elton[80] in correlating measurements by this method with values obtained by the streaming potential and the development of the diffusion theory of the effect[81], it is generally regarded as too erratic in its manifestations for general use. In any case it can give no more information than is more readily obtained by electrophoresis[82]. We shall, therefore, concentrate attention on the streaming potential and electrophoretic methods.

8.2.7.1 Streaming potential

When a liquid is forced through a capillary or a porous plug under a hydrostatic pressure, P, there is generated a potential difference, E_s between the ends of the capillary or plug which is given by:

$$\frac{E_s}{P} = 4\pi\varepsilon_0 \frac{D\zeta}{4\pi\eta\lambda} \tag{8.6}$$

The factor $1/4\pi\varepsilon_0$ has the value 9×10^9 N m^2 C^{-2} in the M.K.S. system but has a value of 1 dyne cm^2 statcoulomb^{-2} in the c.g.s. electrostatic system. Most workers in the field still use a combination of c.g.s. and practical units for quoting results so that some care is required in using equation (8.6). The dielectric constant D is dimensionless in this formulation and the viscosity, η, and conductivity, λ, are usually measured in poise and ohm^{-1} cm^{-1} respectively. With such mixed units the appropriate value for $1/4\pi\varepsilon_0$ is 9×10^{11} V C^{-1} cm and P must be measured in dyne cm^{-2}. Both E and ζ will then be in volts.

Overbeek[83] has given a fairly rigorous derivation of this relation for both capillaries and for plugs of powdered material and the linearity of the plot of E_s v. P has been amply confirmed[84]. At very high Reynolds numbers (>2000) where turbulence is induced in straight capillaries, the relation becomes non-linear[85]. This would presumably occur at much lower Reynolds numbers in porous plugs because non-linearity and turbulence occur more easily in such systems[86]. At the other extreme, Derjaguin[87] claims that at very low flow rates where diffusion becomes comparable in importance with convective transport the relation again becomes non-linear. In order to detect this effect it would be necessary to place the measuring electrodes very near to the ends of the capillary and so far no experimental confirmation has appeared.

Equation 8.6 is derived by balancing the convective movement of charge, caused by the movement of the solution, against the back flow of current under the influence of the developing streaming potential. At low electrolyte concentrations it becomes invalid because a significant fraction of the back flow occurs along the surface of the capillary or pores, rather than through the bulk liquid. The effective conductivity at very low electrolyte concentrations ($<10^{-3}$M) is therefore much higher than would be expected and the corresponding streaming potential is lower than anticipated. A plot of ζ-potential v. concentration of indifferent electrolyte often shows a maximum (in absolute value) at $c.$ 10^{-3}–10^{-4} M as a consequence of failure to take account of this surface conduction effect.

For a single capillary of radius, r, the correction is readily made since the returning current is proportional to[88] $(\pi r^2 \lambda + 2\pi r \lambda_s)$ rather than to $\pi r^2 \lambda$. The corrected equation is, therefore,

$$\frac{E_s}{P} = 4\pi\varepsilon_0 \frac{D\zeta}{4\pi\eta[\lambda + (2\lambda_s/r)]} \qquad (8.7)$$

where λ_s is the surface conductivity (in ohm^{-1}). If λ_s is not known, the true ζ value can be obtained by measurements on capillaries of the same material but of different radii[89] since λ_s can then be eliminated. Strictly speaking the value of λ_s is not a constant but depends on the pore radius as has been shown by Morrison and Osterle[90]. (Figure 8.1.)

In the case of porous plugs the surface conduction problem is more serious. The early workers[91], following a suggestion of Briggs[92], used the conductivity

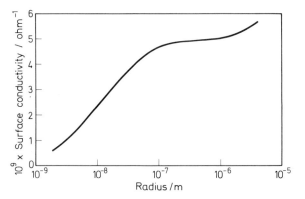

Figure 8.1 Surface conductivity v. tube radius for water in glass combination.
(From Morrison and Osterle[90], by courtesy of Amer. Inst. Phys.)

of the liquid in the plug rather than the bulk value in an effort to circumvent the difficulty but it has since been shown[93, 94] that this will work only if the porous medium can be replaced by a set of parallel capillaries of uniform bore. This should be approximately true for a diaphragm made of spheres of uniform size and Ghosh and his co-workers in a series of papers[95–98] have shown that true ζ-potential values can be obtained even at low concentrations for such systems. The method is tedious since it calls for measurements on plugs made of several different and uniform particle sizes but the results agree well with those on capillaries made from the same material.

A further difficulty with streaming potential (and electro-osmosis) measurements occurs when the radius of the capillary or pores approaches the double layer thickness since equation (8.6) is based on the assumption that the electrostatic potential ψ differs from zero only in a very thin layer near the walls. Bull and Moyer[99, 100] adopted the suggestion of Lens[101] that the potential in very narrow capillaries would resemble that between two concentric cylinders and derived a relation for the dependence of apparent ζ-potential on κr, where r was the capillary radius. A more sophisticated treatment was attempted by Rice and Whitehead[102] more recently, but since they were forced to use the Debye–Hückel approximation for the potential and also assumed a uniform conductivity for the liquid throughout the entire pore volume their theoretical predictions are not likely to be very accurate. Their solution suggested that the direction of the current flow could reverse several times across the diameter of the pore under suitable conditions. A similar analysis was conducted by Oldham, Young and Osterle[103] who used Ohm's law to relate the streaming potential and the current transported and combined this with Poiseuille's law for the fluid flow and the Poisson–Boltzmann equation. Solving the resulting non-linear differential equation with a power series they obtained ζ-potential values which were independent of the size of the capillary radius, which suggests that despite the approximations involved, their treatment was essentially correct.

A rather more realistic treatment has been given by Dresner[104] for the case of good co-ion exclusion. He uses the diffusion coefficient of the counter-ion as the index of ionic mobility which allows a spatial dependence of the conductivity to develop because of the radial concentration gradient in a narrow capillary. The results of the analysis are used to calculate the conditions under which charged membranes may be used for desalination of brackish water.

8.2.7.2 Measurement of streaming potential

As noted above, the most significant change in the apparatus for measuring streaming potential has been the introduction of accurate and stable high impedance electrometers. The basic experimental arrangement[105] remains much the same as that depicted by Alexander and Johnson[106] and used by Gortner[107], though Parriera[108] has recently described an automatic recording version. The electrodes which contain the porous plug should be of expanded platinum or platinum gauze rather than perforated sheet since it is important that the pressure drop should be confined to the plug and not occur partly across the electrode. Some of the early anomalous results at low pressure[109, 110] were probably due to the occurrence of an asymmetry potential between the electrodes. This evidently arises from strains induced in the electrode materials during their fabrication. A procedure for eliminating this effect has been suggested by Hunter and Alexander[111].

Some of the difficulties encountered in the direct measurement of streaming potential could, perhaps, be eliminated by the use of a sinusoidally varying pressure which would induce an a.c. streaming potential. The theory was presented by Cooke[112] and also by Ueda, Watanabe and Tsuji[113] quite some time ago and the preliminary results were promising but much more needs to be done on the method.

For measurements at higher electrolyte concentration when E_s becomes very small it is possible to use the streaming current method[114, 115] which can also be made automatic recording[116] but which also suffers from surface conduction problems at low concentrations.

8.2.7.3 Electrophoresis

(a) *Relaxation and retardation corrections* — The early work on the relation between electrophoretic mobility and ζ-potential of spherical particles was reviewed by Overbeek[117] and by Booth[118, 119] and since then the most significant development has been the publication of the computer solution of the problem[120] incorporating the corrections for retardation and relaxation and using the complete Poisson–Boltzmann expression for the potential around the particle[121], rather than the linearised form of Debye and Hückel. The principal conclusion from the work is that the earlier calculations of Overbeek[122] and Booth[123] were qualitatively sound but somewhat overestimated the size of the correction.

Figure 8.2 shows a comparison of the corrections predicted by the various treatments for different particle sizes and double layer thicknesses (as measured by the parameter κa). The ordinate, E, is a dimensionless variable proportional to the mobility, u:

$$E = \frac{1}{4\pi\varepsilon_0} \frac{6\pi\eta e}{DkT} u \qquad (8.8)$$

The curves are drawn for the case $e\zeta/kT = 5$ (i.e., $\zeta \doteq 125\,\text{mV}$) and for

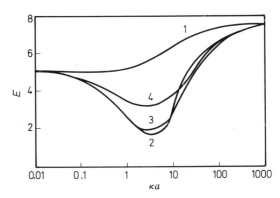

Figure 8.2 Plot of E v. κa for a 1:1 electrolyte given $e\zeta/kT = 5$ and $m_+ = m_- = 0.184$ ($m_\pm = (N_A\varepsilon kT/6\pi\eta)z_\pm/\lambda_\pm$). Curve 1 – Henry; 2 – Overbeek; 3 – Booth; 4 – Wiersema *et al.* numerical results
(From Overbeek and Wiersema[125], by courtesy of Academic Press)

typical values of the ionic mobilities of 1:1 electrolytes ($\lambda_\pm^0 = 70\,\text{ohm}^{-1}\,\text{cm}^2\,\text{mol}^{-1}$). Henry's calculation[124] took account only of the influence of the particle size on the shape of the externally imposed electric field around the particle whilst both Overbeek[122] and Booth[123] also took account of retardation and relaxation effects. It is evident that whilst they were correct in showing that the most difficult area for interpretation was at κa values of c. 6, the more exact treatment reduces their suggested correction to almost half in this region.

The earlier calculations also suggested that at high values of ζ-potential and intermediate values of κa ($1 \leqslant \kappa a \leqslant 10$) the mobility would reach a maximum value and would begin to decrease as ζ increased (Figure 8.3).

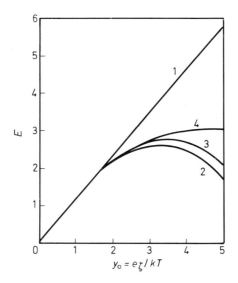

Figure 8.3 Plot of E as a function of $e\zeta/kT$ for a 1:1 electrolyte with $\kappa a = 5$ and $m_+ = m_- = 0.184$. Curve 1 – Henry; 2 – Overbeek; 3 – Booth; 4 – Wiersema et al. numerical results (From Overbeek and Wiersema[125], by courtesy of Academic Press)

This would mean that unique ζ-potential values could not be assigned to a given mobility and it is heartening to see that the computer solution shows E as a monotonic increasing function of y_0 ($=e\zeta/kT$) even though it does appear to approach an upper limit for these κa values. The assumptions which are still involved in this treatment are carefully set out in the original paper[120] and an excellent review of the status of the theory up to about 1967 has now been given by Overbeek and Wiersema[125]. The rest of this section will, therefore, assume a familiarity with that paper.

The results of the computer calculation are given in tabular form for a few values of the ionic mobilities and of the other parameters and interpolation procedures must be used to obtain the maximum accuracy from them. It is apparent from Figure 8.2 that, except for the region $1 \leqslant \kappa a \leqslant 20$, it should be possible to make reasonable corrections with the analytical solutions of Overbeek and Booth. In both of these the mobility is given in the form of a truncated series in ζ whereas the usual requirement is to determine ζ from measured values of the mobility. The early attempt by Stigter and Mysels[126] to invert this series neglected some of the important terms and a more satisfactory relationship was given by Hunter[127]:

$$\frac{e\zeta}{kT} = \frac{E}{X} - \frac{C_3(E/X)^3 + C_4(E/X)^4}{X + 3C_3(E/X)^2 + 4C_4(E/X)^3} \tag{8.9}$$

where E is defined by equation (8.8) and X denotes the function $X_1^*(\kappa a)$ which is presented in graphical form by Henry[124] and corresponds to curve 1 of Figure 8.2 in this review. The coefficients C_3 and C_4 are related to the coefficients of the original power series and are functions of the charges and mobilities of the ions in the double layer. Stigter[128] has shown that equation (8.9) gives results in quite close agreement with the computer values for $\kappa a \approx 1$ (Figure 8.4). This figure also shows that the use of equation (8.9)

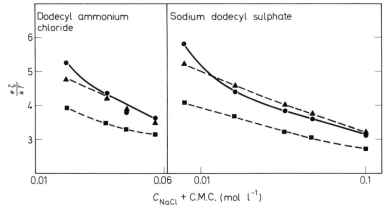

Figure 8.4 ζ-Potential v. ionic strength for detergent micelles in aqueous sodium chloride solutions, as derived from tracer electrophoresis. ▲—equation (8.9); ■—Stigter and Mysels; ●—computer results
(From Stigter[128], by courtesy of Academic Press)

does not give results which are similar to those obtained with the Stigter and Mysels equation, as was claimed by Sengupta and Biswas[129].

Overbeek and Wiersema[125] discuss the effect of various assumptions on the accuracy and utility of a computer solution[126]. They suggest methods by which one may take account of the presence of mixtures of counter-ions and co-ions of different valency. Particles of different shape are also discussed. For insulating particles with $\kappa a \gg 1$ where the von Smoluchowski equation[130] holds:

$$u = 4\pi\varepsilon_0 \, D\zeta/4\pi\eta \tag{8.10}$$

the result is independent of particle shape[131], and no further correction is required for retardation and relaxation (Figure 8.2). For some other shapes the calculations of Henry[124] are relevant though only in the region where the relaxation effect can be neglected.

(b) *Particle conductivity and surface conductivity*—Henry's calculations[124] also took account of the bulk conductivity of the particle though it is now generally recognised that all solid particles behave as insulators due to polarisation effects at the interface[125]. (In effect the difference in potential from one side of the particle to the other is insufficient to allow a suitable charge-transfer process to occur at both surfaces and so the particle does not conduct.) The possibility still remains, however, that some particles may show surface conductivity. So long as this is merely an enhanced conductivity due to the higher concentration of ions in the diffuse double layer, then its effect is taken into account in the theory of the relaxation correction. This is the surface conductivity calculated by Morrison and Osterle[90] (Figure 8.1). Recent measurements by James and Carter[132] show that polystyrene has an unusually high surface conductivity, λ_s. Indeed the value quoted by them ($\lambda_s \doteq 1.3 \times 10^{-7}$ ohm^{-1}) is so high that it should be impossible to obtain meaningful ζ-potential data on such systems. Presumably once again it is the polarisation effects at each end of the particle which prevent surface conduction of this type having its expected influence. James and Carter used an a.c. measuring technique which would overcome the polarisation and allow the full surface conductivity to be manifested. There still remains the possibility that the effective d.c. surface conductivity, though much less than 1.3×10^{-7} ohm^{-1} is still significantly higher than that which would be expected from double-layer considerations. This would make it possible to explain the fact that the mobility of this material when plotted against the concentration of indifferent electrolyte[133] shows a maximum at concentrations of about 10^{-4} M, similar to those found in the early work on streaming potential. Application of the relaxation corrections to the mobility data reduces the height of the maximum but does not eliminate it. Goldfarb et al.[134] found a similar maximum in apparent ζ-potential for suspensions of a paraffin wax stabilised with stearic acid and were able to eliminate it by applying the surface conduction correction suggested by Ghosh and Pal[135]. A similar maximum was observed in suspensions of octadecanol[136]. Although the much larger maxima observed by Sieglaff and Mazur[137] on soap-stabilised polystyrene latices can probably not be entirely ascribed to this effect, it could make a significant contribution. The results show a continuous increase in mobility with increase in KCl concentration at constant soap concentration up to 0.01 M KCl. Although one normally expects surface conduction effects to have disappeared at such concentrations, Ghosh and Bull[138] claim that it is still significant for borosilicate glass particles covered with bovine serum albumin even at 0.01 M NaCl. The persistence of surface conduction effects up to such high electrolyte concentrations is to be expected for values of λ_s as high as those quoted by James and Carter[132] but the accuracy of the results quoted by Ghosh and Bull[138] is questionable in view of the uncertain shape of their particles which were made of crushed glass.

(c) *Concentration suspensions* – The effect of colloid concentration on mobility is briefly discussed by Overbeek and Wiersema[125] who conclude that in view of the uncertainties involved in dealing with concentrated suspensions it is preferable to extrapolate results to infinite particle dilution. The microelectrophoretic method presents no problems in this direction since measurements can only be made at very low particle concentrations.

The methods of moving boundary and mass transport electrophoresis[139] require significantly higher colloid concentrations and Long and Ross[140] suggest that for the mass transport technique the method of Möller, van Os and Overbeek[141] is applicable. In effect the measured ζ-potential of interacting particles is assumed to be equal to the difference between the potential at the shear plane and the potential midway between them.

One of the most difficult questions concerning the interpretation of electrophoretic mobilities is the extent to which the assumptions of the Gouy–Chapman theory limit the validity of the final equations. Since this, and the question of the position of the shear plane, are problems common to all of the electrokinetic procedures they will be deferred for the present.

8.2.7.4 Measurement of electrophoretic mobility

The principal methods of measuring mobility are still the micro-electrophoresis technique and the moving boundary method, as developed by Tiselius[142]. The former is limited to particles of size $>0.2\,\mu m$ although a recent commercial development using a laser light source* is claimed to allow observation of particles as small as 70 nm. Observations have, however, been made on proteins, lipids and other biologically important molecules by this technique[143], after first adsorbing them on a suitable inert surface such as paraffin wax or oil (Nujol). At least for serum albumin there is good correlation between adsorbed and solution protein[144]. The principal advantage of the microscopic method is that measurements can be made at very low particle concentrations using very little material (provided due care is taken against contamination). There is also the possibility of determining the extent of individual variations in ζ-potential from one particle to another rather than merely obtaining a mean value. The natural variation in mobility of mineral particles may have a significant influence on their coagulation and flotation behaviour as it does on their filtration behaviour[145].

One of the problems of the cylindrical microscopic cell introduced by Mattson[146] was the poor optical quality of the particle image[147] caused by the astigmatism introduced by the cell wall acting as a lens system. Although the effect is not too obvious at the stationary level it becomes very obvious at the axis of the cell. The apparent mobility of the particles changes rapidly with depth in the cell in the neighbourhood of the stationary level so that accurate focusing is essential but even so the depth of field of the usual microscope objective is sufficiently large to introduce significant errors. Smith and Lisse[148] attempted to overcome this problem with a two-tube cell in which the radii were so arranged that the stationary level occurred at the axis of the observation tube where the velocity gradient is zero. Unfortunately, the poor focusing at this depth reduces considerably the efficacy of this device. One solution would be to construct a two-tube cell in which the observation tube was a very thin-walled capillary after the manner of the

---

*A very versatile research micro-electrophoresis measuring instrument marketed by Rank Bros. of Bottisham, Cambs., U.K.

van Gils and Kruyt cell[149]. This cell does not suffer from the optical difficulties of the thick walled cell and the by-pass tube could then function as a support for the very fragile observation tube. Another solution has been offered by Hamilton and Stevens[150] who have designed a double-tube flat cell which combines the optical advantages of that shape (very little distortion) with the possibility of measurement at a point where the velocity gradient is zero. Other improvements in the design of the microscopic apparatus have also been suggested[151, 152].

There has been little development in the moving boundary method since the commercial introduction of Schlieren optics, although a modified form of moving boundary procedure, suitable for colloidal surfactants (association colloids), namely tracer electrophoresis, has been introduced by Mysels et al.[153]; it has been applied to soap micelles[126] and to human serum albumin[154]. As mentioned earlier, a cell in which mobility is determined by direct measurement of the mass of colloid transported has been introduced by Long and Ross[139]. They regard their method as valuable for obtaining correlations between ζ-potential and flow properties of concentrated suspensions since both can be measured under comparable conditions of volume concentration. This can at best only lead to qualitative correlations; quantitative relations can be developed between ζ-potentials measured at low particle concentrations and flow behaviour of concentrated suspensions provided that the interaction between the particles is explicitly taken into account[365].

8.2.7.5 Variation of dielectric constant and viscosity in the double layer

The principal limitations of the simple Gouy–Chapman theory stem from its treatment of the ions as point charges immersed in a continuous medium of constant permittivity. We shall take up this matter in more detail below. Suffice it to say at this stage that the calculations of Levine and Bell[155] suggest that (except in the immediate neighbourhood of the surface) these effects are to a large extent self-compensating and the simple theory gives a surprisingly accurate representation of the potential.

There remains, however, the question of the extent to which the electrostatic field influences the viscosity of the liquid in the diffuse double layer. The theory discussed so far assumes that the dispersion medium preserves its bulk viscosity right up to the shear plane, at which point it abruptly rises to infinity and the remaining liquid (and its included ions) is carried along with the particle. Lyklema and Overbeek[156] attempted to remove this restriction by allowing the viscosity to depend on the local electrostatic field as suggested by Andrade and Dodd[157]. Their work has been reviewed extensively by Haydon[158] in his discussion of the variation of D and η in the double layer. Haydon concludes that measured ζ-potentials are limited by the high values of the ratio η/D which occur in the double layer and that the increased values of the ratio are due to an augmented viscosity rather than to a reduction in D. Whilst it is now generally accepted[158] that the values of D in the diffuse part of the double layer are not appreciably less than its bulk value, the possibility of an increased viscosity remains to be established. Li and de Bruyn adduce some evidence for an increased viscosity from their work on the electrokinetic behaviour of quartz[159] and Haydon cites his own data[160] on the relation between ζ and the surface potential, ψ_0, of hydrocarbon droplets covered with surfactant, in support of the Lyklema and Overbeek model[156]. On the other hand, Hunter[161] showed that using the values of viscosity suggested by Lyklema and Overbeek, Haydon's experimental values of ζ should have been impossibly high. Indeed, an upper limit to the effect of the field on the viscosity can be obtained simply from the existence of Haydon's results and this turns out to be of a magnitude comparable to the effect on the dielectric constant. Hunter's analysis[161] allows both effects to be taken into account simultaneously but will only become useful when the parameters for the effect of the field are known with more certainty. At the present time the weight of evidence is that neither the dielectric constant nor the viscosity effect is of great importance[125] and Haydon's data must be attributed to some other process such as the geometrical effect suggested by Hunter[161] or the discreteness of charge effect[165]. This conclusion is confirmed by some calculations made by Stigter[162] on the viscosity and self-diffusion of soap micelles and, more directly, by some neutron scattering measurements[163] of the mobility of water layers contained between vermiculite sheets. Although the electric fields encountered in this latter system are very high there is no evidence for restricted mobility of more than one or two layers of water right at the alumino-silicate surface. The extensive data on augmented viscosities in boundary liquid films collected by Derjaguin[164] must be reassessed in the light of the current views on the nature of 'anomalous' water.

8.3 CHARGE AND POTENTIAL DISTRIBUTION AT THE SOLID/SOLUTION INTERFACE

8.3.1 The potentials involved

It is well known that the total potential difference between two different phases (i.e. the difference between their inner or galvani potentials ϕ) is not accessible to direct experimental measurement[166]. At the same time, it is generally assumed that most of the interesting properties of a colloidal system are determined by this potential difference, and so many attempts have been made to circumvent the difficulty. The general procedure is to set up a plausible model for the distribution of charges and dipoles in the interfacial region and to calculate the resulting variation in electrostatic potential with position. The potential variation can then be used, with various other assumptions, to calculate a number of experimentally accessible quantities which can in turn be used to check the original model. The fewer and more defensible these latter assumptions are, the more directly will such experimental measurements test the proposed model. Unfortunately, the number of situations in which these extra assumptions have gained wide acceptance is severely limited; perhaps Grahame's suggestion[167] (that the inner layer capacity of the mercury/solution interface, in the absence of specific adsorption, is dependent only on the metal charge and independent of electrolyte concentration) could be placed in this category.

The nature of the galvani potential is discussed briefly by Overbeek[168] and more extensively by Lyklema and Overbeek[169]. A very careful description of the relation between the measurable electrostatic parameters of a phase, and those which appear in current models has been given by Parsons[170]. Briefly we may say that the galvani potential of a phase is made up of a part ψ due to the presence of free charges and a part, χ, due to the orientation of dipoles at the interface. The first of these is directly measurable but the second is not, and therein lies the uncertainty in the galvani potential, ϕ.

Models of the double layer usually begin by relating the charge distribution to the potential using an equation such as that of Poisson which concerns itself only with the free charges and introduces the dipoles only in terms of the permittivity of the dielectric. At first sight this would appear to be a calculation of the ψ potential, but in all modern theories of the double layer some account is taken of the special orientation of dipoles at the interface (even if only in terms of a reduced permittivity in the inner region) and, to the extent that this is successful, the calculated potential is a galvani potential. This is important because in most (equilibrium) theories of the double layer the galvani potential is required to establish the thermodynamic criterion for the distribution of ions, i.e. equality of the electrochemical potential throughout the interfacial region[168].

8.3.2 Comparison with mercury/solution interface

The basic features of the structure of the double layer at the mercury/solution interface are now well established[171] and many of the ideas which were developed and tested there have been taken over, with little modification to describe colloidal systems. For soap-stabilised emulsion droplets the surface can be expected to be reasonably smooth or at least regularly indented on the average[172, 173] and the chief differences between this type of surface and the mercury/solution interface would lie in the discrete nature of the surface charge groups and the lower dielectric constant of the drop leading to smaller image forces for adsorbed ions.

By contrast, the parallels which can be drawn between a typical hydrosol interface and that of the mercury/aqueous electrolyte interface are very few and confined almost entirely to the diffuse part of the double layer. A few solid surfaces, such as those on certain clay minerals and on the monodisperse silver iodide particle[174] may be profitably regarded as reasonably planar with an adjacent compact double-layer region and a regular diffuse layer developed at some little distance (c. 0.5–1 nm) from the surface. For the great majority of colloids, however, it is becoming increasingly apparent that the charge on the solid side of the interface is developed over a region of finite depth which may be more or less disordered and may be penetrated by solvent molecules and electrolyte ions. The main reason that the Stern–Gouy–Grahame model of the double layer[175] is still used so widely for the colloid/solution interface is the difficulty of obtaining unequivocal evidence of its limitations, because of the number of adjustable parameters available.

In some cases, such as for the silica/solution interface[159], a suitable pretreatment is said to remove the disturbed layers[176] from the surface but not all workers would agree that the problem can be solved as simply as that. Wright[177] finds that crushed quartz washed with hydrofluoric acid after the manner suggested by Li and de Bruyn[159] still exhibits time-dependent ζ-potentials indicative of a slow development of a gelatinous or disordered surface layer.

Apart from the problem of surface roughness and inhomogeneity we have also to grapple with the uncertainties of assigning an electrostatic potential difference across the interfacial region. At the mercury/solution interface one can impose any potential within quite wide limits and can then independently measure the resulting charge on the metal. Although the absolute value of the potential difference between the interior of the mercury and the interior of the electrolyte is subject to some uncertainty there is no uncertainty about the size of the changes which are imposed when one changes the externally applied potential. On the other hand, at the colloid/solution interface one can alter the potential difference between the bulk of the solution and the solid surface by changing the concentrations of certain ions, but one cannot unequivocally determine the size of these potentials or even the exact size of the alterations in potential. One must always refer to some quasi-thermodynamic model-dependent analysis to relate concentration changes to potentials and potential changes. Even for the silver iodide surface, which is the most widely studied, and most theoretically tractable, there remain some uncertainties in the analysis because there is doubt about the mechanism by which the surface charge and potential are generated[178, 179]. We will discuss these problems (and the more difficult ones associated with oxide surfaces) in more detail in the next section.

Techniques for ensuring cleanliness and reproducibility of surfaces have developed significantly over the past 15 years[180] thus making it possible to develop reasonably sophisticated models of the double layer and to distinguish between them in favourable cases[179]. The model which most faithfully reproduces all of the features of the data and uses the smallest number of arbitrary assumptions and adjustable parameters will obviously be favoured but it can never lay claim to being a unique solution.

8.3.3 Double layer at the solid/solution interface

We will concentrate attention first of all on the silver iodide/solution interface since a considerable amount of work has been done on this system; it may be regarded as the archetypal lyophobic colloid. Since the silver/silver iodide electrode is reversible to the iodide ion it is possible to incorporate this surface in an electrochemical cell and so obtain a direct measure of the phase difference potential ($\Delta\phi$) between the interior of the silver iodide crystal and the interior of the electrolyte[181]. This measurement shares the same uncertainties as those of the mercury solution interface: liquid junctions are involved and the χ-potential must be assumed to be independent of the surface charge[182]. In addition it must be assumed that only the silver and iodide ions are able to influence the surface potential[183]; this means only that certain ions which could influence the potential (like chloride) must be excluded from participation.

The first extensive charge and potential measurements on well defined silver iodide sols in the presence of indifferent electrolyte were made by Lyklema and Overbeek[182] who quoted their results in terms of the capacitance of the double layer. They found that the surface showed some similarities

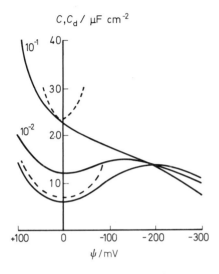

Figure 8.5 Differential capacity C of the double layer on AgI in 10^{-3}, 10^{-2} and 10^{-1} M KF solutions as a function of the potential ψ. Dotted lines: capacity C_d according to the diffuse double layer theory
(From Lyklema and Overbeek[182], by courtesy of Academic Press)

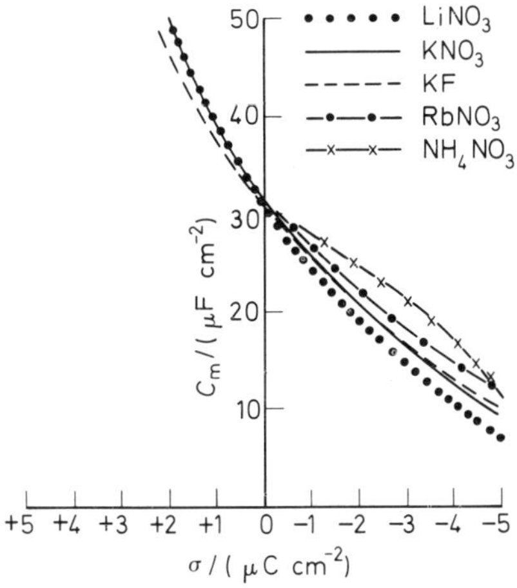

Figure 8.6 Differential capacity of the molecular condenser of the double layer on AgI in solutions of some monovalent electrolytes
(From Lyklema and Overbeek[182], by courtesy of Academic Press)

to the mercury surface and were able to satisfactorily account for their data on the basis of the Stern–Gouy model of the double layer. Thus at low concentrations where the capacity is determined by the behaviour of the diffuse region, the curve follows the hyperbolic cosine function (Figure 8.5) predicted by the Gouy–Chapman theory, at least in the immediate neighbourhood of the point of zero charge (p.z.c.):

$$C_d = 228.5\sqrt{c} \cosh 19.46\, z\psi_0 \qquad (8.11)$$

where C_d is in $\mu F\, cm^{-2}$, c is in g mol l^{-1} and ψ_0 is in volts.

At higher potentials (with respect to the p.z.c.) and at higher concentrations the capacity of the Stern, or compact layer becomes dominant and this shows an almost linear (and very steep) dependence on the surface charge (Figure 8.6). They did not attempt to explain why this should be so but we will return to this point below.

The relationship between the concentration of iodide (or silver) ions in the solution and the phase difference potential ($\Delta\phi$) can be derived as follows.

At equilibrium, the electrochemical potential of the iodide ions must be the same in the bulk solution and at the crystal surface:

$$\mu_s^\circ + RT \ln (a_{I^-})_s + zF\phi_s = \mu_b^\circ + RT \ln (a_{I^-})_b + zF\phi_b \qquad (8.12)$$

A similar expression must also hold at the p.z.c. and if we assume that at this point $\phi_s = \phi_b$ (i.e. we ignore the χ-potential) then we obtain by subtraction[184].

$$\phi_s - \phi_b = \psi_0 = \frac{RT}{F} \ln \frac{(a_{I^-})_b \text{ at p.z.c.}}{(a_{I^-})_b}$$

$$\approx RT/F \ln (C^\circ/C) \qquad (8.13)$$

where C° and C are concentrations of iodide ion and ψ_0 is called the surface potential. Apart from the assumption of ideal behaviour in the last step (which can easily be removed using the Debye–Hückel theory if desired) this analysis assumes that as one changes the bulk activity of the iodide ions there is no concomitant change in the activity of ions in the surface even though some iodide ions are being adsorbed or desorbed. This assumption is valid so long as the silver iodide crystal surface is assumed to consist of large numbers of positive and negative ions. The additional ions required to establish the potential then always remain a small fraction of the total and the assumption of constant activity of potential determining (p.d.) ions at the surface is a justifiable one. Even if the p.d. ions are distributed over a considerable depth in the crystal, as suggested by Grimley[185], this sort of consideration can still be applied: the ions which are adsorbed into the surface phase to establish the potential must still form an insignificant fraction of the total in that phase. The phase difference potential in that case is the difference between the interior of the solid and the interior of the liquid and the surface potential has some intermediate value between the two.

Honig[178] has recently examined the behaviour of a number of alternative models of the AgI/solution interface in which the surface charge extends for some depth into the crystal and is made up of mobile interstitial ions and vacancies (lattice defects). He compares the behaviour of these models with that of Lyklema and Overbeek[182] and dismisses the latter on the grounds that it is not possible to independently determine the capacity of the compact layer. He also regards the required value of the integral capacity of the stern layer at the p.z.c. (~ 31 μF cm^{-2}) as too low since it implies that the dielectric constant must be very small in this region. His estimate of $D \approx 5$ is, however, itself too low and is evidently based on a layer of thickness equal to one ionic radius. A more realistic picture of the inner layer, similar to that proposed for the mercury/solution interface, would give a higher value of D (c. 16) which is quite consistent with the structure proposed in that model[186]. A similar conclusion is reached by Levine, Levine and Smith[179].

Only one of Honig's models is able adequately to reproduce the experimental data and this is one in which there is a diffuse layer in both solution and solid and a fixed charge is present at the boundary. The fixed charge is independent of solution composition so that although the solid and solution are in equilibrium there is a non-equilibrium situation right at the boundary.

The surface potential thus becomes an adjustable parameter with this model since it is no longer related to the phase difference potential as given by equation (8.13). Also the number of lattice defects required for a good fit is c. 0.17% which Honig himself admits is rather high. A more detailed criticism of Honig's work is given by Levine, Levine and Smith[179] who have carried this type of model a good deal further. They show that when a comparison is made between Honig's fixed charge model and the experimentally determined capacitance values, the agreement is much less satisfactory than with the less sensitive comparison of charge versus potential curves which Honig himself used. Better agreement is obtained with Honig's variable charge model (in which the surface charge responds to the solution concentration of p.d. ions) but only after the introduction of a Stern layer which has a capacitance which is a linear function of the charge. By incorporating this additional feature the agreement could be made quite good on both sides of the p.z.c. A similar device had been employed earlier by Levine and Matijevic[187] who were able to reproduce satisfactorily the experimental surface-charge–surface-potential data using a Stern–Gouy model and an inner layer capacitance varying linearly with the surface charge. Considering the shape of the curves shown in Figure 8.6 for this system their success is not very surprising. Their work leaves open the question of how to give a satisfactory physical reason for this linear relationship.

In summary we may say that all that is needed to describe the capacity data for non-specifically adsorbed ions on silver iodide is a Gouy–Chapman diffuse double layer and a compact layer with a capacity which is given by an equation of the type:

$$K = 31 - 1.5\sigma_s \tag{8.14}$$

where K is in $\mu F\ cm^{-2}$ and σ_s is in $\mu C\ cm^{-2}$. The charge on the solid phase may be confined to its surface or it may extend into the crystal and whichever view is adopted seems to be of little importance in most colloid chemical situations. Only the actual position of the p.z.c. seems likely to be affected[179] and that is usually taken as an experimentally determinable quantity.

8.3.4 Specific adsorption and the discreteness of charge effect

For systems in which specific adsorption may occur the situation is much more complex. In this case, either cations or anions or both may be adsorbed into the Stern or compact layer and in this position they are able to interfere with the adsorption of the potential determining ions. Even if the concept of the potential determining ion remains viable, its utility is severely impaired because it becomes difficult if not impossible to measure the point of zero charge in such systems. The titration methods cannot be used to identify the p.z.c. when specific adsorption occurs although one can see which way it is moving with salt content and this indicates which of the ions of the salt is being specifically adsorbed. Using electrokinetic methods, however, one can readily determine the isoelectric point (i.e.p.) in such systems. This is the point at which the ζ-potential is zero and provided one can identify ζ

with the potential at the outer Helmholtz plane (i.e. the diffuse double-layer potential) this gives the point at which the diffuse layer charge is zero.

In order to avoid undue arbitrariness, the models for this type of system usually assume a smeared out surface charge, which does not extend into the solid, and a normal Gouy–Chapman diffuse layer separated by a Stern layer in which one or both ions of the supporting electrolyte may be assumed to be adsorbed. Most of the theoretical developments have concentrated on attempting to make the calculation of the adsorption isotherm for these adsorbed ions as realistic as possible. The first analysis, by Stern himself[188], used a Langmuir type isotherm in which all of the 'specific' chemical effects were embodied in a single constant, called the specific adsorption potential which was independent of the amount of adsorption. Such a treatment ignores (a) the effect of lateral repulsion between the adsorbed ions, (b) the discreteness of charge effect, (c) the influence of the adsorbed ions on the dielectric constant of the material in the compact layer, (d) the possibility that polarisation of the solid by these adsorbed ions and by the potential determining ions could alter the magnitude of the 'chemical' interactions between these ions and the surface material and (e) the effect of the finite size of the ions (and their partial hydration shells) on the availability of sites for adsorption. The use of a Langmuir-type isotherm also implies that the adsorbed ions are immobile which will be true only below a certain temperature[189]. The difference in behaviour between mobile and immobile ions is evident only for extensive adsorption[190] and certainly there are many systems which appear to obey the Langmuir isotherm and yet do not have immobilised adions.

All save the second of the above effects are usually assumed to be negligible for small amounts of adsorption, though Blok[191], and Hunter and Wright[192] have incorporated a correction for the last one. Gellings[193] has made some calculations on the effect of polarisation of an ion on its attraction to the surface but his heavy emphasis on image forces makes the result relevant only to adsorption on metals and other materials of high dielectric constant, e.g. TiO_2. Bell, Mingins and Levine[194] have calculated the magnitude of lateral interactions for a cell model.

Most attention has been paid to the discreteness of charge effect which refers to the fact that the ion, when it is on its adsorption site, does not experience the potential which would have characterised its adsorption site before the ion was adsorbed. The space around the adsorption site must be cleared of all charge before the ion is placed on it and so the potential at the centre of this 'cavity' is smaller, in absolute value, than the average potential before adsorption occurs. The effect was first introduced by Esin and Markov and its application to the mercury/solution interface has been discussed by Parsons, by Ershler and by Grahame (see Haydon[190]). Grahame's claim[195] that the energy required to rearrange the dipoles to make room for the incoming ion is negligible has been challenged more recently by Barlow and MacDonald[196].

The introduction of the discreteness of charge effect into the Stern adsorption isotherm has been accomplished by Levine, Bell, Calvert and Mingins[197, 198] who show that it is able to explain a number of previously puzzling features of colloid behaviour. More recently, Wiese, James and

Healy[199] have invoked the effect to explain the fact that for 2:2 and 3:3 electrolytes on silica, the value of the diffuse double layer potential exhibits a maximum as the surface potential, ψ_0, is increased.

8.3.5 The oxide/solution interface

Because of their very great technological significance, the colloidal chemical behaviour of oxides has been the subject of many investigations. Even though, for example, silica presents difficulties because of slow hydration of the surface layers[176], it is still essential that a detailed understanding of its potential–charge relations be arrived at. As noted earlier, Li and de Bruyn[159] made some study of this system but they did not attempt to set up a detailed model of the interface.

Such a model has been introduced by Levine and Smith[200] along similar lines to those used successfully for the AgI/solution interface. An important difference, however, occurs in the relationship between the surface potential, ψ_0 and the concentration of the p.d. ions, which in this case are H^+ and OH^-. Levine and Smith suggest that the Nernst equation (8.13) is in error in this case essentially because the activity of the p.d. ions at the interface is not independent of the state of charge of the surface. This results from the large fraction of the surface sites that remain uncharged so that the p.d. ions which become adsorbed are a significant fraction of the total number of charges. de Bruyn had earlier entertained the possibility that the Nernst equation might break down for oxide surfaces and Hunter and Wright[192] arrived at the same conclusion from an examination of electrokinetic data on silica and other oxide surfaces. They showed that with a simple Stern–Gouy–Grahame model of the double layer in which the specific adsorption potential of the adsorbed ions was fixed, the observed dependence of ζ-potential on indifferent electrolyte concentration could not be reproduced except by reducing the value of ψ_0 below the value expected from the Nernst equation. Moving the slipping plane into the diffuse part of the double layer did not affect this result.

A really stringent test of a proposed model would require it to simultaneously describe the surface charge–surface potential relationship and at the same time give a quantitative description of the dependence of electrokinetic potential on the concentration of indifferent electrolyte. Preliminary work in this area[201] is not yet very encouraging.

Berube and de Bruyn[202] have given a detailed thermodynamic analysis of the TiO_2/solution interface and although it has been criticised by Ball[203] it leads to a reasonable correlation of the experimental data. The differential capacity is almost what would be predicted from simple Gouy–Chapman theory, indicating that the capacity of the Stern layer is very large and the diffuse layer ions are able to penetrate almost to the plane containing the p.d. ions. The values found for the inner layer capacity are of the order of 150–400 $\mu F\ cm^{-2}$ which would be impossible for a rigid surface containing a smeared out charge. A similar result was found by Blok and de Bruyn[191] for ZnO. Electrokinetic measurements[201] on the other hand, would certainly not suggest that the slipping plane is near the plane of the p.d. ions. The

ζ-potential of TiO_2 is almost constant over a wide range of indifferent electrolyte concentrations which is consistent with a slipping plane near the plane of the p.d. ions[192] but the measured ζ-potential is far too low[201].

An interesting model for the adsorption of hydrolysable metal ions onto oxide surfaces has recently been presented by James and Healy[204] to explain why a positively charged metal ion does not adsorb strongly on a negatively charged surface until a critical pH value is reached. They present evidence that the adsorbed species is a partially hydroxylated metal ion which retains its primary hydration sheath on adsorption. The free energy of adsorption of the species is made up of a simple coulombic term ($ze\psi_x$, where ψ_x is the potential at the centre of the adsorbed ion) and a secondary solvation energy term. Adsorption is opposed by this solvation term because it measures the energy required to remove the secondary hydration sheath and place the ion and its primary layer into a region of low dielectric constant. Only when the effective charge on the ion is lowered by the acquisition of OH^- groups can the adsorption process occur. Despite the approximations involved the model gives a very good description of the adsorption behaviour of a number of transition metal ions.

8.3.6 The diffuse part of the double layer

Up to this point we have assumed that the potential in the diffuse part of the double layer can be reasonably well represented by the solution of the Poisson–Boltzman equation as suggested by Gouy and Chapman. This is generally believed to be true[206], at least for 1 : 1 electrolytes at concentrations below 10^{-2}M. For higher concentrations and valencies the corrections become significant but in these regions the more sophisticated treatments[205] become very complicated; Krylov and Levich[207] used Kirkwood's statistical method to show that for more concentrated solutions (0.1–0.5 M) the rate of decrease of potential with distance increases more rapidly with concentration than would be expected on the basis of Gouy–Chapman theory. Ali-Zade, Martynov and Melamed[208] introduced the image force theory of Onsager and Samaras into the Poisson–Boltzmann equation to generalise the Gouy–Chapman theory and showed that the correction 'was not too serious' for 1:1 electrolytes. Since they were able to perform the calculation only for very low potentials ($e\psi \ll kT$), their suggested correction (10% for 1:1 electrolytes but c. 100% for 2:2 electrolytes) is of doubtful validity. In any case, it is apparent from the work of Levine and Bell[209] that one must take all of the relevant corrections into account simultaneously (volume of ions, variation of permittivity, ion self-atmosphere effects and electrostriction) since they tend, to a considerable extent, to balance one another out.

Detailed discussions of the effects of the more common corrections are contained in the reviews by Haydon[206] and by MacDonald and Barlow[210]. From these works it seems reasonable to assume that in almost all of the situations of interest to colloid scientists, the simple Gouy–Chapman theory is adequate to describe the potential in the region outside the Outer Helmholtz Plane

(O.H.P.)[209, 211] for 1:1 electrolytes. For higher valency types the corrections are so large and so uncertain that it is doubtful whether any useful purpose is served by using the present theory. The evidence from clay mineral systems[212] is that very little diffuse layer development occurs in 2:1 electrolytes, although this may simply be a consequence of the high surface density of charge causing very high adsorption of the divalent metal ion into the compact layer.

8.4 COLLOID STABILITY

8.4.1 Attraction

It is well-known that naked, uncharged colloidal particles undergo coagulation very rapidly if no precautions are taken to stabilise the particles. This is a consequence of the van der Waals attraction between the particles.

8.4.1.1 Calculation of the van der Waals attraction

Ninham and Parsegian[213-217] have made considerable progress in calculating rigorously the van der Waals attraction, between macroscopic bodies of colloidal dimensions. Their elegant approach exploits the powerful methods of quantum electrodynamics. The classical procedure for calculating the attraction focused attention on the London dispersion force, i.e. on the electromagnetic correlations between electronic fluctuations that manifest themselves by a narrow band of frequencies in the ultraviolet. It was therefore held that attraction could be calculated by using the same pair-wise summation method that proved so successful in treating rarefied media, such as dilute gases. But the applicability of this simple technique to condensed media, which involve complicated many-body interactions, has remained a matter for conjecture.

Ninham and Parsegian[214] argue that the origin of the van der Waals force is more complex than commonly assumed. Spontaneous, transient electric polarisations arise at a centre due to molecular distortions and orientations, as well as the motion of electrons. This polarisation acts on the surrounding regions to perturb spontaneous fluctuations elsewhere. The interaction resulting from these perturbations is, of course, such as to lower the energy.

Parsegian and Ninham have performed these more exact calculations for the interactions across a single thin film[214, 215] or across a triple film in air[216] or in water[217]. The virtue of their theory lies in its application of the generalised theory of Lifshitz on the interactions in condensed media. The power of Lifshitz's formalism resides in its correct allowance for all interactions: it implicitly accounts for all many-body interactions; it includes all frequencies of interaction; and any perturbations resulting from the presence of intermediate substances between two interacting substances may be properly estimated.

For a triple film, composed of n-decane in water, the rigorous calculations provide several surprising, not to say revolutionary, conclusions. First, because of the highly polar nature of water, much of the van der Waals interaction derives from polarisations at infrared and microwave frequencies rather than from the ultraviolet. Second, it is wrong to view the van der Waals force between the decane layers as the sum of individual interactions between unit segments of the constituent materials. Indeed, it eventuates that even the very idea of a characteristic 'Hamaker constant' can be misleading for these types of systems; more realistically it should perhaps be termed a Hamaker function. Third, the van der Waals force contains a large temperature-dependent component.

Previously it had been held that the absence of suitable spectroscopic information prevented the application of Lifshitz's theory to colloidal systems[218]. It now seems that these obstacles were largely illusory. What is needed for the calculations are the values of the dielectric susceptibility of the various materials on the imaginary frequency axis. In principle, the full susceptibility function for imaginary frequencies can be derived from the absorption spectrum for real frequencies. In practice, the imaginary susceptibility function is found to a good approximation from knowledge of the major absorption frequencies and indexes of refraction, using a damped oscillator model. The dielectric properties of most common substances are sufficiently well known over the whole range of frequencies to permit numerical estimates to be made with little ambiguity.

The approximate procedure devised by Parsegian and Ninham permits the contributions from each frequency range to be isolated readily. Three ranges are usually identified, namely zero or microwave relaxation frequencies, infrared frequency fluctuations and ultraviolet frequencies. The zero-frequency dielectric-constant contribution contains temperature-dependent and non-additive components. For a triple film in water this is always the most important term. Earlier neglect of this contribution led to the mistaken belief that many-body and temperature-dependent forces are unimportant for colloidal systems.

The correctness of this new approach can be gauged by its prediction of an effective 'Hamaker constant' of $(5.5–6.1) \times 10^{-21}$ J for the n-decane triple film[217]. Experiment suggests a value of 5.6×10^{-21} J [219]. Pair-wise summation of the London dispersion force predicts a value of 0.9×10^{-21} J.

This new theory points to the expectation that the laws relating energy to the distance of separation will sometimes be very different from those predicted by the orthodox summation of the inverse sixth power interactions between infinitesimal volume elements of the interacting bodies. The extension of the theory to other geometrical shapes (e.g. spheres) is eagerly awaited. Moreover, some idea of the magnitude of the temperature dependence of these forces would prove valuable. Finally, we mention that the idea that the attractive and repulsive interactions are completely separable in colloid systems must be applied with caution. Because the nature of the dispersion medium profoundly affects the van der Waals attraction, any changes in this intervening liquid, designed to alter the repulsive component, may also influence the attraction. The effect may be small for some electrostatic systems but may be quite large for some sterically stabilised dispersions.

8.4.1.2 Direct measurement of the van der Waals attraction

The most important recent development in the direct measurement of van der Waals forces between macroscopic objects are the delicate studies of Tabor and Winterton[220-223]. Prior investigations are reviewed by Gregory[224]. In these earlier pioneering measurements the Russian, Dutch and English workers were all forced to work at separations greater than 100 nm. Accordingly only retarded forces were measured. Whilst retarded forces are of interest in colloid stability, especially with larger particles, the normal, unretarded forces are more important in general.

Tabor and Winterton have succeeded in measuring unretarded forces operative over distances as small as 5 nm. This is an order of magnitude improvement in the distance over which the forces can be measured. It has permitted the first experimental investigation of the transition from normal to retarded van der Waals forces.

To achieve such precise measurements one needs smooth surfaces, accurate measurements of their separation and a sensitive method of measuring forces. Smooth surfaces were achieved by using the cleavage face of mica; these were bent into a cylindrical shape and measurements made of the attraction between crossed cylinders. Interferometry permitted the distance of separation to be measured with an accuracy of 0.4 nm. Special precautions were taken to eliminate electrostatic artifacts at these distances. An elastic beam method was used to measure the forces involved.

The existence of a transition from normal to retarded forces was verified experimentally although it occurred at separations of $c.$ 10 nm which is a shorter range than was considered previously. At distances greater than 15 nm both the absolute magnitude and the distance dependence of the measured forces agreed well with the theory of retarded London dispersion forces. Unretarded forces were apparent at distances of less than 10 nm. The measured London–Hamaker constant for mica in air was $c.$ 1×10^{-19} J in good agreement with the value of 1.1×10^{-19} J derived from optical dispersion data.

Unfortunately, these excellent experiments may lull us into a false sense of well-being when it comes to hydrosols. The presence of the water profoundly reduces the attraction and we are far from certain that we can accurately allow for its presence, as the work of Ninham and Parsegian[217] shows. Improvement in the technique by a further order of magnitude will be necessary before direct measurements in the presence of water will become feasible.

8.4.2 Electrostatic stabilisation (DLVO theory)

It is now generally recognised[225] that stability may be conferred on a lyophobic colloidal suspension in a number of ways. The two most significant methods are (a) electrostatic stabilisation and (b) steric stabilisation. The first of these has been extensively studied over the past 30–40 years and the current theoretical ideas are quite well developed and have been subjected to a variety of tests. The second method is historically more ancient (since it encompasses the well known phenomenon of 'protection' of a lyophobic colloid by a lyophilic one) but it has only been subjected to intensive study over the last 10 years or so. The treatment of electrostatic stabilisation is based on the ideas developed by Derjaguin, Landau, Verwey and Overbeek (DLVO) and discussed in detail in the monograph by the latter authors[226] and in Kruyt's textbook[227]. Essentially, stability is conferred on a system if the potential energy of repulsion due to the overlap of approaching electrical double layers is sufficiently large to dominate over attraction at some values of the separation between the particles. The attraction is due to the operation of long-range van der Waals forces and is not easily altered or controlled. The repulsion energy on the other hand can be modified by changing the surface potential on the particles, or the valency or concentration of electrolyte in the solution in which they are immersed. Stability is therefore controlled in many technological applications by the magnitude of the repulsion.

8.4.2.1 Calculation of the repulsion

The early estimates for spheres were valid only for small potentials since they invoked the Debye–Hückel approximation to describe the potential in the double layers of the approaching particles. On the other hand, for flat plates (i.e. larger values of κr) the solution of the complete Poisson–Boltzmann equation can be used and the interaction can be calculated using no further approximations, but assuming that the surface potentials remain constant as the particles approach one another. It is given in terms of elliptic integrals of the second kind and the values of the repulsion potential energy are tabulated for various values of ψ_0 and the separation between the plates, by Verwey and Overbeek[228]. Various analytical approximations, valid under different conditions, are also discussed by those authors. Since that early work the calculation has been extended by Devereux and de Bruyn[229] to the interaction of particles of differing surface potential. Their extensive tabulation of results shows that some surprising effects occur when the two particles differ considerably in potential. For unsymmetrical electrolytes a numerical procedure for calculating the charge and potential distribution between two flat plates has recently been suggested by Bresler[230]. Levine and Bell[231] had undertaken an investigation of this type of system earlier and showed that the influence of the co-ion on the coagulating concentration of the counter-ion depended upon whether or not a Stern layer was incorporated in the model. There seems, therefore, to be little point in pursuing this subtlety at the moment.

Verwey and Overbeek[226] discussed one way in which a Stern layer could be incorporated into the model for interaction and Devereux and de Bruyn[232] have more recently developed some explicit relationships. The result must depend critically on which, if any, of the following factors is considered to remain constant during the interaction: (a) surface potential, (b) surface charge, (c) Stern layer charge or (d) diffuse layer charge. A case could be made out for each possibility, and no doubt for some others as well. Given a suitable choice of parameters the introduction of the Stern layer to the model can permit an explanation of any of the wide variety of phenomena experienced by colloid chemists[233] (antagonism, synergism, superadditivity and stability at high electrolyte concentration). It can even reconcile the viewpoint of the DLVO theory with the ion exchange theories of Tezak and Mirnik, according to Levine and Bell[234].

Restricting ourselves to the interaction of two Gouy–Chapman type double layers still leaves open the question of whether one should assume a constant double layer potential or a constant charge during the interaction. For slow penetration or equilibrium studies the constant potential approximation is appropriate since this corresponds to thermodynamic equilibrium But for rapid and close approach of particles, as occurs in coagulation and rheological situations, Frens[235] has suggested that a constant charge assumption is more appropriate. Wiese and Healy have recently[236] derived a relation which allows comparison of the two assumptions. For two identical spheres of radius r, their equation is:

$$V_R^\sigma = V_R^\psi - 4\pi\varepsilon_0 \tfrac{1}{2}(Dr\psi_0^2)\ln\left[1 - \exp(-2\kappa H_0)\right] \tag{8.15}$$

where H_0 is the separation and the superscripts refer to the quantity which is held constant. This expression is valid only for relatively small potentials (<50 mV say). Fortunately, the difference between the two assumptions is a second-order effect[237] so that for most applications the choice makes little difference. Only when the particles are rapidly forced very close together ($\kappa H_0 \ll 1$) does the difference become very obvious. Honig and Mul[238] have recently presented a number of limiting expressions derived for the case of small separations, large separations, moderate potentials, or moderate charges on the basis of both the constant charge and constant potential assumption.

It should be noted that even without the introduction of a Stern layer it is possible to introduce some specific ion effects into the theory as the work of Sanfeld[239, 240] shows. Using the method of local thermodynamic balance he suggests that the DLVO theory overestimates the repulsion and that the error is greatest[240] for H^+ and less for Li^+ and Na^+ in that order.

The repulsion energy between approaching spheres is more difficult to calculate because of the problem of finding a suitable expression for the potential. McCartney and Levine[241] have improved on Derjaguin's earlier result by using a new integration procedure which gives much greater precision for $\kappa H_0 \geqslant 5$. The results of their analytical expression differ by only about 3% from the numerical calculations. This approach has now been extended[242] to deal with particles of different radii and different surface potential.

8.4.2.2 Tests of the DLVO theory

The theory of stability may be tested either in equilibrium situations or in kinetic situations and both have been used in a variety of ways. The equilibrium situations offer several advantages: the geometry of the interaction is controlled, the rate of approach of the double layers is so slow that thermodynamic equilibrium may be assumed and the magnitude of the interaction can be studied as a function of the separation between the surfaces. In the kinetic methods only the magnitude of the interaction at one particular distance can be estimated.

(a) *Equilibrium studies* – By far the most fruitful studies have been those on the equilibrium thickness of thin soap films as a function of electrolyte concentration. The work of Derjaguin[243], of Scheludko and Exerova[244-246] and also of Lyklema and Mysels[247] has been reviewed by Lyklema[248] who shows that the general agreement between the expected and observed thickness is very good (Figure 8.7). In these measurements a soap film is allowed

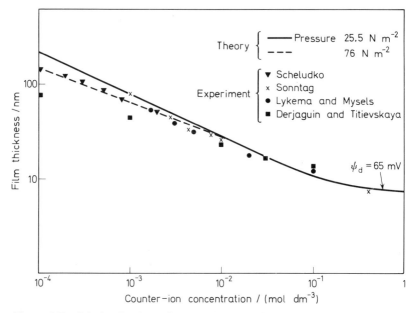

Figure 8.7 Calculated and experimental thicknesses of soap films as a function of ionic strength
(From Lyklema and Mysels[247], by courtesy of the American Chemical Society)

to drain under a known hydrostatic suction until the repulsive interaction between the stabilising soap layer on either side of the film is sufficiently large to prevent any further reduction in thickness. Thinning is, of course, aided by the van der Waals force, and it is not yet certain whether the remaining discrepancies between theory and experiment in this system are due to inaccuracies in the estimates of the repulsion or of the attraction.

At low salt content the films are thinner than expected whilst at high salt content they are rather thicker than expected which suggests that either the attraction falls off less rapidly with thickness than the theory predicts or that the repulsion falls off more rapidly. As noted above the corrections for finite ion size[239, 240] tend to reduce the magnitude of the repulsion but this alone probably does not fully account for the result. Since retardation effects have already been incorporated in Lyklema's theoretical curve one must look elsewhere for an improved estimate of the attraction. Perhaps the approach of Parsegian and Ninham[217] when applied to this system will at least partly remove the anomaly. They show that when the Lifshitz theory is applied to the calculation of the attraction energy its value depends upon the concentration of ions in the solution (since this affects the refractive index of the water). Thus the usual separation between the attractive force and the repulsive force cannot be so readily made. It must be noted, however, that their preliminary results on this problem appear to run in the opposite direction to the experimental values.

Tabor and Roberts[249, 250] have also devised an ingenious method for measuring directly the repulsive forces (usually of the order of 1N cm^{-2}) exerted by charged double layers residing on solid surfaces. Liquid films of the order of 5–20 nm are formed between a soft optically-smooth rubber hemisphere and a flat glass surface. The film thickness is measured optically for different electrolyte concentrations. van der Waals forces are relatively small compared with the applied pressure so that the equilibrium thickness is determined primarily by the double layer repulsion. The results obtained in this way were in good agreement with those of Mysels for black soap films. Experiments in somewhat similar vein were performed by Kitchener and Read[251]. However, in this case a gas bubble was generated electrolytically and pushed against a silica plate. Electrical double layers generated at the air/water interface of the bubble and the silica surface were repulsive and this resulted in the formation of films of finite thickness between the bubble and the plate. Optical measurements of these films were in fair agreement with the theory of wetting films proposed by Langmuir and Frumkin.

Careful equilibrium studies have also been made on the interaction between the clay minerals montmorillonite and vermiculite. These are layer-lattice alumino-silicates which form extended flat sheets of high surface charge density[252]. The sheets are only $c.$ 1 nm in thickness and between each pair of sheets an electrical double layer can develop under suitable conditions. In his early measurements on montmorillonite[253], Norrish determined the separation between the clay sheets as a function of electrolyte concentration, by low-angle x-ray diffraction. Much of the discrepancy between expected and observed separations was subsequently traced to the occurrence of positive charges on the edges of the montmorillonite crystals and the presence of occasional crystals which were not in proper parallel alignment[254, 255]. These tended to bond the crystal packet together and so reduce the spacing. Meanwhile Garrett and Walker[256] showed that vermiculite behaved in a similar fashion and Norrish and Rausell-Colom[255] undertook a similar study of that material. It has the advantage of occurring in much larger sheets so that the problems of edge charge and non-parallel alignment are reduced. Again, the spacings turned out to be smaller than expected, suggesting that either the repulsion was overestimated or the attraction was underestimated by the theory. In these systems it is possible to examine the swelling under an imposed external pressure so that the spacings remain very much smaller and the van der Waals attraction makes a negligible contribution to the balance of forces. Under these conditions, Warkentin and Schofield[364] and Norrish and Rausell-Colom[255] have shown that the relation between repulsion forces and distance is given closely by the DLVO equation (Figure 8.8) which again suggests that the attraction energy is underestimated. In an effort to improve the agreement between experiment and theory, Friend and Hunter[257] measured the ζ-potential of vermiculite suspensions with the intention of using this value rather than the total cation exchange capacity to calculate the repulsion. The measurements revealed some very

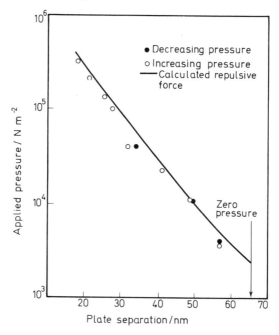

Figure 8.8 Swelling of lithium vermiculite under pressure in 0.03 m LiCl
(From Norrish and Rausell-Colom[255], by courtesy of Clays and Clay Minerals)

peculiar features of this material which could be roughly correlated with the swelling behaviour but which suggested that it was not really an ideal material for testing of double layer theory.

Measurements on Schiller layers of tungstic oxide are said to show good agreement[258] at high concentration where the separation is c. 100 nm but the agreement is poorer at 300–400 nm. These values are much larger than those involved in the other studies and in this region the van der Waals interaction is retarded. It has also been studied carefully for macroscopic objects *in vacuo* and is known to follow the Lifshitz theory[259, 261]. Again the application of the Parsegian and Ninham equations[217] should be attempted here.

A very interesting experiment was described by Watanabe and Gotoh[262] on the interaction between two mercury drops. They brought the drop surfaces together in an electrolyte solution whilst applying a known potential difference to them and observed the correlation between electrolyte concentration and the applied e.m.f. at which coalescence of the drops would just occur. They found very good agreement between their observations and the DLVO theory though it must be noted that this type of experiment can test the magnitude of the interaction energy only at one separation.

Similar measurements on the repulsive force between two thin wires[263] also gave good agreement with DLVO theory up to concentrations of about 0.1 M KCl. At much higher concentrations, however (up to 1 M and again above 2 M) the degree of repulsion is said to increase again and so far no adequate explanation of this effect has been advanced.

Another method of examining interactions is to determine the force which is necessary to detach a particle from a substrate. Detachment can be induced either by a centrifugal effect[264] or by a shearing process[265]. The methods agree with one another [265] and are in general accord with DLVO theory. The desorption of a relatively thick oil film from a flat surface also appears to be in accord with the theory[266], as is the attachment of a mercury drop to a glass sheet[267].

(b) *Kinetic measurements* – The philosophy underlying the kinetic methods of testing the DLVO theory is that we assume that the theory is correct and determine whether experiments yield sensible values for the attractive interaction. The kinetic methods for determining Hamaker constants have been reviewed by Lyklema[268]. The two most important methods are (i) the critical coagulation method and (ii) the rate of coagulation method. In the first procedure, the minimum concentration of electrolyte needed to induce rapid coagulation is determined. The second method involves the measurement of the rate of coagulation both in the presence and absence of a repulsive barrier. This stability ratio method is the one most commonly employed. The results obtained by these two procedures have been reviewed by Lyklema[268], Gregory[224] and Matijevic[269, 270].

Initially, silver iodide sols were studied most frequently as model dispersions. Values of A for silver iodide in the range $(3-9) \times 10^{-20}$ J were commonly obtained from kinetic measurements. More recent studies support these limits[271]. The value calculated for A from dispersion data is 3×10^{-20} J, which agrees to better than a factor of 3 with the experimental results[268, 271].

Silver iodide sols have now been superseded as model dispersions by polymer latices, of which polystyrene latices are studied most frequently. Ironically the agreement between the results of different workers, and between theory and experiment, is by no means impressive. Theoretical estimates of the London–Hamaker constant for polystyrene in water range from $(0.3-5) \times 10^{-20}$ J [269]. Some of the uncertainty arises from the lack of accurate data necessary for the calculations; other variations stem from the different methods of calculation used. The values of A derived from rates of coagulation experiments span the range $(0.1-22) \times 10^{-20}$ J [269]. Admittedly this covers the range of theoretical estimates but it is so large that no firm conclusions can be reached regarding agreement between theory and experiment. In contrast, Matijevic[269, 270] has studied the stability of styrene–butadiene latices using the coagulation concentration method; the most reliable experimental value for A of $(0.3-0.4) \times 10^{-20}$ J was in excellent agreement with the best theoretical value $(0.3 \times 10^{-20}$ J). Greene and Saunders[272] have also measured the Hamaker constant for polystyrene recently and obtained values in the range $(0.2–0.4) \times 10^{-20}$ J.

Recently, Goldstein and Zimm[273] have developed an ultracentrifuge technique to locate the upper and lower bounds of A for polystyrene in water. Again the range $(0.3–0.8) \times 10^{-20}$ J seems reasonable, if rather wide.

Given the confused state of the theoretical and experimental results for the Hamaker constant for polystyrene, it would seem that no strong conclusions may be drawn from these experiments regarding the validity of the DLVO theory vis-à-vis colloid stability. Lyklema[268] reached the same conclusion for metals in water. Even if a disparity were evident between theory and experiment, there would still remain the problem of sheeting the discrepancy home to the repulsive potential or the attractive potential (or both). Thin film experiments, of course, should provide a check on the repulsive potential. Any discrepancy could also arise from the dynamic nature of kinetic experiments; perhaps the better agreement exhibited by the critical coagulation data with theory suggests that the problem arises in the application of the Fuch's theory for diffusion of particles in a force field. What is urgently needed for these studies is a rigorous theoretical estimate of the van der Waals attraction between substances in water. The techniques exploited by Ninham and Parsegian[217] should provide this shortly, although the present indications are that the 'Hamaker constant' is not a constant at all. Interestingly enough Ottewill and Shaw[15] were forced to this conclusion some time ago, as a result of their experiments on polystyrene lattices.

A somewhat different kinetic approach was adopted by Hull and Kitchener[274]. They studied the deposition of polystyrene latex particles onto a rotating disc, the surface of which was also polystyrene. When the disc was charged positively, the deposition rate of negatively charged particles followed the Levich theory for diffusion controlled mass transport. However, when both the surface and the particles were negatively charged, the DLVO theory for sphere/plate interactions seemed broadly applicable. Nevertheless, the rate of deposition was not described precisely by the theory. Hull and Kitchener proposed that this discrepancy arose because deposition occurred mainly on areas especially favoured as a result of their electrical and/or geometrical dispositions.

The foregoing discussion has been concerned primarily with model hydrophobic dispersions. At the other extreme we mention silica dispersions, which exhibit some of the characteristics of hydrophilic dispersions[275–282]. They are sometimes stable at very high concentrations of electrolyte and undergo reversible coagulation. Certainly silica dispersions do not in general obey the DLVO theory but precisely what factors control stability is a moot point. There appears to be good evidence that the silanol groups on the surface may result in strong hydration effects at the interface. It has even been proposed that this tightly bound layer of water contributes significantly to stability[275–282]. Allen and Matijevic[275–278] have suggested that ion exchange processes, which remove protons from the silanol groups, are the primary factors controlling stability, presumably through the concomitant removal of water molecules that were hydrogen bonded to the silanol groups. This is supported by the work of Laskowski and Kitchener[281] who showed that silica can be rendered hydrophobic merely by strong heating to remove the silanol groups. Depasse and Watillon[279], however, believe that acid–base interparticle bonding is of primary importance. Much remains to be done in understanding these complex systems.

8.4.3 Steric stabilisation

A second non-electrostatic method of stabilising colloidal particles is to use non-ionic polymeric stabilisers. Heller[283, 284] has referred to this as 'steric stabilisation', although the classical description 'protective action' (or just 'protection') is also employed[285]. Neither terminology is entirely satisfactory; 'protective action' as used classically may be taken to include electrostatic effects while 'steric stabilisation' has little in common with the steric effects prominent in organic chemistry. The latter arise in the final analysis from the Born repulsion generated as a result of the operation of the Pauli exclusion principle. We note in passing that Born repulsion has not yet apparently been utilised in colloid stabilisation, although this may well come to pass.

Steric stabilisation can lay claim to being one of the oldest unsolved technological problems. It dates back some 4500 years when the early Egyptians devised methods for manufacturing ink and fresco paints[286]. Ink, for example, was prepared by dispersing carbon (lamp black from combustion) in water using natural steric stabilisers, usually proteins such as casein (from milk) or albumin (from egg white). Steric stabilisation is still being exploited in the preparation of paints and inks, as well as in numerous other applications, e.g. the use of non-ionic surfactants in household detergents and pharmaceutical emulsions.

The classic gold-number experiments of Zsigmondy[287] probably represent the first scientific investigation of steric stabilisation. Little theoretical progress proved possible, however, until suitable theories for the thermodynamics of polymer solutions were developed. As these theories are still being elaborated and refined, especially by Flory[288], no definitive theory of steric stabilisation seems likely to appear for some time. There has, however, been considerable activity—both experimental and theoretical—in the field in the past 5 years. Several theories whose predictions can be tested against experiment have already appeared.

8.4.3.1 Theories of steric stabilisation

(a) *Classical thermodynamic approach* – Since colloidal particles coated by non-ionic polymers often exhibit stability, there must exist a repulsive potential energy barrier between two such particles. This barrier counterbalances the van der Waals attraction between the core particles. Repulsion

Table 8.1 Ways of obtaining steric stabilisation

ΔH_R	ΔS_R	$|\Delta H_R|/|\Delta S_R|$	ΔG_R	Stability type
+	+	>1	+	Enthalpic
−	−	<1	+	Entropic
+	−	$\gtreqless 1$	+	Combined

is characterised by the change in the Gibbs free energy of the particles on close approach (ΔG_R). Of course the molecular origins of ΔG_R are of no particular concern in a classical thermodynamic discussion.

The second law of thermodynamics implies that ΔG_R must be suitably positive for stability to be observed. The table summarises the three different ways by which a positive value of ΔG_R may arise from its constituent enthalpy (ΔH_R) and entropy (ΔS_R) changes on close approach[33, 289].

(i) *Enthalpic stabilisation* — If both ΔH_R and ΔS_R are positive and $\Delta H_R > T\Delta S_R$, then a positive value for ΔG_R results. In words this means that the net enthalpy change on close approach, which opposes flocculation, outweighs the net effect of the increase in entropy, which promotes flocculation. This has been termed enthalpic stabilisation[33]. Enthalpic stabilisation has been shown in principle to be characterised by flocculation on heating[289]. Conversely a dispersion that exhibits flocculation on heating is enthalpically stabilised, at least immediately below the critical flocculation temperature (c.f.t.).

(ii) *Entropic stabilisation* — Entropic stabilisation arises when both ΔH_R and ΔS_R are negative but $T|\Delta S_R| > |\Delta H_R|$. The effect of the entropy change opposes flocculation and outweighs the enthalpy term[33, 289, 290]. In principle, entropically stabilised dispersions are characterised by flocculation on cooling. This contrasts with the flocculation on heating of the enthalpically stabilised dispersions.

(iii) *Combined enthalpic–entropic stabilisation* — Both the enthalpy and entropy changes oppose flocculation if ΔH_R is positive and ΔS_R is negative[33, 289, 291]. Dispersions stabilised by this combined enthalpic–entropic mechanism cannot in principle be flocculated at any accessible temperature.

The foregoing classification implies that the generic term steric stabilisation encompasses enthalpic, entropic and combined enthalpic–entropic stabilisation[292].

(b) *Statistical thermodynamic theories* — The classical thermodynamic approach outlined above is independent of any molecular models that might be advanced to describe how ΔG_R arises. It therefore provides little information on how ΔG_R is generated nor does it permit theoretical estimation of its magnitude.

There have been several attempts to calculate by statistical thermodynamics the actual magnitude of ΔG_R. This is, of course, a model dependent procedure. The usual model examined consists of polymer chains irreversibly attached to the particles. The chains project into, and are dissolved by, the dispersion medium (hereafter also termed the solvent). A successful *ab initio* application of this approach should yield a definitive theory of steric stabilisation analogous to the DLVO theory for electrostatic stabilisation.

(i) *Entropy theories* — Mackor[293] was the first to attempt to estimate quantitatively the magnitude of the repulsive potential energy barrier in steric stabilisation. A simple model of a rod shaped molecule adsorbed but freely-jointed at a flat interface was adopted. The magnitude of ΔG_R was calculated solely in terms of the loss in configurational entropy of the rod on close approach of a second uncovered surface (Figure 8.9). The latter condition implies small surface coverages. Only small repulsive potential energies were therefore calculated in this way.

An inflexible rod, or even its subsequent refinement[294, 295] is obviously inadequate to describe the properties of a flexible multi-segment polymer

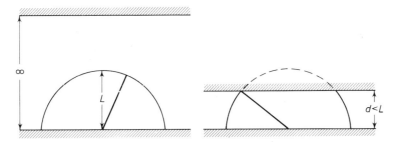

Figure 8.9 Schematic representation of the loss of configurational entropy on close approach of flat plates according to Mackor[293]

chain. Further, high surface coverages are common. Mackor's contribution remains important on two scores. First, it gave rise to the erroneous idea, still current, that all polymers stabilise by an entropic mechanism. This is why the terms 'steric' and 'entropic' stabilisation are often used synonymously[296]. Second, it recently inspired Clayfield and Lumb[297–300] to perform quite elaborate Monte Carlo calculations of the entropy loss of flexible chains when they are compressed. These calculations were performed on macromolecules that are terminally adsorbed, or are in a flattened conformation due to multipoint adsorption. Chains of several hundred links can be handled in this way if a four-choice cubic lattice is constructed so that the bond angle is restricted to 90 degrees.

The results of these calculations will not be described in detail because they seem to be of quite restricted applicability. No allowance is made in these calculations for the solvent quality of the dispersion medium into which the dispersant chains project. Yet experimentally this solvency appears to be absolutely critical in determining stability[33, 289–292]. The computations of Clayfield and Lumb may, however, be applicable to certain athermal dispersion media, although no comparisons with experiment have yet been published.

(ii) *Fischer's solvency theory* — The critical nature of the solvent properties of the dispersion medium in steric stabilisation was first realised by Fischer[301]. The importance of his ideas has been recognised only recently[302]. Fischer noted that when the polymer sheaths surrounding two stable particles interact in a Brownian collision, the chemical potential of the solvent in the interaction zone decreases. A gradient in the chemical potential is accordingly established between the solvent in the interaction zone and that in the external dispersion medium. The net result is that solvent external to the interaction zone may diffuse into that zone and force the polymer chains, and thus the core particles, apart.

An alternative description of this difference in chemical potential, which is actually an excess chemical potential, may be given in terms of the excess osmotic pressure (π_E) in the interaction zone:

$$\pi_E \bar{V}_1 = (\mu_1 - \mu_1^0)_E \qquad (8.16)$$

Here $\bar{V}_1$ is the molar volume of the solvent. Ottewill[302] in his review of protective action has noted that the concept of an excess osmotic pressure is analogous to the disjoining pressure notion of Derjaguin[303]. The absence of a semi-permeable membrane does not invalidate the excess osmotic pressure viewpoint because it merely expresses an alternative way of describing a difference in chemical potential.

Two important conclusions emerged from Fischer's theory. First, the magnitude of the repulsive potential energy is likely to be as high as (100–1000)kT in good solvents. This is more than adequate to impart stability. In addition Fischer related the repulsive potential energy to the second virial coefficient (B) of the polymer in free solution:

$$\Delta G_R \approx 2BRT\langle C_g\rangle^2(\delta V) \qquad (8.17)$$

where $\langle C_g\rangle$ is the mean segment concentration in the stabilising layer and δV is the overlap or interaction volume. Higher virial coefficients are neglected.

According to equation (8.17), a repulsive potential is generated in a dispersion medium of good solvency for the polymer chains because B is positive. In contrast, in a bad solvent, B is negative and the particles are sensitised to flocculation. An interesting case is that of a theta (θ)-solvent in which the polymer chains can telescope one another without prejudice because B, and therefore ΔG_R, is zero. Ottewill and Walker[304] have shown how to express Fischer's formula in terms of the enthalpy (κ_1) and entropy (ψ_1) of dilution parameters of the polymer in free solution. Dilution may be regarded as the reverse of interpenetration and so κ_1 and ψ_1 (with their signs changed) are directly proportional to ΔS_R and ΔH_R.

(iii) *Entropy plus solvency theories* — The assumption that a mean segment density may be used in calculating the repulsion is undoubtedly an approximation. It corresponds to adopting a step function to represent the segment density distribution. What is needed, of course, is the density distribution function relevant to a polymer chain attached to an impenetrable interface. Against this Flory[305] has pointed out that the distance dependence of the repulsive potential energy will not be excessively sensitive to the precise form of the distribution adopted.

It becomes necessary before proceeding further to distinguish between various modes of attachment of the macromolecules to the interface. Attachment is always assumed to be irreversible. A tail is defined as a macromolecule attached terminally at one point. A loop implies attachment at both ends. Recalling the well-known segment density distribution function for unadsorbed macromolecules[306], we may infer intuitively that the function for loops and tails should be positively skewed about some maximum value.

Hesselink[307,308] has calculated the actual segment density function $\rho_1(i, x)$ for monodisperse tails and loops composed of i segments. With these results it becomes possible to specify the distribution functions for other given modes of attachment to the interface. The elegant procedure devised by Hesselink calculates $\rho_1(i, x)$ by summing over all values of k the probability $P_1(i, k, x)$ of finding the kth segment at a distance x from the interface:

$$\rho_1(i, x) = \sum_k P_1(i, k, x) \tag{8.18}$$

The probability $P_1(i, k, x)$ was expressed as the product of the probability of finding the terminal segment of a chain of k segments at x, and the probability of finding the remaining $(i-k)$ segments at $x>0$. For a six-choice cubic lattice, the normalised density distribution function for tails is

$$\rho_1(x) = 6(il^2)^{-1} \int_x^{2x} \exp\{-3t^2/2il^2\}\, dt \tag{8.19}$$

where l is the segment length. That for random loops is

$$\rho_1(x) = 12x(il^2)^{-1} \exp\{-6x^2/il^2\} \tag{8.20}$$

These functions are plotted in Figure 8.10. The root-mean-square distance of the segments from the interface, $\langle x^2 \rangle^{\frac{1}{2}}$, was found to be equal to $([7/18]\,il^2)^{\frac{1}{2}}$ for tails but to be substantially less, $([1/6]il^2)^{\frac{1}{2}}$, for loops. A random loop is accordingly one-third less extended at an interface than is a tail of the same

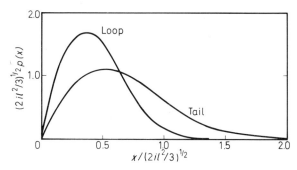

Figure 8.10 The normalised segment density distribution functions for tails and random loops
(From Hesselink[307], by courtesy of the American Chemical Society)

number of segments. These results supersede earlier calculations of Meier which are in error because of the inclusion of configurations that actually penetrated the impenetrable barrier.

If a second impenetrable interface approaches an adsorbed macromolecule the extension of the polymer molecule into the dispersion medium may be restricted. Hesselink[308] has argued that in the period required for a Brownian collision (of the order of, say, 1 μs in water), insufficient time elapses for significant segment adsorption to occur on the approaching interface. The chains do have sufficient time, however, to relax into their equilibrium conformation determined by the restricted region between the interfaces, provided that the macromolecules are not too high in molecular weight (M ⩽ 50 000). The analytical distribution functions for compressed tails and loops between flat plates are complex but they are sketched in Figure 8.11

for various distances of separation (d). Notice that the distribution functions are considerably less skewed than in the absence of the second interface.

Armed with these segment density distribution functions, it should prove possible to calculate the repulsive potential energy in steric stabilisation if a properly formulated theory of the thermodynamics of polymer solutions is available.

It is generally agreed that when two sterically stabilised colloidal particles undergo a Brownian collision, two separate approach stages can be distinguished[290]. These stages are determined by the relative magnitudes of the

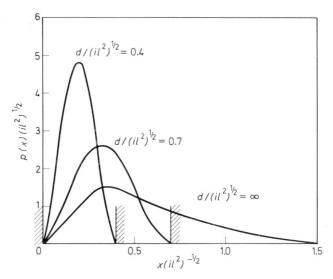

Figure 8.11 The normalised segment density functions for tails on approach of a second impenetrable interface
(From Hesselink[308], by courtesy of the American Chemical Society)

minimum distance of particle separation (H_0) and the contour length of the polymer (L). Interpenetration of the polymers alone is responsible for repulsion if $L \leqslant H_0 < 2L$. The segment density distribution functions corresponding to equations (8.19) and (8.20) are operative in this domain. Closer approach of the particles so that $H_0 < L$ results in macromolecular compression. Therefore both interpenetration and compression contribute to stability. The segment density distribution functions for compressed polymers must now be used for both interpenetration and compression.

Meier[309] combined the entropy approaches of Mackor[293] and Clayfield and Lumb[297-300] with Fischer's[301] solvency concepts to produce a hybrid general theory formulated in terms of the Flory–Huggins theory. He assumed that repulsion could be generated by two phenomena:

(i) The loss of possible chain configurations as the volume available to a chain between the approaching surfaces is reduced. The increase in free energy was calculated solely from the loss in configurational entropy ($-T\Delta S$). This has been referred to as 'the volume restriction effect'[285].

(ii) The possible change in the free energy of mixing of polymer segments and solvent as the segment density increases in the interaction zone. This has been termed 'the osmotic-pressure effect'[285].

Both contributions were evaluated by Meier for tails. The repulsion arising from the volume restriction effect was found by solving the diffusion equation with the relevant boundary conditions. That due to the osmotic pressure effect was estimated using the Flory–Krigbaum theory[310] for the mixing of randomly oriented polymer molecules, whose centres of gravity are fixed in space, with solvent molecules. Of the two factors responsible for repulsion, it was concluded that the volume restriction contribution is the more important if the surface coverage is sparse. Mixing contributions become at least of comparable importance for extensive surface coverage in good solvents.

The conclusions of Meier, which have been challenged on other grounds[290], will not be considered further because, as mentioned above, an incorrect derivation of the segment density distribution function led to an underestimation of the mixing contribution. Hesselink, Vrij and Overbeek[285] (HVO) have corrected this error. They have, in addition, recalculated the loss in configurational entropy of tails and loops resulting from volume restriction. A six-choice cubic lattice was adopted.

The results for tails were identical with those of Meier. The repulsive potential energy according to this scheme is obtained from

$$\Delta G_R = \Delta G_{VR} + \Delta G_M \tag{8.21}$$

where the subscripts VR and M denote volume restriction and mixing respectively. Evaluation of each quantity gave for these terms respectively:

$$\Delta G_R = 2vkTV(i,d) + 2(2\pi/9)^{\frac{3}{2}} v^2 kT(\alpha^2 - 1)\langle r^2 \rangle M(i,d) \tag{8.22}$$

Here α = intramolecular expansion factor for the polymer in free solution, $\langle r^2 \rangle^{\frac{1}{2}}$ = r.m.s. end-to-end length of the polymer in free solution, v = number of molecules per unit surface area, d = distance between the plates, and i = number of segments per tail (or average number per loop). The volume restriction function $V(i,d)$ and the mixing function $M(i,d)$ are quite complicated for loops but for tails, if $d/(il^2)^{\frac{1}{2}} > 1$,

$$V(i,d) = 2(1 - 12d^2/il^2)\exp(-6d^2/il^2) \tag{8.23}$$

and

$$M(i,d) = (3\pi)^{\frac{1}{2}}(6d^2/il^2 - 1)\exp(-3d^2/il^2) \tag{8.24}$$

Numerical values of these functions for both tails and loops have been tabulated[285].

The total potential energy, V_T, is obtained by adding the London attraction and steric repulsion:

$$V_T = \Delta G_R + V_A \tag{8.25}$$

Some general predictions of the HVO theory for tails are succinctly presented in Figure 8.12. This displays the predicted lines of demarcation between stable and unstable dispersions for different molecular weights (M), surface coverage (ω) and edge lengths (k) for flat plates. The intramolecular expansion factor α, which is the abscissa in Figure 8.12, is a measure of the quality of the solvency of the dispersion medium. Clearly the repulsion must decrease as the solvency decreases, i.e., as α decreases.

Some of the more important conclusions to emerge from this theory will now be listed. Reference to Figure 8.12 will highlight these conclusions.

(i) The boundary between stability and instability is strongly dependent upon M. For a decade increase in M, α is predicted to decrease from, say,

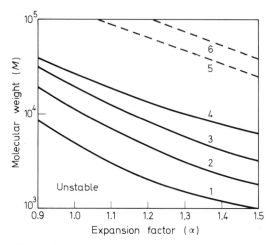

Figure 8.12 The curves of incipient instability for flat plates according to HVO theory. The region below each curve corresponds to instability, that above each curve to stability. The full lines correspond to a Hamaker constant $A = 1 \times 10^{-20}$ J, the dotted lines to $A = 5 \times 10^{-20}$ J. The following values of $\omega/\text{g cm}^{-2}$ and $h/\mu\text{m}$ obtain: curve 1, $\omega = 2 \times 10^{-8}$, $h = 0.1$; curve 2, 5×10^{-9}, 0.1; curve 3, 2×10^{-8}, 0.2; curve 4, 5×10^{-9}, 0.2; curve 5, 2×10^{-8}, 0.2; curve 6, 5×10^{-9}, 0.2.
(From Hesselink et al.[285], by courtesy of the American Chemical Society)

1.6–1.0. This is an enormous change in α, equivalent for most polymers to a temperature change of at least 100–200 K [311]. The flocculation temperatures of two dispersions of plates stabilised by polymeric tails which differ in molecular weights by a factor of ten should thus differ by 100–200 K.

(ii) The points corresponding to $\alpha = 1$ in Figure 8.12 are obviously of no special significance on the demarcation curves. Therefore θ-solvents, to which this value of α corresponds, have no special relationship to incipient instability.

(iii) It follows as a corollary of (ii) that stability is predicted to be observed in dispersion media that are markedly poorer solvents than θ-solvents.

(iv) Incipient instability is strongly dependent upon particle size.

(v) The total potential energy diagram for two sterically stabilised particles (Figure 8.13) exhibits no maximum of the type characteristic of electrostatic stabilisation. Except in the early stages of close approach, repulsion in a good solvent everywhere outweighs the attraction. Nevertheless the potential energy diagram does exhibit a minimum because of the finite cut-off in the repulsion at twice the contour length of the polymer. In a sense this minimum corresponds to the secondary minimum in electrostatic stabilisation. Commonly its depth is extremely small.

Although the foregoing theoretical predictions derive from consideration of tails, comparable qualitative conclusions undoubtedly apply to stabilisation by loops. Moreover, the predictions would not be radically altered were spheres to be considered instead of plates. HVO recognised this by adducing the experimental results for spheres to support their theoretical predictions for plates.

To see this more precisely we recall that Derjaguin[312] developed an

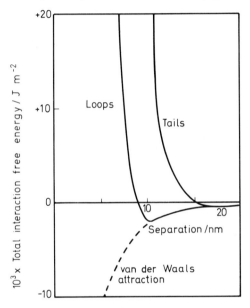

Figure 8.13 The total free energy of interaction as a function of the distance of separation for two flat plates stabilised by equal tails or equal loops. Assumed values are $A = 1 \times 10^{-20}$J, $\alpha = 1.2$, $\omega = 2 \times 10^{-8}$ g cm^{-2} and $M = 6000$
(From Hesselink et al.[285], by courtesy of the American Chemical Society)

ingenious method for calculating the repulsive potential for spheres (ΔG_R^S) of radius a, given the functional dependence of any potential for flat plates:

$$\Delta G_R^S = 2\pi a \int_{H_0}^{\infty} (\Delta G_R^{FP}) \, dd \qquad (8.26)$$

This is strictly valid only if $a/(il^2)^{\frac{1}{2}} \ll 1$. Equation (8.26) implies that the qualitative generalisations (i)–(v) are predicted to hold for spheres as well, although the particle size dependence will be less dramatic.

The theory developed by HVO can be critised for incorporating the now obsolete Flory–Huggins theory. The latter theory assumes that no volume change occurs on mixing segments and solvent and the thermodynamic parameters ψ_1 and κ_1 are taken to be concentration independent. Experimentally this is found to be rarely true. The interaction parameter $\chi_1 (= \frac{1}{2} + \kappa_1 - \psi_1)$ is usually dependent upon segment concentration. The Flory–Prigogine free volume theory[313–317] allows for volume changes on mixing but no application of this sophisticated theory to steric stabilisation has yet appeared. This latter theory also has the advantage that it permits ψ_1 and κ_1 to be predicted from rather more fundamental thermodynamic considerations. Unlike the simpler Flory–Huggins theory, it predicts both positive and negative values for ψ_1 and κ_1.

8.4.3.2 Experimental studies

(a) *Displacement flocculation* — The theoretical predictions arising from the preceding theories may be tested experimentally. Care must be taken to ensure that misleading artifacts are not introduced into these experiments.

Heller[284] has reviewed the early experimental investigations of steric stabilisation. These examined the stabilising properties of poly(ethylene oxide) (POE) chains. One uncertainty that arises in comparing the results of these early studies with current theories, as has been done recently[285], is whether the adsorption could be considered to attach the chains irreversibly onto the surfaces of the particles. POE chains *per se* are only weakly adsorbed on most colloids. Consequently in the stress generated by a Brownian collision the macromolecules are prone to either lateral movement and/or desorption. This results in flocculation, which may be termed 'displacement flocculation'[289].

The tendency towards lateral movement of the chains can be countered by ensuring that the surfaces are totally covered by polymer[33, 289, 290]. Desorption can likewise be eliminated by anchoring the stabilising polymers irreversibly to the surfaces of the colloidal particles[292]. The latter may be achieved by direct chemical combination of the polymer to the surface. Alternatively, the stabilising chains can be covalently bound to a second polymer (the anchor polymer) that is itself strongly attached to the interface by chemical or, more usually, physical forces[33, 289–291].

The need for the presence of an anchor polymer, which has long been recognised industrially in the design of non-ionic surfactants is strikingly illustrated by a comparison of the stabilisation effectiveness of anchored and unanchored POE. Whereas unanchored POE chains generate aqueous dispersions that flocculate in only 0.5 M KCl, latices stabilised by strongly anchored POE are not flocculated at electrolyte concentrations of 3 M KCl[33]. The latter is, of course, a much poorer solvent for POE. Well-anchored dispersions appear to exhibit reversible flocculation because of thermodynamic failure of the stabilising barrier. It may therefore be inferred that unanchored polymer chains permit displacement flocculation to occur.

Even the presence of an anchor polymer does not render a dispersion immune from displacement flocculation. Mathai and Ottewill[318] have demonstrated that at neutral pH silver iodide sols stabilised by low-molecular-weight non-ionics (e.g., n-dodecyl hexaoxyethylene glycol monoether) can be flocculated by small concentrations (10^{-5}–10^{-4} M) of trivalent (La^{3+} or Al^{3+}) and tetravalent (Th^{4+}) cations. Flocculation was attributed to the competitive displacement of the non-ionic surfactant. Their conclusion was supported by electrophoretic measurements that detected a reversal of the particle charge in this electrolyte concentration region. Surface balance studies were also in agreement with the hypothesis of displacement flocculation.

Mathai and Ottewill[318] were unable to exclude bridging by hydrolysed species as a possible flocculation mechanism. Subsequent experiments on latices stabilised by irreversibly anchored POE suggested that bridging was unimportant[9]. These latices were stable near neutral pH in 10^{-6}–2 M solutions of Al^{3+} or La^{3+}.

No quantitative theory for displacement flocculation, which also plays an important role in emulsion stability, has yet been proposed.

(b) *Thermodynamic flocculation* – Polymer latices can be prepared that are not apparently subject to displacement flocculation. These latices, which are specially designed to ensure strong anchoring of the stabilising chains exhibit reversible flocculation[33, 289–292, 319, 320]. This appears to correspond to thermodynamic collapse of the stabiliser. The latices constitute model dispersions whose stability behaviour can be used to screen the various competing theories of steric stabilisation.

The boundary between long-term stability and fast flocculation of sterically stabilised dispersions is very sharp. Experiments show that a temperature change of only 2–3 K is sufficient to induce almost instantaneous flocculation in dispersions that would otherwise exhibit stability for an indefinitely long period of time (e.g. several years)[33]. This is, of course, the critical flocculation temperature (c.f.t.). The addition of non-solvent to the dispersion medium also produces a comparably sharp transition to instability[290]. The volume fraction of non-solvent that must be added to the dispersion medium to induce incipient flocculation is termed the critical flocculation volume (c.f.v.). Apparently the region of slow flocculation commonly exhibited by electrostatically stabilised dispersions is either extremely narrow or even absent in sterically stabilised systems. The sharp transition corresponding to incipient flocculation may be used accordingly to characterise the stability of the dispersion.

To date all sterically stabilised dispersions prepared in water have exhibited flocculation on heating under suitable conditions[33, 291, 319]. This observation implies that such hydrosols are enthalpically stabilised, at least immediately below the c.f.t[290]. The common aqueous stabilisers studied thus far are POE and poly(vinyl alcohol) (PVA). The enthalpy (κ_1) and entropy (ψ_1) of dilution parameters for these polymers are both negative with $|\kappa_1| < |\psi_1|$ if $T < \theta$ [33].

In contrast all non-aqueous dispersions prepared so far have exhibited flocculation on cooling[290]. This corresponds to entropic stabilisation, at least immediately above the c.f.t. In this case ψ_1 and κ_1 are both positive but $\psi_1 > \kappa_1$ if $T > \theta$.

Ottewill[321] has summarised the current situation by tentatively concluding that enthalpic contributions are the most important in aqueous dispersions and entropic contributions in non-aqueous dispersions. It must be recalled, however, that modern theories of polymer solution thermodynamics require that most, if not all, phase diagrams for binary polymer/solvent mixtures display both upper and lower consolute temperatures[288]. Correspondingly most, if not all, dispersions should flocculate both on heating and on cooling if only a sufficient temperature range can be scanned by, e.g. the application of pressure.

Experiments have demonstrated that a given polymer may stabilise by different mechanisms in different dispersion media. POE, e.g. stabilises by an enthalpic mechanism in water but by an entropic mechanism in methanol[290]. Similarly PVA is an enthalpic stabiliser in pure water; addition of dioxan apparently converts it into an entropic stabiliser[291]. This interconversion seems to proceed via a region of combined enthalpic–entropic stabilisation.

All theories which allow for the presence of the solvent in the interaction zone predict a decrease in colloid stability as the quality of the solvency of the dispersion medium diminishes. A considerable body of experimental evidence supports this prediction. It applies whether the decrease in solvency is generated by addition of non-solvent[290] or by a change in temperature[33, 289, 291, 319, 320] This observation naturally fails to discriminate between the validity of the various competing solvency theories; it does, however, exclude from further consideration all the entropy theories that fail to allow for the solvency of the dispersion medium.

The experimental evidence to date suggests that the following factors exert little influence on the transition from stability to instability of sterically stabilised dispersions[33, 290]: the nature of the disperse phase (if polymeric), the anchor polymer (if efficient) and the particle size. These generalisations have led Flory[305] to suggest that perhaps the importance of the London dispersion attraction has been overestimated in these systems in the past. Supplementation of the steric barrier by an electrostatic barrier does not alter the c.f.t. at the high ionic strengths usually required for incipient instability[289].

One of the principal experimental results to emerge from studies of model dispersions is the strong correlation that exists for a given molecular weight between the critical flocculation point and the θ-point. Again this is independent of precisely how the critical flocculation point is reached; both the c.f.v.s and c.f.t.s correlate strongly with their corresponding θ-points[33, 289–291, 319, 320].

This observation is quite discriminating so far as the validity of the various solvency theories is concerned. Curiously enough only Fischer's solvency theory[301], crude and germinal though it is, predicts the observed correlation between the critical flocculation point and the θ-point. Of course this does not prove that Fisher's theory is necessarily correct.

The θ-temperature of a polymer in a given solvent is independent of molecular weight to a good approximation. This is because under θ-conditions polymer segments cannot 'see' thermodynamically other identical polymer segments, a state that is clearly independent of precisely how many segments are involved. It follows that if the correlation of the c.f.t. with the θ-temperature is not an artifact, the c.f.t. should also be independent of the molecular weight of the stabilising moieties. Hydrosols stabilised by strongly anchored POE have been found to exhibit almost identical c.f.t.s despite a three decade variation in the molecular weight of the POE[33, 289]. The strong molecular-weight dependence observed by Heller[284] in his early studies of unanchored POE is apparently an artifact of displacement flocculation. A small variation in the c.f.t. (<10 K) was in fact observed even with properly anchored POE; this was, nevertheless, at least an order of magnitude less than that predicted by the theories of Meier[309] and Hesselink, Vrij and Overbeek[285].

(c) *Molecular models of steric stabilisation* — The cataloguing of the phenomenology of steric stabilisation raises many questions at the molecular level that remain to be answered. Why, for example, do most aqueous dispersions exhibit enthalpic stabilisation? Why does POE function as an enthalpic stabiliser in water but an entropic stabiliser in methanol?

The molecular model to be adopted for common examples of entropic stabilisation seems to be reasonably straightforward. The net entropy change that opposes flocculation is derived from a decrease in the configurational entropy of the chain on interpenetration and/or compression, coupled with a corresponding decrease in the entropy of mixing of polymer segments with solvent[285, 290, 292, 309]. The enthalpy change promoting flocculation may be generated, e.g. by specific dipole–dipole interactions between polar residues in the polymer chain.

The molecular model or models to be associated with enthalpic stabilisation remain equivocal. It is tempting to invoke the traditional colloid science viewpoint that specific interactions, e.g. hydrogen bonding, play a vital role in the stabilisation[322]. This is especially so for POE in water where the evidence for hydrogen bonding between the water molecules and the ether oxygens is quite strong [323, 324]. These strong interactions fall within the traditional, if vague, notion of 'hydration'.

It is in fact possible to explain qualitatively enthalpic stabilisation in terms of hydration. On this model, the enthalpy change opposing flocculation corresponds to the energy required to remove hydrogen-bonded water from the interaction zone[33]. Release of strongly bound water molecules is also accompanied by an increase in entropy, which either may or may not outweigh the decrease in entropy due to both the loss of configurational entropy and entropy of mixing. The first possibility results in enthalpic stabilisation, the second in combined enthalpic–entropic stabilisation.

One of the difficulties associated with such a simple hydration theory is the relative effectiveness of ions of different charge in promoting flocculation[320]. The order of effectiveness for cations is found experimentally to be monovalent > divalent > trivalent. This sequence is just the reverse of that observed for electrostatic stabilisation and, in general terms, the reverse of their order of hydration. The detailed sequence for cations ($Rb^+ = K^+ = Na^+ = Cs^+ > NH_4^+ = Sr^{2+} > Li^+ = Ca^{2+} = Ba^{2+} = Mg^{2+} > Al^{3+}, La^{3+}$) does not follow the classical Hofmeister series. It suggests that the more highly hydrated ions are the least effective flocculants. The orthodox explanation of the Hofmeister series couched in terms of the ions competing with the hydrophilic layer for the water, thus seems inapplicable. Indeed other evidence[33] suggests that the popular idea that flocculation occurs because the stabilising chains are 'precipitated' (whatever that means for single molecules) from solution is quite unfounded.

One explanation for the foregoing results invokes the influence that large hydrated ions exert on the structure of water. The concept of hydration is in fact retained but alternative theories may well be devised which rationalise the experimental results[325].

An additional difficulty confronting the hydration model of enthalpic stabilisation is that recent developments in polymer solution thermodynamics explain the existence of lower consolute temperatures, which presage enthalpic stabilisation, without invoking strong specific interactions. Indeed thus far the theory has been developed solely for weakly interacting segment/solvent systems[288, 313, 314].

The idea is that traditional lattice theories customarily invoke for mixing the approximation $\Delta H = \Delta E$ because mixing is assumed to be ideal and so involves no change in volume. For many polymer–solvent systems this is a reasonable approximation. For other systems, however, the volume change on mixing may be so large that application of the exact relationship $\Delta H = \Delta E + P\Delta V$ actually reverses the expected sign of ΔH. Free volume theories are able to allow for the volume change via equation of state contributions. The relative importance of specific interactions v. equation of state contributions remains unknown for aqueous systems exhibiting enthalpic stabilisation. Both effects seem likely to be important for hydrosols.

8.5 COAGULATION OF HYDROSOLS

An understanding of the principles of colloid stability leads on quite naturally to methods for the coagulation of hydrosols. Some mechanisms for destabilising electrostatically stabilised dispersions have been considered implicitly in the foregoing discussion of stability. Briefly these are (a) a decrease in the absolute value of the surface potential of the particles, (b) an increase in the ionic strength of the dispersion medium by addition of an indifferent electrolyte, leading to compression of the double layer and (c) specific adsorption of counter-ions onto the sol surface. For sterically stabilised dispersions, destabilisation is induced by decreasing the solvency of the dispersion medium for the stabilising moieties.

We now examine some additional destabilising procedures.

8.5.1 Coagulation by hydrolysable cations

Multivalent cations often undergo complicated hydrolysis and oligomerisation reactions in aqueous solutions. The products are usually species of different charge whose adsorptive properties at the solid/liquid interface differ from those of the parent ions. As a result, multivalent ions usually do not fit neatly into the simple pattern predicted by the DLVO theory for the coagulation of negatively charged sols by cations of different charge. The precise pattern followed depends upon the composition and concentration of the various counterion species present and the extent to which these species are specifically adsorbed on the surface of the sol particles. One possible model[204] to describe this adsorption process has been dealt with above (Section 8.3.5). Strong specific adsorption results in a smaller potential drop in the double layer, and hence in a lower stability, than would be expected with an indifferent electrolyte. It has in fact proved possible in favourable cases to run the foregoing reasoning in reverse and to use colloid stability studies to probe the nature of the complex species resulting from cation hydrolysis or complex formation.

The concentration of a multivalent cation necessary to induce coagulation of a negatively charged sol is frequently small (10^{-4}–10^{-7} M) as would be expected from the Shulze–Hardy rule. Obviously the extent and nature of the hydrolysis of the cation is strongly pH dependent. If hydrolysis increases the charge on an ion, it usually results in an increase in its flocculation power; a decrease in charge may be accompanied by a decrease in power. Because hydrolysed species often exhibit quite strong specific adsorption onto the sol particles, they may, if present in suitable amounts, even reverse the sign of the charge on the particles and so produce redispersion.

Matijevic and his collaborators[326–332] have made considerable progress in cataloguing the diverse phenomena that may be observed in the coagulation of dispersions (usually silver halide sols or styrene–butadiene latices) by tri- and tetra-valent cations (e.g. Al^{3+}, La^{3+}, Sc^{3+}, Th^{4+}, Hf^{4+} etc). For example, they studied the coagulation of silver halide sols by aluminium perchlorate. Below pH 4, unhydrolysed aluminium ions ($[Al(H_2O)_6]^{3+}$) were the effective coagulating ions. But above c. pH 5 the coagulating species were the hydrolysis products of Al^{3+}. Experiments[332] with styrene–butadiene latices suggested the charge reversal generated in that system at pH 5–6 was due to the adsorption of the oligomeric polynuclear complex $[Al_8(OH)_{20}]^{4+}$.

With the silver halide sols, the coagulation effectiveness of the Al^{3+} was decreased if sulphate anions were present[331]. This is a good example of the antagonism of one ion towards the coagulation effectiveness of another. The increase in the critical concentration of aluminium salt necessary to induce incipient coagulation appeared to arise from the formation of the $AlSO_4^+$ complex. Indeed this system seems to be a remarkably simple one: only one complex of lower charge is formed and little specific adsorption of the species occurs onto the sol. As a result, it is possible to use coagulation experiments to determine the formation constant of $AlSO_4^+$.

A second example of antagonism with aluminium ions is provided by fluoride ions. Aluminium ions complex fluoride ions so that species of lower charge are formed[332]. Consequently the aluminium ions are less effective flocculants in the presence of F^-. It was also inferred that the fluoride complexes were not as strongly adsorbed as complexes formed in the absence of fluoride ions because the former were unable to reverse the charge on the sol. These results may be important in the treatment of fluoridated water.

It is clear from these studies that adsorption of a complex species onto a sol or latex need not be determined by electrostatic interactions. Thus hafnium in solution at pH = 4 exists as the neutral complex $[Hf(OH)_4]$ [75]. This is strongly adsorbed onto surfaces of latices presumably because of the hydroxyl groups. This does not, of course, prove that electrostatic effects play no role when charged species are present, only that charge is not mandatory for adsorption. However, as $Th(OH)_2^+$ seems to be more efficient than $[Al_8(OH)_{20}]^{4+}$ in reversing charge, it does perhaps suggest that the charge is frequently not the primary factor controlling adsorption[326].

Hahn and Stumm[333, 334] studied the coagulation of silica dispersions by hydrolysed aluminium ions. They argue that coagulation proceeds by (a) hydrolysis and oligomerisation reactions leading to hydroxyl complexes, (b) the highly-charged cations so formed adsorb on the sol surface and reduce the surface charge density leading to instability and (c) Brownian motion, coupled with velocity gradients from agitation or convection lead to coagulation. Apparently the rate determining step is the last process.

8.5.2 Mechanical and surface coagulation

Some aqueous dispersions, e.g., α-FeOOH sols, which exhibit an indefinitely long 'shelf-life' when left undisturbed are found to coagulate quite rapidly when subject to mechanical agitation (moderate stirring or shaking). Heller[335-339] has shown that such coagulation obeys quantitatively a theory that he developed for the kinetics of coagulation at surfaces and interfaces. Surface coagulation occurs on passage of gas bubbles through a sol or latex. Heller assumed that coagulation at an interface obeys second-order kinetics and that a Langmuirian adsorption isotherm of primary particles into the interface might be operative. Highly stable dispersions, whether stabilised electrostatically or sterically, do not apparently exhibit mechanical flocculation. A critically low stability seems necessary for this phenomenon to be observed. Additives, which by preferential adsorption at the interface reduce the population density of the particles at the surface, also render dispersions immune to coagulation.

The central question raised by mechanical and surface coagulation is why particles that are stable in the bulk become unstable at the interface. No definitive answer has yet appeared. However, Heller[335] has proposed several possible explanations: the asymmetry of the particle double layers at the interface; the lower dielectric constant of the liquid at or near the surface; the distribution of the electrolyte between the surface and the bulk.

Surface coagulation may play an important role in the clotting of blood[340] and is presumably also involved in mineral separations by flotation.

8.5.3 Flocculation by polymers

There has been an upsurge of interest in the flocculation of electrostatically stabilised hydrosols by macromolecules. This interest has been generated by the considerable importance of polymeric flocculation in water treatment, mineral processing, etc.

The water soluble polymers used as flocculants include both natural polymers, such as starches, proteins and gums, and synthetic materials, such as non-ionic polymers and polyelectrolytes. Both anionic and cationic polyelectrolytes, either singly or mixed, have been exploited, sometimes aided by the presence of simple inorganic electrolytes. The concentration of solids in the dispersion may vary from only a few p.p.m. (e.g. certain river waters) to upwards of 10–30% solids (e.g. clay suspensions).

The mechanism or mechanisms of flocculation by polymers have proved to be somewhat controversial. Extensive observations have led to the postulation of several different mechanisms, which are not necessarily mutually exclusive. These fall into three main categories, namely charge neutralisation polymer bridging and mutual dehydration[341].

(a) *Charge neutralisation* — Charge neutralisation covers flocculation mechanisms resulting from neutralisation of the electrostatic repulsion between particles. It includes both compression of the electrical double layers due to added polyelectrolyte (i.e. an ionic-strength effect) and the loss of part or whole of the electrical double layer by specific interaction of the surface charge with the polyelectrolyte or other added ions (i.e. a surface charge-density effect).

The molecular weight might be expected to play a relatively minor role in this type of mechanism in the sense that what is critical is the total number of charged groups present on the polyelectrolyte.

Particle flocculation can be viewed kinetically as resulting primarily from a change in the fraction of collisions leading to aggregation. The fraction of 'sticky' collisions between particles is increased from virtually zero to effectively unity. The collision frequency remains almost unaltered.

(b) *Polymer bridging* — This covers several postulated mechanisms by which the flocculant interferes with the free movement of the particles, either by adsorption of a number of particles on a polymer network, or by the formation of long bridges between particles which otherwise might not meet so readily by diffusion or free, or forced, convection. For the bridging mechanism to be operative long unbranched polymer chains should be advantageous. The molecular weight of the flocculant would be expected to be critically important. Bridging occurs at low surface coverages because only then is free surface readily available on which adsorption of loops of molecules attached to other particles may occur.

Kinetically polymer bridging can be regarded primarily as increasing the effective collision frequency of the particles. No decrease in the repulsive potential energy between the particles is in principle necessary for the formation of bridges, although collisions are in fact 'sticky'.

(c) *Mutual dehydration* — This represents a group of postulated flocculation mechanisms whereby a hydrophilic colloid is flocculated by a non-ionic macromolecule. Interaction between the two components results in an insoluble product. This phenomenon, which has scarcely been studied, is likely to be important with sterically stabilised dispersions. To date only classical hydrophilic colloids where there are no discrete particles but only extended highly hydrated chains have been mainly considered. The molecular weight is probably only of minor importance in this type of flocculation.

Because most naturally occurring colloids are negatively charged, cationic polyelectrolytes have been most thoroughly investigated as polymeric flocculants. However, Smellie and La Mer[342] showed as early as 1951 that anionic polyelectrolytes may flocculate negatively charged hydrosols provided, e.g. suitable complexing cations are present. Overbeek[343] has pointed out that the adsorption of negatively charged groups onto negatively charged surfaces is a common phenomenon in colloid science. For example, where an anionic surfactant is adsorbed onto a neutral surface it imparts a negative potential to the surface. Despite this charge, adsorption continues until the equilibrium value is reached.

All that is needed for polyelectrolyte adsorption onto a surface of the same sign is for the coulombic repulsion to be counteracted by another interaction of chemical or electrical origin. Mechanisms for the attachment of polyanions onto anionic surfaces that have commonly been considered[344] are van der Waals forces, hydrogen bonding, anion interchange with adsorbed anions (such as OH^-), or the interaction with cations on the colloid surface.

Sommerauer, Sussman and Stumm[344] have investigated the flocculation of negatively charged silver bromide sols by the following negatively charged polyelectrolytes: poly(acrylic acid), hydrolysed polyacrylamide and poly(styrene sulphonate). They concluded that flocculation of polymer/sol systems of like charges occurs only if an appropriate concentration of a suitable electrolyte (e.g. a calcium salt) was present. Nevertheless, flocculation occurs at electrolyte concentrations much smaller than those required in the absence of the polymer. This type of flocculation is obviously a sensitisation phenomenon. The counter-ions (in this case Ca^{2+}) may be regarded as acting as electrostatic bridges. They promote the adsorption of the polyelectrolyte onto the surface through complex formation with its functional groups at, or near, the surface. The polyelectrolyte chains themselves form bridges between the colloidal particles. It is this bridging that induces flocculation. The counterions thus sensitise the particles to flocculation by a mechanism of forming 'bridges-within-bridges'. Obviously no reduction in the repulsive potential energy between particles is in principle necessary for bridging flocculation.

Nemeth and Matijevic[345] studied the stability of silver halide sols in the presence of gelatin of the same charge. The gelatin sensitised the particles to flocculation apparently by reducing the effective surface charge density of the particles.

The main technological interest in polymer flocculation still rests, however, with polymer/sol systems of opposite sign. There has been considerable patent activity, which will not be reviewed here, in the preparation of suitable polyelectrolytes. Typical are the studies of Fanta et al.[346, 347] who have prepared cationic graft copolymers of starch. One graft copolymer of starch and 10–20% poly(2-hydroxy-3-methacryloyloxypropyltrimethyl ammonium chloride) compared favourably with the ubiquitous high molecular weight polyacrylamide as a flocculating agent for diatomaceous silica and nonmagnetic iron ore dispersions[347].

One of the most interesting, and certainly the most controversial, contributions to appear recently is that of Ries and Meyers[348–350]. They studied the flocculation of silica dispersions and polystyrene latices (both negatively charged) by a commercial high molecular weight cationic polyelectrolyte described as a polyamine sulphate, Primafloc C-7). Zeta-potential measurements showed that incremental additions of polyelectrolyte gradually reduced the magnitude of the ζ-potential of the particles. Maximum flocculation was observed near $\zeta = 0$, suggesting that a reduction in ζ-potential may play a role in flocculation. Further addition of polyelectrolyte reversed the sign of the ζ-potential and ultimately led to re-dispersion. Re-dispersion is a consequence of a combination of electrostatic and steric (in this case entropic[351]) effects. If incipient charge reversal was produced with a relatively low molecular weight cationic polymer, large flocs were still observed on addition of a high molecular weight anionic flocculant.

Ries and Meyers also published shadowed electron micrographs of both silica and polystyrene particles flocculated by the polyelectrolyte. These exhibited fibre-like structures that extended radially from the surfaces of the particles. The fibres, which presumably were composed of polyamine, ranged in width from 30 nm down to 2 nm. The latter size is the expected diameter of a single polyelectrolyte chain subject to some folding or coiling.

It is scarcely surprising that the electron micrographs of Ries and Meyers have been greeted with some scepticism. Artifacts abound in electron microscopy and conceivably the fibrous bridges were introduced during specimen preparation. The micrographs would have been more convincing had it been demonstrated that no comparable bridging was evident when stable dispersions of coated particles were dried down on the electron microscope grid.

Ries and Meyers would have rendered their case less vulnerable had they cited the previous studies of Richardson[352] and Rochow[353] on polyacrylamide (PAM). Very careful electron microscopy, which permits single polymer molecules to be observed, strongly suggests that fibrillar aggregates of PAM exist in aqueous solutions. Indeed, it had even been proposed as early as 1965 by Audsley and Fursey[354] that such aggregates play a central role in the flocculation of kaolin by natural polysaccharides especially galactomannan. Fibrillar bridges were also observed by electron microscopy in their studies.

Narkis and Rehbun[355] have proposed that PAM aggregates could be due to incomplete disentanglement of the highly tangled network of molecules that result when highly mobile monomers undergo rapid random chain growth. Shyluk[356, 357] has investigated the ageing of freshly-prepared high intrinsic viscosity solutions of PAM in water. A rapid decrease in viscosity was observed in, say, 24 h at 65 °C, consistent with the disaggregation theory. The ageing of the PAM solutions is paralleled by a loss in the potency of the polyelectrolyte as a polymeric flocculant for kaolin. However, it must be pointed out that the entanglement theory seems unlikely to explain the aggregation of naturally occurring polysaccharides which are formed, not by rapid chain growth, but by condensation reactions.

Ries and Meyers[348-350] concluded that both charge neutralisation and interparticle bridging are important in flocculation by polymers and seem likely to function simultaneously in certain systems. Gregory[358] reached similar conclusions with two polyamine flocculants (Primafloc C-3 and C-7) interacting with monodisperse polystyrene lattices. Flocculation studies, coupled with electrophoretic measurements, indicated that polymer bridging was important for higher molecular weight polyelectrolytes ($M > 10^6$), as was originally proposed by Ruehrwein and Ward[359]. Charge neutralisation became the dominant mechanism for flocculation with low molecular weight flocculants ($M \approx 10^4$).

Williams and Ottewill[271] have investigated the flocculation of positively charged silver bromide sols by a series of negatively charged poly(acrylic acid) (PAA) samples. These ranged in molecular weight from 1×10^4 to

5×10^6. The technique used was similar to that devised to study the flocculation of the same sol by simple electrolytes. Of course, as La Mer[361] pointed out, the DLVO theory could scarcely be expected to describe flocculation by polyelectrolytes. Nevertheless, flocculation of the sol occurred at a well-defined concentration of PAA (usually in the range 10^{-5}–10^{-3}% w/v). Higher concentrations of PAA were less effective in producing flocculation, presumably because of a combination of electrostatic and steric (in this case entropic[351]) stabilisation. The critical concentration of PAA decreased monotonically with increasing molecular weight. These results are in accord with the concept of bridging flocculation. However, they may be in conflict with the suggestion of La Mer and Healy[362] that a maximum should be observed in the plot of critical flocculation concentration versus molecular weight.

Williams and Ottewill observed no flocculation when negatively charged PAA was added to negatively charged silver iodide sols. This is in accord with Ries and Meyers'[348–350] observations that direct addition of an anionic polymer flocculant to a negatively charged sol may render the ζ-potential of the particles considerably more negative. Moreover, it does not conflict with the ideas of Sommerauer et al.[344] that suitable cations must be present to permit the bridges-within-bridges mechanism to be operative. It may be inferred that the bridge formation that occurred between the negatively charged PAA and positively charged silver iodide sol originates in the coulombic attraction between the particle surfaces and the polyelectrolyte.

Two further points should be noted. First Yarar and Kitchener[363] have observed apparently the selective flocculation of mixtures of minerals (e.g. quartz–galena mixtures) by polymers. This observation could have considerable technological importance. Second Hesselink, Vrij and Overbeek[285] have examined theoretically the equilibrium conformation and the potential energy diagram for polymer chains bridging colloidal particles.

Walles[341] has reconsidered the question of the particle collision frequency for monodisperse particles in the presence and absence of polymer bridges of thickness h and subjected only to Brownian motion. By close analogy with von Smoluchowski's equation for the kinetics of rapid coagulation the ratio of the collision frequencies in the presence and absence of bridges was shown to be given by

$$Z_h/Z_0 = (2+(h/a))^2(1.6c^{-\frac{1}{3}}-2)^4/4(1.6c^{-\frac{1}{3}}-2-(h/a))^4 \qquad (8.27)$$

where a = particle radius and c = weight concentration of the dispersion. Numerical computations based on equation (8.27) show that the collision frequency may be increased more than 10^4–10^6 under suitable conditions. Further at sol concentrations of 0.01–1% solids, the flocculant should have a length 4–10 times the particle radius in order to function as an effective bridge. At these sol concentrations doubling of the molecular weight of the polymer can increase by a factor of 10–100 the initial flocculation rate. These results were used to rationalise the experimental finding that for bauxite red mud liquor the settling rate in the presence of poly(sodium styrene sulphonate) increased as the sixth power of the molecular weight (M between $(4–8) \times 10^6$).

One final point on terminology. La Mer[342] advocated that a differentiation be made between 'flocculation' and 'coagulation'. The term coagulation was to be reserved for aggregation resulting from compression of the double layer; flocculation was to refer to the formation of loosely structured flocs as a consequence of polymer bridging. Although Overbeek[343] criticised this distinction on historical grounds, the differentiation between the two terms seems to be gaining popularity. However, the distinction is by no means clear-cut: since lower molecular weight flocculants appear to function by both compression of the double layer and bridging, it is a subjective matter as to whether they induce flocculation or coagulation. Moreover, aggregation of purely sterically stabilised dispersions is not covered by La Mer's dictum. To complicate matters even further, coagulation commonly has a different meaning in emulsion science: it means aggregation of the emulsion droplets followed by coalescence. Such considerations imply that no great reliance can be placed on the use of the terms 'coagulation' and 'flocculation' in the literature.

References

1. Vanderhoff, J. W., van den Hul, H. J., Tausk, R. J. M. and Overbeek, J. Th. G. (1970). *Clean Surfaces*, 15, (New York: Marcel Dekker)
2. Thiele, H. and von Levern, H. S. (1965). *J. Colloid Sci.*, **20**, 679
3. Sinclair, D. and La Mer, V. K. (1949). *Chem. Rev.*, **44**, 245
4. Ottewill, R. H. and Woodbridge, R. F. (1961). *J. Colloid Sci.*, **16**, 58
5. Petres, J., Dezelic, G. and Tezak, B. (1966). *Croat. Chem. Acta*, **38**, 277
6. Takuyama, K. (1958). *Bull. Chem. Soc. Japan*, **31**, 950
7. Herak, M. J., Kratohvil, J. and Herak, M. M. (1958). *Croat. Chem. Acta*, **30**, 221
8. Watillon, A. and Dauchot, J. (1968). *J. Colloid Interface Sci.*, **27**, 507
9. Frens, G. and Overbeek, J. Th. G. (1969). *Kolloid-Z. Z. Polymer*, **233**, 922
10. Stöber, W., Fink, A. and Bohn, E. (1968). *J. Colloid Interface Sci.*, **26**, 62
11. Flachsbart, H. and Stöber, W. (1969). *J. Colloid Interface Sci.*, **30**, 568
12. Demchak, R. and Matijevic, E. (1969). *J. Colloid Interface Sci.*, **31**, 257
13. Dezelic, N., Petres, J. J. and Dezelic, G. (1970). *Kolloid-Z. Z. Polymere*, **242**, 1142
14. Shaw, J. N. and Ottewill, R. H. (1965). *Nature (London)*, **208**, 681
15. Ottewill, R. H. and Shaw, J. N. (1966). *Discuss. Faraday Soc.*, **42**, 154
16. Ottewill, R. H. and Shaw, J. N. (1967). *Kolloid-Z. Z. Polymere*, **215**, 161
17. Ottewill, R. H. and Shaw, J. N. (1967). *Kolloid-Z. Z. Polymere*, **218**, 34
18. Shaw, J. N. and Marshall, M. C. (1968). *J. Polymer Sci.*, *A-1*, **6**, 449
19. Shaw, J. N. (1969). *J. Polymer Sci. C*, **27**, 237
20. van den Hul, H. J. and Vanderhoff, J. W. (1968). *J. Colloid Interface Sci.*, **28**, 336
21. van den Hul, H. J. and Vanderhoff, J. W. (1970). *Brit. Polymer J.*, **2**, 121
22. McCann, G. D., Bradford, E. B., van den Hul, H. J. and Vanderhoff, J. W. (1971). *J. Colloid Interface Sci.*, **36**, 159
23. Davidson, J. A. and Collins, E. A. (1969). *J. Colloid Interface Sci.*, **29**, 456
24. Kotera, A., Furusawa, K. and Takeda, Y. (1970). *Kolloid-Z. Z. Polymere*, **239**, 677
25. Priest, W. J. (1952). *J. Phys. Chem.*, **56**, 1077
26. Napper, D. H. and Parts, A. G. (1962). *J. Polymer Sci.*, **61**, 113
27. Fitch, R. M., Prenosil, M. B. and Sprick, K. J. (1966). *Polymer Preprints*, **4**, 707
28. Johnson, G. A., Lecchini, S. M. A., Smith, E. G., Clifford, J. and Pethica, B. A. (1966). *Discuss. Faraday Soc.*, **42**, 120
29. Fitch, R. M. (1971). *Polymer Colloids*, (New York: Plenum Press)
30. Valko, E. I. (1971). *J. Polymer Sci. B*, **9**, 637
31. Osmond, D. W. J. and Walbridge, D. J. (1970). *J. Polymer Sci.*, **30**, 381
32. Walbridge, D. J. and Waters, J. A. (1966). *Discuss. Faraday Soc.*, **42**, 294
33. Napper, D. H. (1970). *J. Colloid Interface Sci.*, **32**, 106
34. Netschey, A., Napper, D. H. and Alexander, A. E. (1969). *J. Polymer Sci. B*, **7**, 829

35. Orr, C. and Dallavalle, J. M. (1959). *Fine Particle Size Measurements*, (New York: MacMillan)
36. Irani, R. R. and Callis, C. F. (1963). *Particle Size Measurements, Interpretation and Application*, (New York: Wiley)
37. Cadle, R. D. (1965). *Particle Size*, (New York: Reinhold)
38. Kerker, M. (1969). *The Scattering of Light*, 351, (New York: Academic Press)
39. Epenschied, W. F., Kerker, M. and Matijevic, E. (1964). *J. Phys. Chem.*, **68**, 3093
40. Watterson, J. G. (1971). *J. Macromol. Sci., Chem. A*, **5**, 1007
41. Rowell, R. L. and Levit, A. B. (1970). *J. Colloid Interface Sci.*, **34**, 585
42. Wallace, T. P. and Kratohvil, J. P. (1969) *Polymer Preprint Amer. Chem. Soc., Div. Polymer Chem.*, **10**, 343
43. O'Toole, J. T. (1969). *J. Polymer Sci. C*, **27**, 171
44. Saidel, G. M. and Katz, S. (1969). *J. Polymer Sci. C*, **27**, 149
45. Gardon, J. L. (1968). *J. Polymer Sci. A-1*, **6**, 687
46. Marinkovic, V. and Peterlin, A. (1956). *J. Stefan Inst. Report*, **3**, 225
47. Alexander, A. E. and Robb, I. D. (1968). *Kolloid-Z. Z. Polymere*, **228**, 64
48. Kleinschmidt, A. (1955). *Kolloid-Z.*, **142**, 72
49. Sheppard, E. and Tcheurekdjian, N. (1968). *Kolloid-Z. Z. Polymere*, **225**, 162
50. Sheppard, E. and Tcheurekdjian, N. (1968). *J. Colloid Interface Sci.*, **28**, 481
51. Finnigan, J. A., Jacobs, D. J. and Marsden, J. C. (1971). *J. Colloid Interface Sci.*, **37**, 102
52. Chu, B. and Schoenes, F. J. (1968). *J. Colloid Interface Sci.*, **27**, 424
53. Gucker, F. T., Chiu, G., Osbourne, E. C. and Tuma, J. (1968). *J. Colloid Interface Sci.*, **27**, 395
54. Gucker, F. T. and Lin, H. M. (1971). *J. Colloid Interface Sci.*, **35**, 139
55. Phillips, D. T., Wyatt, P. J. and Berkman, R. M. (1970). *J. Colloid Interface Sci.*, **34**, 159
56. Gucker, F. T. and Egan, J. J. (1961). *J. Colloid Interface Sci.*, **16**, 68
57. Maron, S. H. and Pierce, P. E. (1969). *J. Polymer Sci. C*, **27**, 183
58. Lange, H. (1968). *Kolloid-Z. Z. Polymere*, **223**, 24
59. Rowell, R. L., Wallace, T. P. and Kratohvil, J. P. (1968). *J. Colloid Interface Sci.*, **26**, 494
60. Livesey, P. J. and Billmeyer, F. W. (1969). *J. Colloid Interface Sci.*, **30**, 447
61. Brill, O. L., Weil, C. G. and Schmidt, P. W. (1968). *J. Colloid Interface Sci.*, **27**, 479
62. Jennings, B. R., Plummer, H., Closs, W. J. and Jarrard, H. G. (1969). *J. Colloid Interface Sci.*, **30**, 134
63. Jennings, B. R., Brown, B. L. and Plummer, H. (1970). *J. Colloid Interface Sci.*, **32**, 606
64. Bryant, F. D. and Latimer, P. (1969). *J. Colloid Interface Sci.*, **30**, 291
65. Coulter, W. H. (1956). *Proc. of National Electronics Conf. Chicago*, **12**, 1034
66. Matthews, B. A. and Rhodes, C. T. (1968). *J. Colloid Interface Sci.*, **28**, 71
67. Spielman, L. and Goren, S. L. (1968). *J. Colloid Interface Sci.*, **26**, 175
68. Cooper, W. D. and Parfitt, G. D. (1968). *Kolloid-Z. Z. Polymere*, **223**, 160
69. Matthews, B. A. and Rhodes, C. T. (1969). *J. Pharm. Pharmacol.*, **20**, 2045
70. Walstra, P. and Oortwijn, H. (1969). *J. Colloid Interface Sci.*, **29**, 424
71. Matthews, B. A. and Rhodes, C. T. (1970). *J. Colloid Interface Sci.*, **32**, 332
72. Matthews, B. A. and Rhodes, C. T. (1970). *J. Colloid Interface Sci.*, **32**, 339
73. Princen, L. J. and Kwolek, W. F. (1965). *Rev. Sci. Instrum.*, **36**, 646
74. Wales, M. and Wilson, J. N. (1961). *Rev. Sci. Instrum.*, **32**, 1132
75. Stryker, L. J. and Matijevic, E. (1969). *J. Colloid Interface Sci.*, **31**, 39
76. Bier, M. (ed.), (1959). *Electrophoresis*, Vol. I. (New York: Academic Press)
77. Bier, M. (ed), (1967). *Electrophoresis*, Vol. II. (New York: Academic Press)
78. Sennett, P. and Olivier, J. P. (1965). *Ind. Eng. Chem.*, **57**, 33
79. MacKenzie, J. M. W. (1971). *Minerals Science and Engineering, (July)*, 25
80. Pearce, J. B. and Elton, G. A. H. (1960). *J. Chem. Soc.*, 2186: (1956), 22
81. Derjaguin, B. V. and Dukhin, S. S. (1959). *Doklady Akad. Nauk. SSSR*, **129**, 1328
82. de Groot, S. R., Mazur, P. and Overbeek, J. Th. G. (1952). *J. Chem. Phys.*, **20**, 1825
83. Overbeek, J. Th. G. (1952). *Colloid Science*, Vol. I, 204 (H. R. Kruyt, editor) (Amsterdam: Elsevier)
84. Bull, H. B. (1934). *Kolloid-Z.*, **66**, 20
85. Stewart, P. R. and Street, N. (1961). *J. Colloid Sci.*, **16**, 192
86. Philip, J. R. (1960). *J. Hyd. Div. Prov. Amer. Soc. Civil Eng.*, **86**, Hy5, 179
87. Derjaguin, B. V. (1960). *Colloid J. (USSR)*, **22**, 155
88. Overbeek, J. Th. G. (1952). Ref. 83, 206

89. Rutgers, A. J. (1940). *Trans. Faraday Soc.*, **36**, 69
90. Morrison, F. A. Jr. and Osterle, J. E. (1965). *J. Chem. Phys.*, **43**, 2111
91. Bull, H. B. and Gortner, R. A. (1931). *J. Phys. Chem.*, **35**, 307
92. Briggs, D. R. (1928). *J. Phys. Chem.*, **32**, 641
93. Overbeek, J. Th. G. and Wijga, P. W. O. (1946). *Rec. Trav. Chim.*, **65**, 556
94. Overbeek, J. Th. G. and van Est, W. T. (1953). *Rec. Trav. Chim.*, **72**, 97
95. Ghosh, B. N. (1954). *J. Indian Chem. Soc.*, **31**, 273
96. Ghosh, B. N. (1955). *Naturwissenschaften*, **42**, 121
97. Ghosh, B. N., Choudhury, B. K. and De, P. K. (1954). *Trans. Faraday Soc.*, **50**, 955
98. Ghosh, B. N. and Pal, P. K. (1961). *Trans. Faraday Soc.*, **57**, 116
99. Bull, H. B. and Moyer, L. S. (1936). *J. Phys. Chem.*, **40**, 9
100. Bull, H. B. (1943). *Physical Biochemistry*, 1st edn, 160, (New York: John Wiley)
101. Lens, J. (1933). *Proc. Roy. Soc. (London)*, **A139**, 596
102. Rice, C. L. and Whitehead, R. (1965). *J. Phys. Chem.*, **69**, 4017
103. Oldham, J. B., Young, F. J. and Osterle, J. F. (1963). *J. Colloid Sci.*, **18**, 328
104. Dresner, L. (1963). *J. Phys. Chem.*, **67**, 1635
105. Lewis, A. F., Nammari, M. Z. and Myers, R. R. (1964). *Lab. Practice (January)*, 50
106. Alexander, A. E. and Johnson, P. (1948). *Colloid Science*, 341, (Oxford: The University Press)
107. Gortner, R. A. (1938). *Outlines of Biochemistry*, 2nd edn, 160, (New York: John Wiley)
108. Parriera, H. C. (1965). *J. Colloid Sci.*, **20**, 1
109. Martin, W. McK. and Gortner, R. A. (1930). *J. Phys. Chem.*, **34**, 1509
110. Edelberg, R. and Hazel, F. (1949). *J. Electrochem. Soc.*, **96**, 13
111. Hunter, R. J. and Alexander, A. E. (1962). *J. Colloid Sci.*, **17**, 781
112. Cooke, C. E. Jr. (1955). *J. Chem. Phys.*, **23**, 2299
113. Ueda, S., Watanabe, A. and Tsuji, F. (1951). *Memoirs College of Agr., Kyoto Univ.*, **60** *(March)*, 1, 8
114. Neale, S. M. (1946). *Trans. Faraday Soc.*, **42**, 473
115. Neale, S. M. and Peters, R. H. (1946). *Trans. Faraday Soc.*, 478
116. Pravdic, V. (1963). *Croat. Chem. Acta*, **35**, 233
117. Overbeek, J. Th. G. (1950). *Advan. Colloid Sci.*, **3**, 97
118. Booth, F. (1948). *Nature (London)*, **161**, 83
119. Booth, F. (1953). *Progr. Biophys. and Biophys. Chem.*, **3**, 175
120. Wiersema, P. H., Loeb, A. L. and Overbeek, J. Th. G. (1966). *J. Colloid Interface Sci.*, **22**, 78
121. Loeb, A. L., Wiersema, P. H. and Overbeek, J. Th. G. (1961). *The Electrical Double Layer Around a Spherical Colloid Particle*, (Cambridge, Mass: M.I.T. Press)
122. Overbeek, J. Th. G. (1943). *Kolloid Beihefte*, **54**, 287
123. Booth, F. (1950). *Proc. Roy. Soc. (London)*, **A203**, 514
124. Henry, D. C. (1931). *Proc. Roy. Soc. (London)*, **A133**, 106
125. Overbeek, J. Th. G. and Wiersema, P. H. (1967). in ref. 77., Chap. 1
126. Stigter, D. and Mysels, K. J. (1955). *J. Phys. Chem.*, **59**, 45
127. Hunter, R. J. (1962). *J. Phys. Chem.*, **66**, 1367
128. Stigter, D. (1967). *J. Colloid Interface Sci.*, **23**, 379
129. Sengupta, M. and Biswas, D. N. (1969). *J. Colloid Interface Sci.*, **29**, 536
130. von Smoluchowski, M. (1903). *Bull. Acad. Sci. Cracovie*, 184. See ref. 83, 202
131. Morrison, F. A. Jr. (1970). *J. Colloid Interface Sci.*, **34**, 210
132. James, A. M. and Carter, M. N. A. (1969). *J. Colloid Interface Sci.*, **29**, 696
133. Friend, J. P. (1970). *Ph.D. Thesis*, University of Sydney
134. Goldfarb, J., Johnson, G. A., Lagos, A. and Sepulveda, L. (1962). *J. Colloid Sci.*, **17**, 690
135. Ghosh, B. N. and Pal, P. K. (1961). *Trans. Faraday Soc.*, **57**, 116
136. Hollingshead, S., Johnson, G. A. and Pethica, B. (1965). *Trans. Faraday Soc.*, **61**, 577
137. Sieglaff, C. L. and Mazur, J. (1962). *J. Colloid Sci.*, **17**, 66
138. Ghosh, S. and Bull, H. B. (1963). *J. Colloid Sci.*, **18**, 157
139. Long, R. P. and Ross, S. (1965). *J. Colloid Sci.*, **20**, 438
140. Long, R. P. and Ross, S. (1968). *J. Colloid Interface Sci.*, **26**, 434
141. Möller, J. H. M., van Os, G. A. J. and Overbeek, J. Th. G. (1961). *Trans. Faraday Soc.*, **57**, 325
142. Overbeek, J. Th. G. (1952). *Colloid Science*, Vol. I, 214, (H. R. Kruyt, editor) (London: Elsevier)

143. Douglas, H. W. and Shaw, D. J. (1957). *Trans Faraday Soc.*, **53**, 512
144. Benhamou, N., Baruch, M., Guastalla, J. and de Mende, S. (1962). *J. Chim. Phys.*, **59**, 289
145. Hunter, R. J. and Alexander, A. E. (1963). *J. Colloid Sci.*, **18**, 846
146. Mattson, S. (1933). *J. Phys. Chem.*, **37**, 223
147. Hall, E. S. (1964). *Nature (London)*, **203**, 1371
148. Smith, M. E. and Lisse, M. W. (1936). *J. Phys. Chem.*, **40**, 399
149. van Gils, G. E. and Kruyt, H. R. (1936). *Kolloid Beihefte*, **45**, 60; Troelstra, S. A. and Kruyt, H. R. (1942). *Kolloid-Z.*, **101**, 182
150. Hamilton, J. D. and Stevens, T. J. (1967). *J. Colloid Interface Sci.*, **25**, 519
151. Gittens, G. T. and James, A. M. (1960). *Analyt. Biochem.*, **1**, 478
152. Neihof, R. (1969). *J. Colloid Interface Sci.*, **30**, 128
153. Hoyer, H. W., Mysels, K. J. and Stigter, D. (1954). *J. Phys. Chem.*, **58**, 385
154. Mysels, E. K. and Mysels, K. J. (1961). *J. Amer. Chem. Soc.*, **83**, 2049
155. Levine, S. and Bell, G. M. (1966). *Discuss. Faraday Soc.*, **42**, 69
156. Lyklema, J. and Overbeek, J. Th. G. (1961). *J. Colloid Sci.*, **16**, 501
157. Andrade, E. N. daC. and Dodd, C. (1947). *Proc. Roy. Soc. (London)*, **A187**, 296; (1951). **204**, 449
158. Haydon, D. A. (1964). In *Recent Progress in Surface Science*, p.140 et seq. (J. F. Danielli, K. G. A. Pankhurst, and A. C. Riddiford, editors) (London: Academic Press)
159. Li, H. C. and de Bruyn, P. L. (1966). *Surface Sci.*, **5**, 203
160. Haydon, D. A. (1960). *Proc. Roy. Soc. (London)*, **A258**, 319
161. Hunter, R. J. (1966). *J. Colloid Interface Sci.*, **22**, 231
162. Stigter, D. (1964). *J. Phys. Chem.*, **68**, 3600
163. Hunter, R. J., Stirling, G. and White, J. W. (1971). *Nature (Phys. Sci.)*, **230**, 192; Olejnik, S., Stirling, G. C. and White, J. W. (1970). *Discuss. Faraday Soc. (Preprint)*
164. Derjaguin, B. V. (1966). *Discuss. Faraday Soc.*, **42**, 109
165. Levine, S., Mingins, J. and Bell, G. M. (1963). *J. Phys. Chem.*, **67**, 2095
166. Guggenheim, E. A. (1929). *J. Phys. Chem.*, **33**, 842; (1930). **34**, 1540
167. Grahame, D. C. (1947). *Chem. Rev.*, **41**, 441
168. Overbeek, J. Th. G. (1952). in *Colloid Sci.*, Vol. I. 124, (H. R. Kruyt, editor) (London and Amsterdam: Elsevier)
169. Lyklema, J. and Overbeek, J. Th. G. (1959). *Electrophoresis*, Vol. I. Chapter I, (M. Bier, editor). (New York: Academic Press)
170. Parsons, R. (1954) in *Modern Aspects of Electrochemistry*, Vol. I, 103, (J. O'M. Bockris, editor) (London: Butterworths)
171. Delahay, P. (1965). *The Double Layer and Electrode Kinetics*, (New York: Interscience)
172. Stigter, D. (1964). *J. Phys. Chem.*, **68**, 3603
173. Levine, S., Bell, G. M. and Pethica, B. A. (1964). *J. Chem. Phys.*, **40**, 2304
174. Ottewill, R. H. and Woodbridge, R. F. quoted by Haydon, D. A. (1964). in *Recent Progress in Surface Science*, (Danielli, Pankhurst and Riddiford, editors), Vol. I. 182, (New York: Academic Press)
175. Bockris, J. O'M, Devanathan, M. A. V. and Müller, K. (1963). *Proc. Roy. Soc. (London)* **A274**, 55
176. van Lier, J. A., de Bruyn, P. L. and Overbeek, J. Th. G. (1960). *J. Phys. Chem.*, **64**, 1675
177. Wright, H. J. L. (1972). *Ph.D. Thesis*, University of Sydney
178. Honig, E. P. (1969). *Trans. Faraday Soc.*, **65**, 2248
179. Levine, P. L., Levine, S. and Smith, A. L. (1970). *J. Colloid Interface Sci.*, **34**, 549
180. Goldfinger, G. (1970). *Clean Surfaces*, (New York: Marcel Dekker)
181. Overbeek, J. Th. G. (1952). ref. 24, 159
182. Lyklema, J. and Overbeek, J. Th. G. (1961). *J. Colloid Sci.*, **16**, 595
183. Lyklema, J. (1961). *Trans. Faraday Soc.*, **59**, 418
184. Verwey, E. J. W. and Overbeek, J. Th. G. (1948). *Theory of Stability of Lyophobic Colloids*, (Amsterdam: Elsevier)
185. Grimley, T. B. (1950). *Proc. Roy. Soc. (London)*, **A201**, 40
186. Levine, S., Bell, G. M. and Smith, A. L. (1969). *J. Phys. Chem.*, **73**, 3534
187. Levine, S. and Matijevic, E. (1967). *J. Colloid Interface Sci.*, **23**, 188
188. Stern, O. (1924). *Z. Elektrochem.*, **30**, 508
189. MacDonald, J. K. and Barlow, C. A. jun. (1965). *Canad. J. Chem.*, **43**, 2985
190. Haydon, D. A. (1964). ref. 174, 109
191. Blok, L. and de Bruyn, P. L. (1970). *J. Colloid Interface Sci.*, **32**, 518, 527, 533

192. Hunter, R. J. and Wright, H. J. L. (1971). *J. Colloid Interface Sci.*, **37**, 564
193. Gellings, P. J. (1962). *J. Appl. Chem.*, **12**, 113
194. Bell, G. M., Mingins, J. and Levine, S. (1966). *Trans. Faraday Soc.*, **62**, 949
195. Grahame, D. C. (1958). *Z. Elektrochem.*, **62**, 264
196. Barlow, C. A., jun. and MacDonald, J. R. (1965). *J. Chem. Phys.*, **43**, 2575
197. Levine, S., Bell, G. M. and Calvert, D. (1962). *Canad. J. Chem.*, **40**, 518
198. Levine, S., Mingins, J. and Bell, G. M. (1965). *Canad. J. Chem.*, **43**, 2834
199. Wiese, G. R., James, R. O. and Healy, T. W. (1971). *Discuss. Faraday Soc., Surface Chemistry of Oxides*, (Preprint)
200. Levine, S. and Smith, A. L. (1971). *Discuss. Faraday Soc., Surface Chemistry of Oxides*, (Preprint)
201. Wright, H. J. L. (1972). *Ph.D. Thesis*, University of Sydney
202. Berube, Y. G. and de Bruyn, P. L. (1968). *J. Colloid Interface Sci.*, **27**, 305; **28**, 92
203. Ball, B. (1969). *J. Colloid Interface Sci.*, **30**, 424
204. James, R. O. and Healy, T. W. (1972). *J. Colloid Interface Sci.*, March
205. Buff, F. B. and Stillinger, F. H., jun. (1963). *J. Chem. Phys.*, **39**, 1911
206. Haydon, D. A. (1964). see ref. 174
207. Krylov, V. S. and Levich, V. G. (1963). *Russian J. Phys. Chem.*, **37**, 1224
208. Ali-Zade, P. G., Martynov, G. A. and Melamed, V. G. (1963). *Doklady Akad. Nauk. SSSR*, **151**, 601
209. Levine, S. and Bell, G. M. (1966). *Discuss. Faraday Soc.*, **42**, 69
210. MacDonald, J. K. and Barlow, C. A. (1963). *1st Australian Conference on Electrochemistry*, (London: Pergamon)
211. Grahame, D. C. (1947). *Chem. Rev.*, **41**, 441
212. Aylmore, L. A. G. and Quirk, J. P. (1960). *Nature (London)*, **187**, 1046
213. Parsegian, V. A. and Ninham, B. W. (1969). *Nature (London)*, **224**, 1197
214. Ninham, B. W. and Parsegian, V. A. (1970). *Biophys. J.*, **10**, 646
215. Parsegian, V. A. and Ninham, B. W. (1970). *Biophys. J.*, **10**, 664
216. Ninham, B. W. and Parsegian, V. A. (1970). *J. Chem. Phys.*, **52**, 4578
217. Parsegian, V. A. and Ninham, B. W. (1971). *J. Colloid Interface Sci.*, **37**, 332
218. Overbeek, J. Th. G. (1966). *Discuss. Faraday Soc.*, **42**, 7
219. Haydon, D. A. and Taylor, J. (1968). *Nature (London)*, **217**, 739
220. Tabor, D. and Winterton, R. H. S. (1968). *Nature (London)*, **219**, 1120
221. Tabor, D. and Winterton, R. H. S. (1969). *Proc. Roy. Soc.*, **A312**, 435
222. Tabor, D. (1969). *J. Colloid Interface Sci.*, **31**, 364
223. Tabor, D. (1971). *Chem. and Ind. (London)*, p. 969
224. Gregory, J. (1969). *Advan. Colloid Interface Sci.*, **2**, 396
225. Overbeek, J. Th. G. (1966). *Discuss. Faraday Soc.*, **42**, 7
226. Verwey, E. J. W. and Overbeek, J. Th. G. (1948). *Theory of the Stability of Lyophobic Colloids*. (Amsterdam: North Holland)
227. Kruyt, H. R. (1952). *Colloid Science*. Vol. 1. Chapters VI, VII and VIII, by Overbeek, J. Th. G. (Amsterdam: Elsevier)
228. Ref. 226. Table XI
229. Devereux, O. F. and de Bruyn, P. L. (1963). *Interaction of Plane-Parallel Double Layers*. (Cambridge, Mass.: M. I. T. Press)
230. Bresler, E. (1970). *J. Colloid Interface Sci.*, **33**, 278
231. Levine, S. and Bell, G. M. (1960). *Canad. J. Chem.*, **38**, 1346
232. Devereux, O. F. and de Bruyn, P. L. (1964). *J. Colloid Sci.*, **19**, 302
233. Barboi, V. M., Glazman, Yu, M. and Dykman, I. M. (1961). *Colloid J. (USSR)*, **23**, 317, 321
234. Levine, S. and Bell, G. M. (1962). *J. Colloid Sci.*, **17**, 838
235. Frens, G. (1968). *Ph. D. Thesis*. State University of Utrecht
236. Wiese, G. and Healy, T. W. (1970). *Trans. Faraday Soc.*, **66**, 490
237. Jones, J. E. and Levine, S. (1969). *J. Colloid Interface Sci.*, **30**, 241
238. Honig, E. P. and Mul, P. M. (1971). *J. Colloid Interface Sci.*, **36**, 258
239. Sanfeld, A. and Defay, R. (1966). *J. Chim. Phys.*, **63**, 577
240. Sanfeld, A., Devillez, C. and Terlinck, P. (1970). *J. Colloid Interface Sci.*, **32**, 33
241. McCartney, L. N. and Levine, S. (1969). *J. Colloid Interface Sci.*, **30**, 345
242. Bell, G. M., Levine, S. and McCartney, L. N. (1970). *J. Colloid Interface Sci.*, **33**, 335

243. Derjaguin, B. V. and Titievskaya, A. S. (1957). *Proc. Int. Cong. Surface Activity*, Vol. I, p. 211. (London: Butterworths)
244. Scheludko, A. (1957). *Kolloid-Z.*, **155**, 39
245. Scheludko, A. and Exerova, D. (1960). *Kolloid-Z.*, **168**, 24
246. Scheludko, A. (1958). *Doklady Akad. Nauk. SSSR*, 905
247. Lyklema, J. and Mysels, K. J. (1965). *J. Amer. Chem. Soc.*, **87**, 2539
248. Lyklema, J. (1967). *Pontif. Acad. Sci. Scripta Varia*, **31**, 221
249. Roberts, A. D. and Tabor, D. (1968). *Nature (London)*, **219**, 1122
250. Roberts, A. D. and Tabor, D. (1968). *Wear*, **11**, 163
251. Kitchener, J. A. and Read, A. D. (1969). *J. Colloid Interface Sci.*, **30**, 391
252. van Olphen, H. (1963). *An Introduction to Clay Colloid Chemistry* (New York: Interscience)
253. Norrish, K. (1954). *Discuss. Faraday Soc.*, **18**, 120
254. van Olphen, H. (1962). *J. Colloid Sci.*, **17**, 660
255. Norrish, K. and Rausell-Colom, J. A. (1963). *Clays and Clay Minerals*, **10**, 123
256. Garrett, W. G. and Walker, G. F. (1962). *Clays and Clay Minerals*, **9**, 557
257. Friend, J. P. and Hunter, R. J. (1970). *Clays and Clay Minerals*, **18**, 275
258. Furusawa, K. and Hachisu, S. (1968). *J. Colloid Interface Sci.*, **28**, 167
259. Prosser, A. P. and Kitchener, J. A. (1956). *Nature (London)*, **178**, 1339
260. Devereux, O. F. and de Bruyn, P. L. (1962). *J. Chem. Phys.*, **37**, 2147
261. Black, W., de Jongh, J. G. V., Overbeek, J. Th. G. and Sparnaay, M. J. (1960). *Trans. Faraday Soc.*, **56**, 1597
262. Watanabe, A. and Gotoh, R. (1963). *Kolloid-Z.*, **191**, 36
263. Derjaguin, B. V., Voropayeva, T. N., Kabanov, B. N. and Titiyevskaya, A. S. (1964). *J. Colloid Sci.*, **19**, 113
264. Krupp, H., Walter, G., Kling, W. and Lange, H. (1968). *J. Colloid Interface Sci.*, **28**, 170
265. Visser, J. (1970). *J. Colloid Interface Sci.*, **34**, 26
266. Hamilton, W. C. and Jennings, W. G. (1968). *J. Colloid Interface Sci.*, **26**, 478
267. Usiu, S. and Yamasaki, T. (1969). *J. Colloid Interface Sci.*, **29**, 629
268. Lyklema, J. (1967). *Pontif. Acad. Scient. Scripta Varia*, **31**, Contribution 7
269. Force, C. G. and Matijevic, E. (1968). *Kolloid-Z. Z. Polymere*, **224**, 51
270. Force, C. G., Matijevic, E. and Kratohvil, J. P. (1968). *Kolloid-Z. Z. Polymere*, **223**, 31
271. Williams, D. J. A. and Ottewill, R. H. (1971). *Kolloid-Z. Z. Polymere*, **243**, 141
272. Greene, B. W. and Saunders, F. L. (1970). *J. Colloid Interface Sci.*, **33**, 393
273. Goldstein, B. and Zimm, B. H. (1971). *J. Chem. Phys.*, **54**, 4408
274. Hull, M. and Kitchener, J. A. (1969). *Trans. Faraday Soc.*, **65**, 3093
275. Allen, H. A. and Matijevic, E. (1969). *J. Colloid Interface Sci.*, **31**, 287
276. Allen, H. A. and Matijevic, E. (1970). *J. Colloid Interface Sci.*, **33**, 420
277. Matijevic, E., Mangravite, F. J. and Cassell, E. A. (1971). *J. Colloid Interface Sci.*, **35**, 560
278. Allen, H. A. and Matijevic, E. (1971). *J. Colloid Interface Sci.*, **35**, 66
279. Depasse, J. and Watillon, A. (1970). *J. Colloid Interface Sci.*, **33**, 430
280. Harding, R. D. (1971). *J. Colloid Interface Sci.*, **35**, 172
281. Laskowski, J. and Kitchener, J. A. (1969). *J. Colloid Interface Sci.*, **29**, 670
282. Watillon, A. and Gerard, P. (1964). *Proc. 4th Int. Congr. Surface Activ.*, 1261
283. Heller, W. and Pugh, T. L. (1954). *J. Chem. Phys.*, **22**, 1778
284. Heller, W. (1966). *Pure Appl. Chem.*, **12**, 249
285. Hesselink, F. Th., Vrij, A. and Overbeek, J. Th. G. (1971). *J. Phys. Chem.*, **75**, 2094
286. (1968). *Encyclopaedia Brittanica*, **12**, 257; **17**, 38
287. Zsigmondy, R. (1901). *Z. Anal. Chem.*, **40**, 697
288. Patterson, D. (1969). *Macromolecules*, **2**, 672
289. Napper, D. H. and Netschey, A. (1971). *J. Colloid Interface Sci.*, **37**, 528
290. Napper, D. H. (1968). *Trans. Faraday Soc.*, **64**, 1701
291. Napper, D. H. (1969). *Kolloid-Z. Z. Polymere*, **234**, 1149
292. Napper, D. H. (1970). *Ind. Eng. Chem. (Prod. Res. and Development)*, **9**, 467
293. Mackor, E. L. (1951). *J. Colloid Sci.*, **6**, 492
294. Mackor, E. L. and van der Waals, J. H. (1952). *J. Colloid Sci.*, **7**, 535
295. Ash, S. G. and Findenegg, G. H. (1971). *Trans. Faraday Soc.*, **67**, 2122
296. Elworthy, P. H., Florence, A. T. and Rogers, J. A. (1971). *J. Colloid Interface Sci.*, **35**, 23, 34
297. Clayfield, E. J. and Lumb, E. C. (1966). *Discuss Faraday Soc.*, **42**, 314

298. Clayfield, E. J. and Lumb, E. C. (1966). *J. Colloid Sci.*, **22**, 269
299. Clayfield, E. J. and Lumb, E. C. (1966). *J. Colloid Sci.*, **22**, 285
300. Clayfield, E. J. and Lumb, E. C. (1968). *Macromolecules*, **1**, 133
301. Fischer, E. W. (1958). *Kolloid-Z.*, **160**, 120
302. Ottewill, R. H. (1967). *Nonionic Surfactants*, 627 (New York: Marcel Dekker)
303. Derjaguin, B. V., Voropayeva, T. N., Kabanov, B. N. and Titiyevskaya. (1964). *J. Colloid Sci.*, **19**, 113
304. Ottewill, R. H. and Walker, T. (1968). *Kolloid-Z. Z. Polymere*, **227**, 108
305. Flory, P. J. (1971). Personal communication
306. Flory, P. J. (1953). *Principles of Polymer Chemistry*. Chap. X. (Ithaca: Cornell University Press)
307. Hesselink, F. Th. (1969). *J. Phys. Chem.*, **73**, 3488
308. Hesselink, F. Th. (1971). *J. Phys. Chem.*, **75**, 65
309. Meier, D. J. (1967). *J. Phys. Chem.*, **71**, 1861
310. Flory, P. J. and Krigbaum, W. R. (1950). *J. Chem. Phys.*, **18**, 1086
311. Flory, P. J. (1953). *Principles of Polymer Chemistry*, 600. (Ithaca: Cornell University Press)
312. Derjaguin, B. V. (1934). *Kolloid-Z.*, **69**, 155
313. Eichinger, B. E. and Flory, P. J. (1968). *Trans. Faraday Soc.*, **64**, 2035, 2053, 2061, 2068
314. Flory, P. J., Ellenson, J. L. and Eichinger, B. E. (1968). *Macromolecules*, **1**, 279
315. Flory, P. J. and Höcker, H. (1971). *Trans. Faraday Soc.*, **67**, 2258
316. Höcker, H. and Flory, P. J. (1971). *Trans. Faraday Soc.*, **67**, 2270
317. Höcker, H., Shih, H. and Flory, P. J. (1971). *Trans. Faraday Soc.*, **67**, 2275
318. Mathai, K. G. and Ottewill, R. H. (1970). *Kolloid-Z. Z. Polymere*, **236**, 147
319. Napper, D. H. (1969). *J. Colloid Interface Sci.*, **29**, 168
320. Napper, D. H. (1970). *J. Colloid Interface Sci.*, **33**, 384
321. Ottewill, R. H. (1969). *Ann. Rept. Progr. Chem. (Chem. Soc. London) A*, **66**, 183
322. Glazman, Y. M. (1966). *Discuss. Faraday Soc.*, **42**, 255
323. Rösch, M. (1956). *Kolloid-Z.*, **147**, 78
324. Horne, R. A., Almeida, J. P., Day, A. F. and Yu, N. T. (1971). *J. Colloid Interface Sci.*, **35**, 77
325. Erlander, S. R. (1970). *J. Colloid Interface Sci.*, **34**, 53
326. Matijevic, E. (1967). *Principles and Application of Water Chemistry*, 328. (New York: John Wiley)
327. Stryker, L. J. and Matijevic, E. (1968). *Advan. in Chem.*, **79**, 44
328. Matijevic, E., Levit, A. B. and Janauer, G. E. (1968). *J. Colloid Interface Sci.*, **28**, 10
329. Matijevic, E. and Force, C. G. (1968). *Kolloid-Z. Z. Polymere*, **225**, 33
330. Stryker, L. J. and Matijevic, E. (1969). *Kolloid-Z. Z. Polymere*, **233**, 912
331. Stryker, L. J. and Matijevic, E. (1969). *J. Phys. Chem.*, **73**, 1484
332. Matijevic, E., Kratohvil, S. and Stickels, J. (1969). *J. Phys. Chem.*, **73**, 564
333. Hahn, H. H. and Stumm, W. (1968). *Advan. in Chem.*, **79**, 91
334. Hahn, H. H. and Stumm, W. (1968). *J. Colloid Interface Sci.*, **28**, 134
335. Heller, W. and Peters, J. (1970). *J. Colloid Interface Sci.*, **32**, 592
336. Peters, J. and Heller, W. (1970). *J. Colloid Interface Sci.*, **33**, 578
337. De Lauder, W. B. and Heller, W. (1971). *J. Colloid Interface Sci.*, **35**, 308
338. Heller, W. and De Lauder, W. (1971). *J. Colloid Interface Sci.*, **35**, 60
339. Heller, W. and Peters, J. (1971). *J. Colloid Interface Sci.*, **35**, 300
340. Clark, H. G., Ikenberry, L. D. and Mason, R. G. (1970). *Clean Surfaces*, 60. (New York: Marcel Dekker)
341. Walles, W. E. (1968). *J. Colloid Interface Sci.*, **27**, 797
342. La Mer, V. K. (1964). *J. Colloid Sci.*, **19**, 291
343. Overbeek, J. Th. G. (1966). *Discuss. Faraday Soc.*, **42**, 276
344. Sommerauer, A., Sussman, D. L. and Stumm, W. (1968). *Kolloid-Z. Z. Polymere*, **225**, 147
345. Nemeth, R. and Matijevic, E. (1968). *Kolloid-Z. Z. Polymere*, **225**, 155
346. Fanta, G. F., Burr, R. C., Russell, C. R. and Rist, C. E. (1971). *J. Appl. Polymer. Sci.*, **15**, 1889
347. Fanta, G. F., Burr, R. C., Russell, C. R. and Rist, C. E. (1970). *J. Appl. Polymer Sci.*, **14**, 2601
348. Ries, H. E. and Meyers, B. L. (1968). *Science*, **160**, 1449

349. Ries, H. E. and Meyers, B. L. (1969). *Abstracts of Papers, 158th Meeting, Amer. Chem. Soc., New York*
350. Ries, H. E. (1970). *Nature (London)*, **226,** 72
351. Silberberg, A., Eliassaf, J. and Katchalsky, A. (1957). *J. Polymer. Sci.*, **23,** 259
352. Richardson, M. J. (1964). *Proc. Roy. Soc. (London)*, **A279,** 50
353. Rochow, T. G. (1961). *Analyt. Chem.*, **33,** 1810
354. Audsley, A. and Fursey, A. (1965). *Nature (London)*, **208,** 753
355. Narkis, N. and Rehbun, M. (1966). *Polymer*, **7,** 507
356. Shyluk, W. P. and Stow, F. S. (1969). *J. Appl. Polymer Sci.,* **13,** 1023
357. Shyluk, W. P. and Smith, R. W. (1969). *J. Polymer Sci., A-2*, **7,** 27
358. Gregory, J. (1969). *Trans. Faraday Soc.*, **65,** 2260
359. Ruehrwein, R. A. and Ward, D. W. (1952). *Soil Sci.*, **73,** 485
360. Williams, D. J. A. and Ottewill, R. H. (1971). *Kolloid-Z. Z. Polymere,* **243,** 141
361. La Mer, V. K. (1966). *Discuss. Faraday Soc.*, **42,** 248
362. La Mer, V. K. and Healy, T. W. (1963). *Rev. Pure Appl. Chem.*, **13,** 112
363. Yarar, B. and Kitchener, J. A. (1970). *Inst. Mining Met. Trans. Section C.,* **79,** C23
364. Warkentin, B. P. and Schofield, R. K. (1960). *Clays and Clay Minerals,* **7,** 343
365. Friend, J. P. and Hunter, R. J. (1971). *J. Colloid Interface Sci.*, **37,** 548